Student Solutions Manual

to accompany

Chemistry: The Practical Science

Paul Kelter, *University of Illinois at Urbana-Champaign*
Michael Mosher, *University of Nebraska at Kearney*
Andrew Scott, *Perth College, UHI Millennium Institute*

Scott Darveau

University of Nebraska at Kearney

HOUGHTON MIFFLIN COMPANY **Boston** **New York**

Publisher and Editor-in-Chief: Charles Hartford
Development Editor: Rebecca Berardy Schwartz
Assistant Editor: Amy Galvin
Editorial Associate: Henry Cheek
Senior Project Editor: Charline Lake
Manufacturing Coordinator: Susan Brooks
Senior Marketing Manager: Laura McGinn
Marketing Assistant: Kris Bishop

Printed in the U.S.A.

ISBN-10: 0-618-73623-9
ISBN-13: 978-0-618-73623-2

1 2 3 4 5 6 7 8 9-VHO-10 09 08 07 06

TABLE OF CONTENTS

WORD TO THE STUDENT

This manual contains the complete solutions to the in-chapter practice problems and to every blue-numbered end-of-chapter question. As a convenience, the bulleted points from "The Bottom Line" in the text are reprinted at the beginning of each solution manual chapter. You may use these summary ideas as a quick reminder of the main concepts in the chapter.

As you study chemistry, it is best that you not become reliant on this manual in your attempts to solve the problems. These solutions should be used only as a final check of your answers or as a "last resort" when you may encounter difficulty. Try other problems that are similar in nature, reread the chapter, and rework the in-chapter exercises before resorting to looking up an answer here. Most (if not all) chemistry instructors will emphasize that a lot of practice working problems is crucial to develop the understanding not only of the problem-solving process, but also of the concepts found in the book. Students who "solve" their assigned problems by starting with the solution manual often do not develop the skills needed to succeed in understanding the material, let alone succeed on their exams or quizzes.

Every attempt has been made to use the significant figure rules that were presented in the textbook. In some cases, it is desirable to carry at least one extra digit beyond that allowed in the rules throughout a calculation to prevent cumulative rounding errors. These extra digits are most often shown as a subscript. If your answer differs slightly from the values shown here, repeating the calculation without rounding should result in a matching answer.

There are two cases where an additional concept is explained and used in solving problems. The first of these occurs in Chapter 12 and involves the use of simple line structures to represent molecules. A brief primer on line structure is presented at the beginning of the chapter. The second case occurs in Chapter 18 and introduces the idea of a millimole (10^{-3} mol), which is often easier to use and avoids the repeated conversion of volumes from milliliters to liters.

I would like to thank Estelle Lebeau, Linda Bush, Bette A. Kreuz, Richard Hartmann, Tsun Mei Chang and Annina Carter for their keen eyes and minds in ensuring the accuracy of these solutions. I would also like to thank Rebecca Berardy Schwartz, Liz Hogan and the rest of the staff at Houghton Mifflin Company, whose professionalism made this endeavor possible.

Chapter 1: The World of Chemistry

The Bottom Line

- Our universe and everything in it are made of chemicals. Chemicals are involved in all of the changes that affect and sustain us. (Section 1.1)
- Atoms are the most fundamental particles of the chemical world. A substance containing only one kind of atom is called an element. The particles of chemistry—atoms and molecules—are incredibly tiny compared to the objects we see in the everyday world. (Section 1.2)
- Molecules are composed of two or more atoms chemically bonded together. A chemical compound is any substance that contains different elements chemically bonded together. (Section 1.2)
- Chemical changes occur when chemicals undergo reactions in which new chemical products are formed from the initial chemical reactants. Physical changes occur when chemicals undergo changes in their state. (Section 1.2)
- Scientists find out about nature, and learn how to change it, using the scientific method. (Section 1.3)
- The International System (SI) defines the fundamental base units, and a variety of derived units, that are used to measure physical quantities. (Section 1.4)
- The calculations in chemistry include the conversion of units by using factors that relate identical quantities. This can be done via a method known as dimensional analysis. (Section 1.5)
- Precision and accuracy are important to the discussion of chemistry. They relate the uncertainties in measurements. The number of significant digits a number contains is related to the precision of that number. (Section 1.6)
- Our understanding of chemistry can be used to solve problems in many fields, including medicine, agriculture, pollution control, and nanotechnology. In addition, we can broaden our understanding of many issues, including global warming, materials, meeting our energy needs, and life itself. (Section 1.7)

Solutions to Practice Problems

1.1. a. The rusting of a copper statue results in the formation of copper oxide, a new chemical, from the copper; therefore, this process is a chemical change.

b. The yellowing of an old newspaper involves oxidation of the fibers in the paper, a chemical change.

c. Boiling water on the stove involves only a change in state, from liquid to gas. It is a physical change.

d. Shaping a piece of wood into a post involves mechanical removal of some of the wood. The post and the shavings left behind are still wood; this is a physical change.

e. Sharpening a lawnmower blade is much like creating the wooden post; some of the original material is removed mechanically. This process is a physical change.

f. Developing a photograph involves changes that permanently affect the chemicals in the photographic paper. This is a chemical change.

1.2. a. Hot tea is a homogeneous mixture. In the absence of leftover tea leaves, the tea is a transparent mixture the components of which can be distinguished only at the molecular level; it is the same throughout.

b. A milkshake is a heterogeneous mixture of particles of fat and protein in liquid. It is possible to separate the components mechanically and to see those particles as separate. It is not the same throughout.
c. A root beer float is heterogeneous; it has regions that are different from one another (ice cream versus root beer, and so on).
d. Gasoline is a homogeneous mixture; it is the same throughout.
e. Bronze is a homogeneous mixture; the different atoms of the metal are thoroughly mixed, and it is the same throughout.
f. A chef's salad is heterogeneous; there are definitely different items within.

1.3. Answers to this problem may vary. The solution that follows is only one possibility.

Step 1: Formulating a question. The question is provided: "There is a dark liquid in my cup. What is the liquid and how did it get there?"

Step 2: Finding out what is already known about your question. You could ask people nearby if they knew what was in your cup or whether they had themselves poured the liquid in the cup. You may have your answer to the question after completing this step. Similar findings occur in scientific investigations in that a specific problem may have already been solved.

Step 3: Making observations. Your first observations may be the odor you smell from the cup, the actual color and deepness of color of the liquid, and the thickness of the liquid in the cup. As a final step, you may wish to taste the liquid, which should reveal the true identity of the liquid. [Note that practicing chemists NEVER taste anything unknown in the laboratory. It might also be unwise to taste a mysterious liquid even in a social setting.]

Step 4: Creating a hypothesis. Here you would begin to posit an explanation of how the liquid got into your cup: "My friend Anne poured the coffee into my cup." Or "My nephew emptied his juice cup into my cup." In order to pose the proper question, you need to suggest an explanation of both what is in the cup and how it got there.

Step 5: Designing and performing experiments. You could create several experiments that could range from asking people in the vicinity whether if they saw what had happened to checking for fingerprints. If this is a repeated occurrence, you could hold a "stakeout."

Depending on what you find, you may need to change your hypothesis. For example, Anne might not have been at school or work that day and thus couldn't have done it. If it is a repeated occurrence and the same thing happened each time, you might have a reasonable theory that could be stated, "My nephew empties his grape juice into my cup each time he is here." Otherwise, if a reasonable explanation cannot be found, you might be limited to stating a law such as "Each Tuesday, coffee appears in my cup."

1.4. Converting to Celsius:

$$t_C = (t_F - 32)/1.8$$

$$t_C = (220 - 32)/1.8$$

$$t_C = 104°C$$

Converting from Celsius to Kelvin:

$$t_K = t_C + 273$$

$$t_K = 104 + 273$$

$$t_K = 377\text{ K}$$

1.5. The viscosity of a liquid does not depend on whether you have a single milliliter or 1000 cubic meters; therefore, viscosity is an intensive property.

1.6. To solve this problem, we need to calculate the volume of 81.76 g of aluminum. Because

$$Density = \frac{mass}{volume}, \text{ we know that } Volume = \frac{mass}{density}.$$

$$\text{Therefore, Volume} = \frac{81.76\ \text{g}}{2.7\ \text{g}/\text{cm}^3} = 3.0 \times 10^1\ \text{cm}^3$$

The volume of water displaced is the same as the volume of aluminum: $3.0 \times 10^1\ \text{cm}^3$.

1.7. a. We have the density of water, which can be used to convert from volume to mass. In order to complete the problem, we need to convert gallons to milliliters. The necessary conversion factors are 1 gal = 3.785 L and 1 L = 1000 mL. Using dimensional analysis:

$$6.50\ \text{gal} \times \frac{3.785\ \text{L}}{1\ \text{gal}} \times \frac{1000\ \text{mL}}{1\ \text{L}} \times \frac{1.00\ \text{g}}{1\ \text{mL}} = 24{,}600\ \text{g water}$$

b. Using 1 gal = 3.785 L:

$$58.6\ \text{L} \times \frac{1\ \text{gal}}{3.785\ \text{L}} = 15.5\ \text{gal}$$

c. Using \$1.34 US/gallon and \$1.30 CN = \$1.00 US:

$$11.2\ \text{gal} \times \frac{\$1.34\ \text{US}}{\text{gal}} \times \frac{\$1.30\ \text{CN}}{\$1.00\ \text{US}} = \$19.51\ \text{CN}$$

1.8. To calculate square meters from square yards, we need to use 3 ft = 1 yd, 12 in = 1 ft, 2.54 cm = 1 in., and 100 cm = 1 m. We will also need to square each factor to match the units of area.

$$1200\ \text{yd}^2 \times \left(\frac{3\ \text{ft}}{1\ \text{yd}}\right)^2 \times \left(\frac{12\ \text{in}}{1\ \text{ft}}\right)^2 \times \left(\frac{2.54\ \text{cm}}{1\ \text{in}}\right)^2 \times \left(\frac{1\ \text{m}}{100\ \text{cm}}\right)^2 = 1003\ \text{m}^2$$

To get in^2 we can stop the calculation after the second conversion:

$$1200\ \text{yd}^2 \times \left(\frac{3\ \text{ft}}{1\ \text{yd}}\right)^2 \times \left(\frac{12\ \text{in}}{1\ \text{ft}}\right)^2 = 1{,}555{,}200\ \text{in}^2$$

1.9. a. 1.3090 has five significant figures. Since the value is greater than 1, all digits to the right of the decimal place are significant.

b. 3450 has an ambiguous number of significant figures: either three or four, depending on whether the number is 3.45×10^3 or 3.450×10^3

c. 0.0020 has two significant figures. The first two zeroes right of the decimal place are placeholders and not significant. The number of significant digits is clear when you write the number as 2.0×10^{-3}.

d. 2.000 has four significant figures. Since the value is greater than 1, all digits to the right of the decimal place are significant.

The following calculation must be broken into steps: first the addition in the numerator and then the division. (The subscripted numbers are not significant.) The result of the addition is only good to the tenths position because of the presence of the number 2.3, so it has only two significant figures. After the division, the number of significant figures remains at two.

$$\frac{2.3 + 0.88}{79.4} = \frac{3.1_8}{79.4} = 0.040$$

Solutions to Student Problems

1. "Chemistry is concerned with the systematic study of the matter of our universe. This study involves the composition, structure, and properties of matter." Since chemistry is a systematic study (following the scientific method) of our surroundings, it is a science.

3. The answers to this problem are infinitely variable. Some possible answers include carpet, plastic soda bottles, computers, foods, inks, building materials, ceramic tile, sunglasses, ANYTHING!

5. If 1 cup of water (no more, no less) is added to the mix, the recipe will work; too much water and the batter will be thin and watery; too little water and there might not be enough liquid to fully mix and bind the ingredients. Likewise, precise amounts are required for chemical reactions and processes to ensure the correct products or processes are achieved.

7. An element is composed of only a single type of atom and cannot be further simplified by chemical or physical processes; a compound is composed of more than one type of atom (or element) and can be separated into its component elements. Oxygen, carbon, and sodium are three examples of elements. Water, salt, and rust (iron oxide) are three examples of compounds. (There are many other possible elements and compounds that could be named.)

9. Elemental oxygen exists as a diatomic molecule (it contains two atoms of oxygen). It is an element because there is only one type of atom involved. It is a molecule because more than one atom constitutes this natural form. It is not a compound because there is not two or more different types of atoms involved.

11. Mixtures and compounds both comprise more than one type of substance (or atom). Both can therefore be separated into simpler substances: the mixture separated by mechanical means into its component parts, the compound by chemical means into its component elements.

13. a. Hydrogen gas is elemental.
b. Table salt is a compound made from sodium and chlorine.
c. Glucose is a compound made from carbon, hydrogen, and oxygen.
d. Neon is an elemental gas.
e. Copper sulfate is a compound made from sulfur, copper, and oxygen.
f. Titanium is an elemental metal.

15. a. Lake water could be either homogeneous or heterogeneous, depending on the conditions. A clear lake on a calm day is likely to be a homogeneous mixture (discounting the fish and other animals), but a muddy lake is a heterogeneous mixture.
b. Yellow notebook paper is a heterogeneous mixture of dyes, cellulose, and possible clay or other binders.
c. Marble is a heterogeneous mixture of different rock types.
d. Soda is a homogeneous mixture of water, sugar, and other flavorings.
e. Milk is a heterogeneous mixture of fat and protein particles in a solution of other substances.
f. Dirt is a heterogeneous mixture of clays, sand, and organic matter (worms?).

17. A few possible ways to separate homogeneous mixtures are distillation, reverse osmosis, electrodialysis, and forcing changes in state to remove the major component (boiling or freezing).

19. a. The chocolate is only changing shape; this is a physical process.
b. The combustion of the wood used to generate the heat is a chemical process.
c. The drying of clothes happens, and liquid water is evaporated (it only changes state, not identity); this is a physical process.
d. The melting of the snow again is simply a change of state, a physical process.

21. Air is a mixture because it consists of different molecules of gases: oxygen, nitrogen, argon, and others.

23. By stating that we should "Eat natural food, not chemicals," the writer implies that natural food is not made up of chemicals. All the food we eat is exactly that—chemicals. All of everyday matter is composed of chemicals! What the writer was trying to convey is that we should eat natural food, not synthetically or artificially produced chemicals (or additives) or those foods produced with pesticides, artificial fertilizers, and the like.

25. You would make observations to gather data about what is not working or why it is not working. On the basis of those observations, you could develop a working hypothesis consistent with your observations that might explain what had happened. You would then develop experiments that would test your hypothesis. Depending on the outcome of the tests, you will have either solved the problem or ruled out your hypothesis, which would make a new hypothesis and appropriate experiments necessary.

27. Anything that involves observation or analysis of experimental data can be open to interpretation and can influence what hypotheses or further experiments might be developed. Finding out what is already known is helpful, and well-established theories should be the least ambiguous because they have been the most widely tested and refined.

29. A hypothesis is a possible explanation for some observations, usually with little or no testing. With much testing, a hypothesis may eventually become a theory, which carries much more weight scientifically than a hypothesis. What is commonly referred to in lay terms as a theory or hunch is actually the scientific equivalent of a hypothesis.

31. Conflicting results can often be attributed to some variables not being the same in the two studies. When humans are involved, these variables can include weight, fitness, family history, and diet, among many others.

33. Some of the important questions: Are the fish dying only in town? Are there places in the river where they are not dying? Are there contaminants in the water known to be lethal to fish? How do the levels of the contaminants vary with proximity to the industrial area, to the farmland, or to the town? Chemical analysis of the water at various locations, surveys of fish populations, and analysis of contaminants in the fish themselves are all tests that should be conducted.

35. 1 terameter (10^{12} m) > 1 kilometer (10^{3} m) > 1 millimeter (10^{-3} m) > 1 nanometer (10^{-9} m)

37. a. $100\text{ kg} \times \dfrac{1000\text{ g}}{1\text{ kg}} = 1\times10^5\text{ g}$

b. $25.9\text{ m} \times \dfrac{1\text{ km}}{1000\text{ m}} = 2.59\times10^{-2}\text{ km}$

c. $t_F = 1.8\,t_C + 32$

$t_F = 1.8\,(25) + 32$

$t_F = 77°\text{F}$

d. $3.20\text{ mg} \times \dfrac{1\text{ g}}{1000\text{ mg}} = 3.20\times10^{-3}\text{ g}$

e. $9.11\text{ nm} \times \dfrac{1\text{ m}}{10^9\text{ nm}} \times \dfrac{10^{12}\text{ pm}}{1\text{ m}} = 9.11\times10^3\text{ pm}$

f. $t_C = (t_F - 32)/1.8$

$t_C = (98.6 - 32)/1.8$

$t_C = 37.0°\text{C}$

39. a. $8.7\text{ kg} \times \dfrac{1000\text{ g}}{1\text{ kg}} \times \dfrac{1000\text{ mg}}{1\text{ g}} = 8.7\times10^6\text{ mg}$

b. $25.9\text{ dm} \times \dfrac{1\text{ m}}{10\text{ dm}} = 2.59\text{ m}$

c. $t_F = 1.8\,t_C + 32$

$t_F = 1.8\,(190) + 32$

$t_F = 374°\text{F}$

d. $3.20\text{ dL} \times \dfrac{1\text{ L}}{10\text{ dL}} \times \dfrac{1\text{ kL}}{1000\text{ L}} = 3.20\times10^{-4}\text{ kL}$

e. $9.11\text{ s} \times \dfrac{10^9\text{ ns}}{1\text{ s}} = 9.11\times10^9\text{ ns}$

f. $t_C = (t_F - 32)/1.8$

$t_C = (350 - 32)/1.8$

$t_C = 177°\text{C}$

41. We can create a new conversion constant, 100 cm^3 = 1 container, to complete the problem using dimensional analysis:

$$2.00\text{ L} \times \frac{1000\text{ cm}^3}{1\text{ L}} \times \frac{1\text{ container}}{100\text{ cm}^3} = 20\text{ containers}$$

43. a. $327\text{ km} \times \dfrac{1000\text{ m}}{1\text{ km}} = 3.27\times10^5\text{ m}$

b. $327\text{ km} \times \dfrac{1000\text{ m}}{1\text{ km}} \times \dfrac{1000\text{ mm}}{1\text{ m}} = 3.27\times10^8\text{ mm}$

c. $327\text{ km}\times\frac{1000\text{ m}}{1\text{ km}}\times\frac{10^6\ \mu\text{m}}{1\text{ m}}=3.27\times10^{11}\ \mu\text{m}$

d. $327\text{ km}\times\frac{1000\text{ m}}{1\text{ km}}\times\frac{10^9\text{ nm}}{1\text{ m}}=3.27\times10^{14}\text{ nm}$

45. To do any density calculation, we need to remember that density = mass / volume.

$$\text{Density}=\frac{\text{mass}}{\text{volume}}=\frac{62.56\text{ g}}{4.60\text{ mL}}=13.6\text{ g/mL}$$

47. In order to solve for the final density, we need to know both the total volume and the total mass of the solution. We can use the initial density and volume to get the initial mass:

$$\text{Mass}=\text{density}\times\text{volume}=1.37\text{g/mL}\times250.0\text{ mL}=342.5\text{ g}$$

The total mass = 342.5 g + 30.0 g = 372.5 g. The total volume is 250.0 mL + 24.6 mL = 274.6 mL, so the new density is

$$\text{Density}=\frac{\text{mass}}{\text{volume}}=\frac{372.5\text{ g}}{274.6\text{ mL}}=1.36\text{ g/mL}$$

49. a. Velocity is distance / time. The fundamental units give velocity as m/s.
b. Acceleration is distance /(time)^2. Using fundamental units gives m/s^2 for acceleration.
c. Volume has units of (distance)^3 and is m^3 in fundamental units.
d. Specific heat has units of energy / (mass × temperature). In fundamental units, it is J/kg•K.

51. The ruler with four divisions between successive numbers will provide more significant figures, because an estimate will not needed be until the measurement is much smaller. In other words, an estimate is not needed until after a good measurement to the nearest quarter of a division is accurately read, while the other rulers accurately read only to the nearest half of a division or full division.

53. The maximum temperature is (75 + 3)°C = 78°C. The minimum temperature is (75 – 3)°C = 72°C. Converting these to Fahrenheit:

$$t_F=1.8\,t_C+32 \qquad t_F=1.8\,t_C+32$$
$$t_F=1.8\,(72)+32 \qquad t_F=1.8\,(78)+32$$
$$t_F=162°\text{F} \qquad t_F=172°\text{F}$$

55. We can create a conversion using 1 atom = 77 pm × 2 = 154 pm to use in the calculation (We double the radius of the atom to get the diameter.)

$$1\text{ cm}\times\frac{1\text{ m}}{100\text{ cm}}\times\frac{10^{12}\text{ pm}}{1\text{ m}}\times\frac{1\text{ atom}}{154\text{ pm}}=6.5\times10^7\text{ atoms}$$

57. a. $0.07\text{ s}\times\frac{1000\text{ ms}}{1\text{ s}} = 7 \times 10^1\text{ ms}$

b. $4\text{ half-lives}\times\frac{0.07\text{ s}}{1\text{ half-life}}\times\frac{10^6\ \mu\text{s}}{1\text{ s}} = 2.8 \times 10^5\ \mu\text{s}$

59. The red tomato is denser and sinks only because its density is greater than that of water, while the green tomato floats because it is less dense than water. Two tomatoes with the same mass can have different densities because they have different volumes. Since density is mass / volume, the red tomato must have a smaller volume than the green tomato, even though they have the same mass.

61. The density of the water would decrease. We can write the formula $\text{density} = \frac{\text{mass}}{\text{volume}}$. As the volume increases with no change in the mass, the density goes down. (The larger value in the denominator decreases the overall value when the numerator is unchanged.)

63. This problem gives us two new conversions: 10 pinches = 2 smidgeons and 10 "just a bits" = 1 smidgeon. Therefore,

$$2\text{ pinches}\times\frac{2\text{ smidgeons}}{10\text{ pinches}}\times\frac{10\text{ "just a bits"}}{1\text{ smidgeon}} = 4\text{ "just a bits"}$$

65. In order to convert from mph to kph, we could use the following plan:

$$\frac{\text{miles}}{\text{hr}} \xrightarrow{\text{mi to ft}} \frac{\text{ft}}{\text{hr}} \xrightarrow{\text{ft to in}} \frac{\text{in}}{\text{hr}} \xrightarrow{\text{in to cm}} \frac{\text{cm}}{\text{hr}} \xrightarrow{\text{cm to m}} \frac{\text{m}}{\text{hr}} \xrightarrow{\text{hr to min}} \frac{\text{m}}{\text{min}} \xrightarrow{\text{min to s}} \frac{\text{m}}{\text{s}}$$

67. Converting the density:

$$\frac{22.6\text{ kg}}{1\text{ L}}\times\frac{1000\text{ g}}{1\text{ kg}}\times\frac{1\text{ L}}{1000\text{ cm}^3} = 22.6\text{ g/cm}^3$$

Calculating the mass (using the density as a conversion factor):

$$0.50\text{ L}\times\frac{22.6\text{ kg}}{1\text{ L}} = 11\text{ kg}$$

69. a. $\frac{35\text{ mi}}{\text{hr}}\times\frac{1.609\text{ km}}{1\text{ mi}} = 56\text{ kph}$

b. $\frac{22.4\text{ L}}{\text{s}}\times\frac{1000\text{ cm}^3}{1\text{ L}} = 2.24\times10^4\text{ cm}^3/\text{s}$

c. $\frac{733\text{ mi}}{\text{gal}}\times\frac{1.609\text{ km}}{1\text{ mi}}\times\frac{1\text{ gal}}{3.785\text{ L}} = 312\text{ km/L}$

d. It is important to realize that in the following conversion, the temperature unit refers to a temperature change, not an absolute temperature. For temperature changes, 1°C = 1.8°F.

$$\frac{4.184\text{ g}}{°\text{C}}\times\frac{1°\text{C}}{1.8°\text{F}}\times\frac{1\text{ kg}}{1000\text{ g}}\times\frac{2.205\text{ lb}}{1\text{ kg}} = 5.125\times10^{-3}\text{ lb/°F}$$

71. a. $\frac{186,000\text{ mi}}{\text{s}} \times \frac{60\text{ s}}{1\text{ min}} = 1.12 \times 10^7\text{ mi/min}$

b. $\frac{186,000\text{ mi}}{\text{s}} \times \frac{60\text{ s}}{1\text{ min}} \times \frac{60\text{ min}}{1\text{ hr}} \times \frac{24\text{ hr}}{1\text{ day}} = 1.61 \times 10^{10}\text{ mi/day}$

c. $\frac{186,000\text{ mi}}{\text{s}} \times \frac{60\text{ s}}{1\text{ min}} \times \frac{60\text{ min}}{1\text{ hr}} \times \frac{24\text{ hr}}{1\text{ day}} \times \frac{365.25\text{ day}}{\text{yr}} = 5.87 \times 10^{12}\text{ mi/yr}$

73. a. $2.0\text{ oz} \times \frac{1\text{ lb}}{16\text{ oz}} = 0.12\text{ lb}$

b. $4.0\text{ qt} \times \frac{0.9464\text{ L}}{1\text{ qt}} = 3.8\text{ L}$

c. $160\text{ lb} \times \frac{1\text{ kg}}{2.205\text{ lb}} = 72.6\text{ kg}$

d. $96\text{ in} \times \frac{2.54\text{ cm}}{1\text{ in}} \times \frac{1\text{ m}}{100\text{ cm}} = 2.4\text{ m}$

e. $13\text{ ft} \times \frac{12\text{ in}}{1\text{ ft}} \times \frac{2.54\text{ cm}}{1\text{ in}} = 4.0 \times 10^2\text{ cm}$

f. $32\text{ mi} \times \frac{1.609\text{ km}}{1\text{ mi}} = 51\text{ km}$

75. In years: $50\text{ min} \times \frac{1\text{ hr}}{60\text{ min}} \times \frac{1\text{ day}}{24\text{ hr}} \times \frac{1\text{ yr}}{365.25\text{ day}} = 9.5 \times 10^{-5}\text{ yr} = 95 \times 10^{-6}\text{ yr} = 95\ \mu\text{yr}$

In centuries:

$$50\text{ min} \times \frac{1\text{ hr}}{60\text{ min}} \times \frac{1\text{ day}}{24\text{ hr}} \times \frac{1\text{ yr}}{365.25\text{ day}} \times \frac{1\text{ century}}{100\text{ yr}} = 9.5 \times 10^{-7}\text{ century} = 0.95\ \mu\text{century}$$

In decades:

$$50\text{ min} \times \frac{1\text{ hr}}{60\text{ min}} \times \frac{1\text{ day}}{24\text{ hr}} \times \frac{1\text{ yr}}{365.25\text{ day}} \times \frac{1\text{ decade}}{10\text{ yr}} = 9.5 \times 10^{-6}\text{ decade} = 9.5\ \mu\text{decade}$$

77. $\frac{197\text{ g}}{6.022 \times 10^{23}\text{ atoms}} = 3.27 \times 10^{-22}\text{ g/atom} = 327 \times 10^{-24}\text{ g/atom} = 327\text{ yg/atom}$

$1\text{ yg} = 1\text{ yoctogram} = 1 \times 10^{-24}\text{ g}$

79. Per second: $\frac{200\text{ Mbucks}}{5\text{ yr}} \times \frac{\$1 \times 10^6}{\text{Mbuck}} \times \frac{1\text{ yr}}{365.25\text{ days}} \times \frac{1\text{ day}}{24\text{ hr}} \times \frac{1\text{ hr}}{60\text{ min}} \times \frac{1\text{ min}}{60\text{ s}} = \$1.27/\text{s}$

Per Game: $\frac{200\text{ Mbucks}}{5\text{ yr}} \times \frac{\$1 \times 10^6}{\text{Mbuck}} \times \frac{1\text{ yr}}{365.25\text{ days}} \times \frac{1\text{ day}}{24\text{ hr}} \times \frac{3\text{ hr}}{\text{game}} = \$13,700/\text{game}$

81. The volume of the room is 12 ft × 14 ft × 9.0 ft = 1512 ft^3. We need to covert this volume to cm^3 and use the density to convert the volume to mass.

$$1512\,\text{ft}^3 \times \left(\frac{12\,\text{in}}{1\,\text{ft}}\right)^3 \times \left(\frac{2.54\,\text{cm}}{1\,\text{in}}\right)^3 \times \frac{1.20\,\text{mg}}{\text{cm}^3} \times \frac{1\,\text{g}}{1000\,\text{mg}} = 5.14 \times 10^4\,\text{g} = 51.4\,\text{kg}$$

83. This set of throws is neither accurate nor precise. The throws are scattered fully to both sides of the horseshoe pit and are not evenly spread around the post.

85. a. Student 1: $\text{ave} = \dfrac{(17.516\,\text{g} + 17.888\,\text{g} + 19.107\,\text{g})}{3} = 18.170\,\text{g}$

Student 2: $\text{ave} = \dfrac{(15.414\,\text{g} + 16.413\,\text{g} + 14.408\,\text{g})}{3} = 15.412\,\text{g}$

Student 3: $\text{ave} = \dfrac{(13.893\,\text{g} + 13.726\,\text{g} + 13.994\,\text{g})}{3} = 13.871\,\text{g}$

b. The measurements by student 3 are the most precise, with all measurements away from the average by less than 0.2 g; the other two students vary by about 1 g.

c. Student 2's average of 15.412 g is the closest to the true mass of 15.384 g and therefore is the most accurate.

d. Assuming that they all used the same balance to weigh the copper shot, the main source of error is human error. If they used different balances, there could be instrumental error as well.

87. a. 12.000000 g is not exact: eight significant figures

b. 3125 students is an exact figure (no partial students allowed) with an infinite number of significant figures

c. 12.2 L is not exact: three significant figures

d. 12 L is not exact: two significant figures

e. 1 g is not exact: one significant figure

f. 42 test tubes is exact (no fractions of test tubes allowed) with an infinite number of significant figures

89. a. $0.7\underline{00}$ cm

b. $0.1\underline{0}1$ kg

c. $1\underline{00.0}$ cm

d. 100 m (ambiguous, one, two or none of the zeroes could be significant)

e. $0.01\underline{0}1\underline{0}$ g

91. a. 6.07×10^{-15} has three significant figures (two nonzero numbers, plus a zero between)

b. 0.003840 has four significant figures (three nonzero numbers, plus a zero at the end of a number containing a decimal point)

c. 17.00 has four significant figures (two nonzero numbers, plus two zeroes at the end of a number containing a decimal point)

d. 8×10^8 has one significant figure (one nonzero number)

e. 463.8052 has seven significant figures (six nonzero numbers, plus a zero between)

f. 1406.20 has six significant figures (five nonzero numbers, plus one zero between, plus one zero at the end of a number containing a decimal point)

g. 0.0007 has only one significant figure (one nonzero number; the zeroes before the 7 are not significant)

h. 1600.0 has five significant figures (two nonzero numbers, plus three zeroes at the end of a number containing a decimal point)
i. 0.0261140 has six significant figures (five nonzero numbers, plus one zero at the end of a number containing a decimal point; the zeroes before the 2 are not significant)
j. 1.250×10^{-3} has four significant figures (three nonzero numbers, plus one zero at the end of a number containing a decimal point)

93. a. $3.44 + 6.2 = 9.6_4 = 9.6$
b. $12.57 - 3.998 = 8.57_2 = 8.57$
c. $2.534 + 1.23 + 2.0500 = 5.81_4 = 5.81$
d. $12.54 \times 5.0 = 62._7 = 63$
e. $84 \times 100 = 8_{400} = 8 \times 10^3$ (assuming 100 has only one significant figure)
f. $45.6 \div 2.4 = 19$
g. $(754 + 0.8) \div 1.3 = (754._8) \div 1.3 = 5.8_{06} \times 10^2 = 5.8 \times 10^2$
h. $(49.53 \times 1.20) + 12 = (59.4_{36}) + 12 = 71._{436} = 71$
i. $(35.865 \div 84.2) + 2.3890 = (0.42_{595}) + 2.3890 = 2.81_{495} = 2.81$

95. Converting all values to the same units: 10.0 cm + 91 cm + 4.0 cm = $105._0$ cm= 105 cm. (Because of the value 91 cm, the calculation is significant only to the ones position.)

97. $$\text{Density} = \frac{\text{mass}}{\text{volume}} = \frac{14.53\,\text{g}}{1.43\,\text{cm}^3} = 10.16\ \text{g/cm}^3 = 10.2\,\text{g/cm}^3$$

99. a. *Nano-* refers to the metric multiplier 10^{-9}, implying technology on very small scales.
b. $$10\,\text{nm} \times \frac{1 \times 10^{-9}\ \text{m}}{1\,\text{nm}} \times \frac{100\,\text{cm}}{1\,\text{m}} = 1 \times 10^{-6}\ \text{cm}$$

101. One advantage is that genetic engineering of corn has greatly increased the yield per acre. The disadvantages of some engineering include the decrease in genetic diversity of the corn and the risk that pesticide-resistant insects or herbicide-resistant weeds will evolve. If you do not include the current lab-based genetic engineering, genetic selection of crops to enhance desired properties is as old as agriculture itself.

103. The responses here could be extremely varied. Possible answers include improved materials, clean water, and life-saving drugs.

105. One of the main energy problems that the United States faces is continuing growth of energy demands while oil, gas, and coal reserves are nearing depletion. Because these reserves affect the global energy market, this problem affects developing countries as well. The United States, with its resources, is in a much better position than poorer countries to tackle the problem.

107. All aspects of daily life are likely to be affected. New drugs and materials will affect our quality of life; the generation of power using new materials in solar cells and fuel cells is likely to be very important.

109. Chemical changes: the incorporation of carbon dioxide and oxygen to form glucose in photosynthesis and the intake of nitrogen (from nitrates or other nitrogen-containing molecules) for protein and DNA formation. Physical changes: evaporation and condensation of water in the water cycle that provides moisture to the plants.

111. Observationally, the sugar appears to break up and disappear into the water as it dissolves; chemically, the molecules in the sugar crystal disperse among the molecules of the water.

113. a. Any measurement has some uncertainty, no matter how precise the measurement, while an exact number is infinitely precise.

b. $6\,\text{packs} \times \dfrac{6\,\text{cans}}{1\,\text{pack}} = 36\,\text{cans}$

c. The 7200 mL contained in the cans is not an exact number, because the volume placed in the cans or the volumes of the cans themselves could vary. The 36 cans is an exact value.

115. Two extensive properties of seawater are its volume and mass. Two intensive properties of seawater are its density and temperature.

117. The original sources are

Miller, S. 1953. "A production of amino acids under possible primitive earth conditions." *Science* 117:528–529.

Miller, S. and H. Urey. 1959. "Organic compound synthesis on the primitive earth." *Science* 130:245–251.

You should be very careful if you conduct Internet searches for this material. Much of what is posted is biased and often patently wrong in that the scientific method has not been properly used. This topic is very controversial, and both sides are passionate in their arguments. Although minor parts of the evolutionary theory are still being actively researched and scientists are still working to explain all the details, this is not an indication that the theory itself is wrong. The vast scientific consensus is that biological evolution is real and deserves its status as a theory in the strictest scientific sense.

Chapter 2: Atoms—A Quest for Understanding

The Bottom Line

- All matter is composed of atoms. (Section 2.1)
- The law of conservation of mass states that the mass of the chemicals at the start of a reaction is equal to the mass of the chemicals at the end of the reaction. (Section 2.1)
- The law of definite composition states that any particular chemical is always composed of its components in a fixed ratio, by mass. (Section 2.1)
- The law of multiple proportions states that when the same elements can produce more than one compound, the ratio of the masses of the element that combine with a fixed mass of another element corresponds to a small whole number. (Section 2.2)
- Dalton's atomic theory stated that every substance is made of atoms; atoms are indestructible; atoms of any one element are identical; atoms of different elements differ in their masses; and chemical changes involve rearranging the attachments between atoms. (Section 2.2)
- The law of combining volumes states that when gases combine, they do so in small whole-number ratios, provided that all the gases are at the same temperature and pressure. (Section 2.2)
- Atoms are composed of electrons, protons, and neutrons. The most common isotope of the hydrogen atom is the only exception. It contains only electrons and protons. (Section 2.3)
- The modern model of the atom indicates a tiny, but dense, positively charged nucleus surrounded by a diffuse electron cloud. (Section 2.3)
- Nuclei can contain both positively charged protons and neutral neutrons. Isotopes of the same element differ in their number of neutrons. All atoms of the same element have the same atomic number (the same number of protons). (Section 2.4)
- Atoms are electrically neutral and contain equal numbers of protons and electrons. Ions are charged particles and are formed from atoms or groups of atoms transferring electrons to other atoms or groups of atoms. Cations are positively charged ions, and anions are negatively charged ions. (Section 2.4)
- The atomic mass of an element is the weighted average of all the isotopes of that element. (Section 2.5)
- The periodic table of the elements lists all the known elements, arranged into periods and groups in a manner that reflects the chemical characteristics that particular elements share. (Section 2.6)
- A chemical formula makes use of the atomic symbols and subscripts, as needed, to represent the atoms in a chemical compound. (Section 2.7)
- Molecules are distinct substances made up of two or more atoms linked together by sharing electrons between their nuclei, rather than by the transfer of electrons from one atom to another.We call bonds that are formed by the sharing of electrons between atoms covalent bonds. (Section 2.8)
- There are systematic rules for naming compounds and listing their formulas. (Section 2.9)

Solutions to Practice Problems

2.1. We can use the ratio of hydrogen to oxygen established in the exercise to find the new amount:

$$\frac{7.99\,\text{g oxygen}}{1.00\,\text{g hydrogen}} = \frac{?\,\text{g oxygen}}{24.5\,\text{g hydrogen}}$$

$$\frac{7.99\,\text{g oxygen}}{1.00\,\text{g hydrogen}} \times 24.5\,\text{g hydrogen} = ?\,\text{g oxygen}$$

$$= 196\,\text{g oxygen}$$

Because the mass has to be conserved, the amount of water produced is the total of the oxygen and hydrogen used: 196 g oxygen + 24.5 g hydrogen = 220. g water.

2.2. ${}^{14}_{7}\text{N}$ has an atomic number (Z) of 7 and atomic mass number (A) of 14. The number of protons is the same as the atomic number. In a neutral atom, the number of electrons is the same as the number of protons, and the number of neutrons (n) is the difference between the mass number and atomic number ($n = A - Z$). So ${}^{14}_{7}\text{N}$ has 7 protons, 7 neutrons, and 7 electrons.

Name	Symbol	Protons	Neutrons	Electrons
nitrogen-14	${}^{14}_{7}\text{N}$	7	7	7
neon-20	${}^{20}_{10}\text{Ne}$	10	10	10
titanium-48	${}^{48}_{22}\text{Ti}$	22	26	22
carbon-11	${}^{11}_{6}\text{C}$	6	5	6
lithium-7	${}^{7}_{3}\text{Li}$	3	4	3
phosphorus-31	${}^{31}_{15}\text{P}$	15	16	15

2.3. To get the average atomic mass, we multiply the isotopic masses by their fractional abundance:

$$\text{atomic mass} = (63.9296011 \times 0.4863) + (65.9260368 \times 0.2790) + (66.9271309 \times .0410) + (67.9248476 \times 0.1875) + (69.925325 \times 0.0062)$$

$$= 65.40\,\text{amu}$$

2.4.

Symbol	Protons	Neutrons	Electrons	Charge
${}^{52}_{24}\text{Cr}^{6+}$	24	28	18	+6
${}^{39}_{19}\text{K}^{1+}$	19	20	18	+1
${}^{79}_{35}\text{Br}^{1-}$	35	44	36	−1

2.5. Since cesium is found in Group IA, it will form ions with a charge of +1. Chlorine is in Group VIIA, so it will form an ion with charge −1. To form an electrically neutral compound, there must be one of each ion. The formula will be CsCl.

2.6. Sulfur tetrafluoride = SF_4
Carbon tetrachloride = CCl_4
Diphosphorus pentoxide = P_2O_5
PCl_3 = phosphorus trichloride
N_2O = dinitrogen monoxide
OF_2 = oxygen difluoride

2.7.

Name	Formula
magnesium chloride	$MgCl_2$
lithium fluoride	LiF
sodium bromide	NaBr
lithium oxide	Li_2O

2.8.

Formula	Name
$CuCl_2$	copper(II) chloride
CrO_3	chromium(VI) oxide
NiO	nickel(II) oxide
PdS_2	palladium(IV) sulfide

2.9. This compound contains K^+, leaving MnO_4^-, which is the permanganate ion; therefore, $KMnO_4$ is potassium permanganate. Ammonium dichromate contains ammonium, NH_4^+, and dichromate, $Cr_2O_7^{2-}$; balancing the charges gives $(NH_4)_2Cr_2O_7$.

Solutions to Student Problems

1. The slow formation of the crystals on the branches from apparently clear air must indicate the presence of very small particles that are slowly adding to the crystal until their numbers are great enough to be seen.

3. Taking the word *dormitory* apart yields the following letters: d, i, m, o, o, r, r, t, y. These letters could be recombined to form the words: *dim, dorm, door, toy, dot, rot, trim, try, moor, it,* and others. Each of these new words has different meaning and function than the original word *dormitory*.

5. The Law of Conservation of Mass means that atoms can't spontaneously appear or disappear in a chemical reaction; they have to go somewhere.

7. Just like Democritus, the researcher trying to discover the age of the earth has to ask what processes are happening and whether each process has always been the same. The researcher must look for similarities in current and past processes to discover a common theme that can offer a way to determine the age of the earth.

9. To find how much water is lost in the second sample, we can use the information from the first as a conversion factor:

$$23.4\,\text{g hydrate} \times \frac{3.6\,\text{g water}}{14.7\,\text{g hydrate}} = 5.7\,\text{g water}$$

In order to complete the problem, we used the Law of Definite Composition by assuming that the ratio of water to hydrate would always be constant, and we used the Law of Conservation of Mass by assuming that mass lost could not simply disappear and must therefore be the water.

11. a. Lavoisier's careful work in determining the mass changes in chemical reactions led to his discovery of the Law of Conservation of Mass.
b. If mass was conserved, the mass of water would have to be the difference: 250.0 g – 2.2 g = 247.8 g of water.

13. To fill out the table, we can use the ratio as a conversion factor:

$$17.0\,\text{g copper} \times \frac{1.00\,\text{g oxygen}}{3.97\,\text{g copper}} = 4.28\,\text{g oxygen}$$

$$11.5\,\text{g oxygen} \times \frac{3.97\,\text{g copper}}{1.00\,\text{g oxygen}} = 45.7\,\text{g copper}$$

Grams of Copper	Grams of Oxygen
17.0	4.28
45.7	11.5

15. If the ratio of 65 g Cu to 16 g O (4.1:1.0) represents a 1:1 ratio of atoms, you would expect a mass ratio of 8.1:1.0 for a 2 Cu: 1 O compound or a 2.0:1.0 mass ratio for a 1 Cu: 2 O compound.

17. The calcium carbonate is 40 g / 100 g × 100% = 40.% calcium. In 10.0 g of calcium chloride there are (10.0 – 6.40) = 3.6 g of calcium. This compound is 3.6 g / 10.0 g ×therefore% = 36% calcium. The calcium carbonate has a higher percentage of calcium and is then the better calcium source.

19. The ratio of the oxygen that combines with 2 g of hydrogen in water to the oxygen that combines with 2 g of hydrogen in hydrogen peroxide is 16:32, which is 1:2. This ratio is purely a small whole-number ratio, which agrees with the Law of Multiple Proportions.

21. Since the charge on the proton and that on the electron are equal in magnitude, the charge on the proton is +2. The neutron is a chargeless particle, so the neutron charge of 0 is unchanged.

23. J. J. Thomson's model: Rutherford's Model:

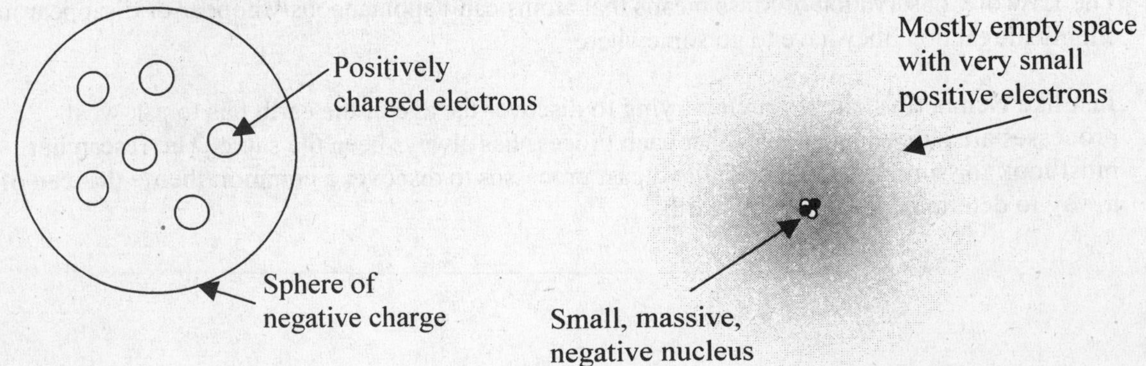

25.

Isotope	Protons	Neutrons	Electrons	Charge
Carbon-12	6	6	6	0
Aluminum-27	13	14	10	+3
Chlorine-35	17	18	18	–1

27. a. The number of protons determines the identity an atom.
b. A neutron has about the same mass as a proton.
c. Changing either the number of neutrons or the number of electrons will not change the identity of an atom.

29. 1 proton (1.6726×10^{-24} g) + 1 electron (9.1094×10^{-28} g) = 1.6735×10^{-24} g
1 proton + 1 neutron (1.6749×10^{-24} g) + 1 electron = 3.3484×10^{-24} g

31. 1 proton (+1 change) + 1 electron (–1 charge) = 0 charge
1 proton (+1 change) + 1 neutron (0 charge) + 1 electron (–1 charge) = 0 charge

33. 1 proton + 1 electron = 1.6735×10^{-24} g; 1 proton + 11 electrons = 1.6826×10^{-24} g

The percent increase is $\frac{\left(1.6826\times10^{-24}\text{ g} - 1.6735\times10^{-24}\text{ g}\right)}{1.6735\times10^{-24}\text{ g}}\times 100\% = 0.54\%$

35. a. $^{56}_{26}Fe$ has 30 neutrons, which is more than $^{52}_{24}Cr$ (only 28).
b. $^{128}_{52}Te$ has 76 neutrons, which is more than $^{127}_{53}I$ (only 74).
c. $^{59}_{27}Co$ has 32 neutrons, which is more than $^{59}_{28}Ni$ (only 31).
d. $^{20}_{10}Ne$ has 10 neutrons, which is more than $^{4}_{2}He$ (only 2).

37. In order of increasing number of protons (increasing atomic number):
$_{1}H$, $_{5}B$, $_{9}F$, $_{13}Al$, $_{20}Ca$, $_{26}Fe$.

39. In order of increasing number of electrons, which is equal to number of protons (atomic number) in a neutral atom: $_{1}H$, $_{5}B$, $_{9}F$, $_{13}Al$, $_{20}Ca$, $_{26}Fe$.

41. The first element that does not have the same number of neutrons and protons is the first element, hydrogen! It has 1 proton and no neutrons.

43. If an atom had two more protons than electrons, it would have a +2 charge and be a cation.

45.

Symbol	Protons	Electrons	Neutrons
$^{24}_{11}Na$	11	11	13
$^{131}_{53}I$	53	53	78
$^{60}_{27}Co$	27	27	33
$^{51}_{24}Cr$	24	24	27
$^{32}_{15}P$	15	15	17

47. Oxygen-16 would have a mass of 16/12 × 6 = 8.000000 amu.
Hydrogen-1 would have a mass of 1/12 × 6 = 0.5000000 amu.

49. The mass of 1 atom of carbon-12: $\frac{12\,\text{amu}}{1\,\text{atom}} \times \frac{1.6605\times10^{-24}\ \text{g}}{1\,\text{amu}} = 1.9926\times10^{-23}\ \text{g/atom}$

The mass of 1 atom of nitrogen-14: $\frac{14\,\text{amu}}{1\,\text{atom}} \times \frac{1.6605\times10^{-24}\ \text{g}}{1\,\text{amu}} = 2.3247\times10^{-23}\ \text{g/atom}$

The mass of 1 atom of fluorine-19: $\frac{19\,\text{amu}}{1\,\text{atom}} \times \frac{1.6605\times10^{-24}\ \text{g}}{1\,\text{amu}} = 3.1550\times10^{-23}\ \text{g/atom}$

51. The total fractional abundance of the two isotopes must equal 1.00; therefore, $A_{113} + A_{115} = 1$, where A represents the abundance of each isotope. We calculate the average mass according to the equation $M_{113}A_{113} + M_{115}A_{115} = M_{\text{In}} = 114.82\text{amu}$, where M is the mass of the individual isotopes. We use the two equations (two unknowns each) with some algebra to solve for the abundances:

$$
\begin{aligned}
&A_{113} + A_{115} = 1 \text{ so } A_{113} = 1 - A_{115} \\
&M_{113}A_{113} + M_{115}A_{115} = 114.82\text{amu} \\
&M_{113}(1 - A_{115}) + M_{115}A_{115} = 114.82\text{amu} \\
&112.9043(1 - A_{115}) + 114.9041A_{115} = 114.82 \\
&112.9043 - 112.9043A_{115} + 114.9041A_{115} = 114.82 \\
&(114.9041 - 112.9043)A_{115} = 114.82 - 112.9043 \\
&1.9998A_{115} = 1.9157 \\
&A_{115} = 0.958 \\
&A_{113} = 1 - A_{115} = 1 - 0.958 = 0.042
\end{aligned}
$$

So indium-113 has an abundance of 4.2%, and indium-115 has an abundance of 95.8%.

53. We calculate the atomic mass by summing the product of each isotope's abundance and mass. Notice on the mass spectrum that the peak at 33 amu is very small and the peak at 36 is imperceptible at this scale.

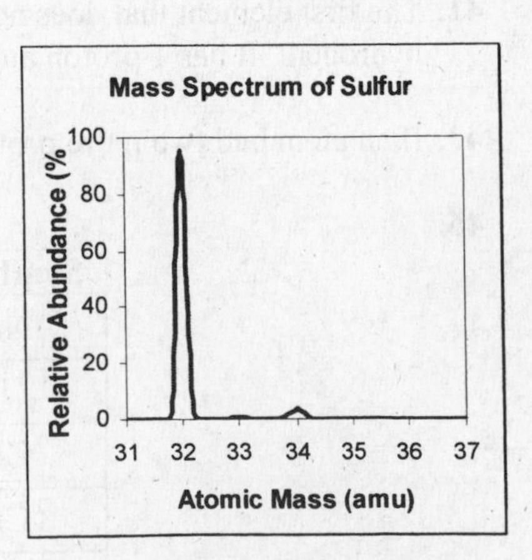

$$
\begin{aligned}
\text{Atomic mass} &= A_{32}M_{32} + A_{33}M_{33} + A_{34}M_{34} + A_{36}M_{36} \\
&= (0.95\times31.97) + (0.0076\times32.97) \\
&\quad + (0.0422\times33.97) + (0.00014\times35.97) \\
&= 32.06\,\text{amu}
\end{aligned}
$$

55. We calculate the atomic mass by summing the product of each isotope's abundance and mass. Notice on the mass spectrum that the peak at 136 amu and the peak at 138 are imperceptible at this scale.

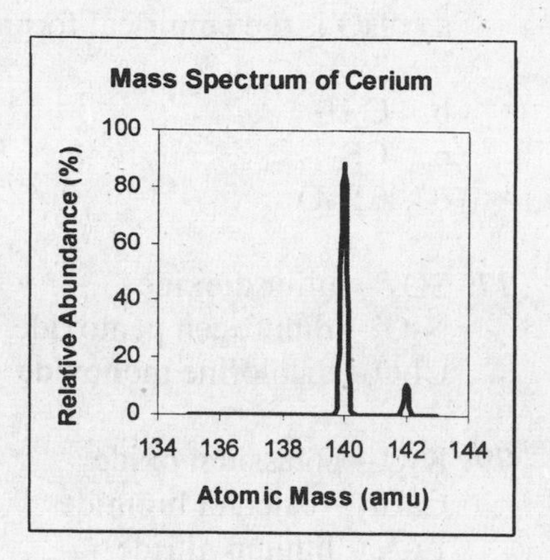

$$\begin{aligned}\text{Atomic mass} &= A_{136}M_{136} + A_{138}M_{138} + A_{140}M_{140} + A_{142}M_{142} \\ &= (0.0019\times 135.907) + (0.0025\times 137.906) \\ &\quad + (0.8848\times 139.905) + (0.1108\times 141.909) \\ &= 140.1\,\text{amu}\end{aligned}$$

57. a. Be – beryllium
b. Mn – manganese
c. Kr – krypton

59. Five elements are classified as halogens (F, Cl, Br, I, At).

61. sulfur – chalcogens (Group VIA)
iodine – halogens (Group VIIA)
helium – noble gases (Group VIIIA)
beryllium – alkaline earth metals (Group IIA)
francium – alkali metals (Group IA)

63. Metals are the elements that are usually shiny: gold, silver, and lead are metals. Additionally, silicon, as a metalloid, may be shiny.

65. When deciding which atom is likely to become a cation, it is good to note that metals are usually cation formers:
Ca or Br: Ca is the metal; it will form the cation.
S or Al: Al is the metal; it will form the cation.
Cl or Al: Al is the metal; it will form the cation.
Sr or N: Sr is the metal; it will form the cation.
I or Be: Be is the metal; it will form the cation.

67. LiCl; $BeCl_2$; NaCl; $CaCl_2$; $AlCl_3$

69. $MgBr_2$, Mg^{2+}, and Br^-
$FeCl_3$, Fe^{3+}, and Cl^-
KI, K^+, and I^-
Na_2S, Na^+, and S^{2-}

71. a. Ionic compounds
b. Molecules
c. Ionic compounds

73. Molecular compounds form between nonmetals. The following are not molecular compounds: K_2O and WO_2.

75. Empirical formulas have the simplest possible ratio of atoms in the formula.

a. HO is the empirical formula (it has the same ratio of atoms as H_2O_2, but 1:1 is simpler than 2:2).
b. C_2H_5
c. CF_2
d. CH_2O

77. SO_2 – sulfur dioxide
N_2O_5 – dinitrogen pentoxide
Cl_2O – dichlorine monoxide
PCl_3 – phosphorus trichloride
CCl_4 – carbon tetrachloride

79. K_2O – potassium oxide
$CaBr_2$ – calcium bromide
Li_3N – lithium nitride
$AlCl_3$ – aluminum chloride
BaS – barium sulfide

81. $CaBr_2$ – calcium bromide
$Fe(NO_3)_3$ – iron(III) nitrate
$CaSO_4$ – calcium sulfate
NH_4Cl – ammonium chloride
NaCl – sodium chloride

83. copper(II) hydroxide – $Cu(OH)_2$
chromium(III) oxide – Cr_2O_3
sulfur hexachloride – SCl_6
carbon tetraiodide – CI_4
aluminum hydroxide – $Al(OH)_3$
magnesium sulfate – $MgSO_4$
sodium sulfite – Na_2SO_3
ammonium hydroxide – NH_4OH
boron tribromide – BBr_3
sodium acetate – $NaCH_3COO$

85. $Mn_2(SO_4)_5$; $MnCl_5$; $Mn(NO_2)_5$; $Mn_2(CO_3)_5$; $Mn(HSO_3)_5$

87. MnC_2O_4; $Cu_2C_2O_4$; $Fe_2(C_2O_4)_3$; $Mn_2(C_2O_4)_5$; $Ti(C_2O_4)_2$

89. $(NH_4)_2CO_3$ – ammonium carbonate
$Cu(HSO_3)_2$ – copper(II) bisulfite
$KMnO_4$ – potassium permanganate
$Mg(CN)_2$ – magnesium cyanide
$NaHCO_3$ – sodium bicarbonate
$Ca(OH)_2$ – calcium hydroxide
Na_3PO_4 – sodium phosphate
$LiClO_3$ – lithium chlorate

91. $V(NO_3)_5$ has charge on V = +5
$TiSO_4$ has charge on Ti = +2
$W(C_2O_4)_3$ has charge on W = +6
AgOH has charge on Ag = +1
$Ru(HCO_3)_3$ has charge on Ru = +3

93. "$Fe(NO_3$" is not enough information to identify the salt, because it could either be $Fe(NO_3)_2$ or $Fe(NO_3)_3$. By using the compound in a chemical reaction, you should be able to determine its identity. The two forms of iron nitrate also have different physical properties that could be used to identify the salt.

95. The major objections to Dalton's original Atomic Theory have been that atoms do appear to be divisible (they can emit radiation) and there do appear to be atoms of the same element that are not identical (isotopes). We could rewrite Dalton's Atomic Theory, which would serve most of chemistry, as follows:

- Every substance is made of atoms.
- Atoms are divisible and consist of electrons, protons, and neutrons. The proton and neutron occupy the center of the atom, the nucleus; and the electrons occupy the space around the nucleus.
- All atoms of any one element have the same number of protons and have the same chemical properties.
- The average atomic masses of different elements are different, and the masses of individual isotopes are related to the total number of protons, neutrons, and electrons in the atom.
- A chemical reaction rearranges the attachments between atoms in a compound.

97. The Law of Combining Volumes would apply in both cases, in that both predict small whole-number ratios. The Law of Definite Composition also applies to both cases for the same reason. The important fact that Dalton did not account for was that elemental hydrogen and chlorine exist as diatomic molecules. Because hydrogen chloride contains one atom of each, two molecules of hydrogen chloride are formed from one molecule each of hydrogen and chlorine.

99. You would need to know the number of balls per box and the total price of the box.

101. ^{63}Cu has 29 protons, 29 electrons, and 34 neutrons. The total mass of the protons, electrons, and neutrons is:

$$29(1.6726\times10^{-27}\text{ kg})+29(9.1094\times10^{-31}\text{ kg})+34(1.6749\times10^{-27}\text{ kg})=1.05478\times10^{-25}\text{ kg}$$

The mass of the copper atom is 62.9296011 amu $\times$ 1.66053873 $\times$ 10^{-27} kg = 1.04497 $\times$x 10^{-25} kg. The mass lost is (1.05478 $\times$ 10^{-25} kg – 1.04497 $\times$ 10^{-25} kg) = 9.81 $\times$ 10^{-28} kg. This mass was lost as energy.

103. One condition that might have changed that would affect radiocarbon dating is the relative amount of carbon-13 available to be incorporated into the plants. If the amount in the past was greater than what is measured now, the amount of carbon 13 still present would be higher than expected, making the artifact appear not as old as it is. On the other hand, if the amount of carbon-13 in the past was lower than the current amount, the artifact would appear to be older than it is.

Chapter 3: Introducing Quantitative Chemistry

The Bottom Line

- Quantities in chemistry are crucial. Different amounts of the same chemical can have very different effects on chemical systems such as living things, environmental systems, and industrial processes. (Chapter opening)
- The formula mass (formula weight) of a chemical is the total mass of all the atoms present in its formula, in atomic mass units (amu) or in grams per mole. (Section 3.1)
- The average molecular mass (molecular weight) of a molecule is the total mass of the molecule, in atomic mass units (amu) or in grams per mole. (Section 3.1)
- The mole is the basic counting unit of chemistry—the chemist's "dozen"—and 1 mol of any chemical contains Avogadro's number (6.022×10^{23}) of molecules, atoms, or formula units (which entity to use depends on the chemical concerned). (Section 3.2)
- We use the mole to convert between molecules and grams of a substance. (Section 3.3)
- The percent, by mass, of an element in a compound is called its mass percent. (Section 3.4)
- The empirical formula for a compound indicates the simplest whole-number ratio in which its component atoms are present. The molecular formula indicates the actual number of each type of atom in one molecule of the compound. (Section 3.5)
- A chemical equation uses chemical formulas to indicate the reactants and products of a reaction and uses numbers before the formulas to indicate the proportions in which the chemicals involved react together and are formed. (Section 3.6)
- Stoichiometry is the study and use of quantitative relationships in chemical processes. (Section 3.7)
- The limiting reagent in a reaction is the one that is consumed first, causing the reaction to cease despite the fact that the other reactants remain "in excess." (Section 3.7)
- The percentage yield of a reaction equals the actual yield expressed as a percentage of the theoretical yield:

$$\text{Percentage yield} = \frac{\text{actual yield}}{\text{theoretical yield}} \times 100\%$$

  (Section 3.7)
- Chemistry is a quantitative science. The practice of chemistry in the real world demands mastery of the quantitative skills introduced in this chapter. (Section 3.8)

Solutions to Practice Problems

3.1. a. Codeine = $C_{18}H_{21}NO_3$

18 atoms of carbon @ 12.01 amu	=	216.18
21 atoms of hydrogen @ 1.008 amu	=	21.17
1 atom of nitrogen @14.01 amu	=	14.01
3 atoms of oxygen @ 16.00 amu	=	48.00
formula mass	=	299.4 amu

b. Magnesium sulfate = $MgSO_4$

1 atom of magnesium @ 24.31 amu	=	24.31
1 atom of sulfur @ 32.07 amu	=	32.07
4 atoms of oxygen @ 16.00 amu	=	64.00
formula mass	=	120.38 amu

c. Trinitrotoluene = $C_7H_5N_3O_6$

7 atoms of carbon @ 12.01 amu	=	84.07
5 atoms of hydrogen @ 1.008 amu	=	5.040
3 atoms of nitrogen @ 14.01 amu	=	42.03
6 atoms of oxygen @ 16.00 amu	=	96.00
formula mass	=	227.14 amu

3.2. $$100.0\,g \times \frac{1\,amu}{1.6605 \times 10^{-24}\,g} \times \frac{1\,atom}{197\,amu} = 3.06 \times 10^{23}\text{ atoms}$$

3.3. We can use Avogadro's number to create a conversion constant between moles and formula units:

$$2.88 \times 10^{20}\text{ formula units} \times \frac{1\,mol}{6.022 \times 10^{23}\text{ formula units}} = 4.78 \times 10^{-4}\,mol$$

We can also use Avogadro's number to convert from moles to molecules:

$$1.5\,mol \times \frac{6.022 \times 10^{23}\text{ molecules}}{1\,mol} = 9.0 \times 10^{23}\text{ molecules}$$

3.4. Table salt, NaCl, has a molar mass of (22.99 +35.45) = 58.44 g/mol. We can use the molar mass as a conversion constant in the calculation: 1 mol = 58.44 g.

$$26\,mol \times \frac{58.44\,g}{1\,mol} = 1.5 \times 10^3\,g$$

$$0.0025\,mol \times \frac{58.44\,g}{1\,mol} = 0.15\,g$$

For water, H_2O, the molar mass is 2(1.008) + 16.00 = 18.02 g/mol.

$$26\,mol \times \frac{18.02\,g}{1\,mol} = 4.7 \times 10^2\,g$$

$$0.0025\,mol \times \frac{18.02\,g}{1\,mol} = 0.045\,g$$

For aspartame, $C_{14}H_{18}N_2O_5$, the molar mass is 14(12.01) + 18(1.008) + 2(14.01) + 5(16.00) = 294.3 g/mol.

$$26\,mol \times \frac{294.3\,g}{1\,mol} = 7.7 \times 10^3\,g$$

$$0.0025\,mol \times \frac{294.3\,g}{1\,mol} = 0.74\,g$$

3.5. Just as in Practice 3.4, the molar mass of the compound serves as a conversion constant in the calculations. Lactic acid, $C_3H_6O_3$, has a molar mass of 3(12.01) + 6(1.008) + 3(16.00) = 90.08 g/mol.

$$5.3\,g \times \frac{1\,mol}{90.08\,g} = 0.059\,mol$$

$$0.0022\,g \times \frac{1\,mol}{90.08\,g} = 2.4 \times 10^{-5}\,mol$$

Sulfuric acid, H_2SO_4, has a molar mass of 2(1.008) + 1(32.07) + 4(16.00) = 98.09 g/mol.

$$5.3\,\text{g} \times \frac{1\,\text{mol}}{98.09\,\text{g}} = 0.054\,\text{mol}$$

$$0.0022\,\text{g} \times \frac{1\,\text{mol}}{98.09\,\text{g}} = 2.2 \times 10^{-5}\,\text{mol}$$

3.6. A conversion between grams and molecules (and vice versa) takes two steps: One is a conversion between grams and moles, and the second is a conversion between moles and molecules. For water (18.02 g/mol):

$$1.50\,\text{g} \times \frac{1\,\text{mol}}{18.02\,\text{g}} \times \frac{6.022 \times 10^{23}\,\text{molecules}}{1\,\text{mol}} = 5.01 \times 10^{22}\,\text{molecules}$$

$$5.33 \times 10^{29}\,\text{molecules} \times \frac{1\,\text{mol}}{6.022 \times 10^{23}\,\text{molecules}} \times \frac{18.02\,\text{g}}{1\,\text{mol}} = 1.59 \times 10^{7}\,\text{g}$$

Since there are about 1000 g of water per liter, there are roughly 16,000 L of water in the above sample. You would need a swimming pool to contain that volume.

3.7. Potassium chloride, KCl, has a formula mass of 39.10 + 35.45 = 74.55 g/mol. Of that, the potassium is 39.10 g/mol, so the percentage of potassium in the sample is

$$\text{Mass percent} = \frac{39.10\,\text{g/mol K}}{74.55\,\text{g/mol KCl}} \times 100\% = 52.45\%\,\text{K}$$

3.8. As in the exercise, 100 g of compound is an excellent starting point. In 100 g of compound, there will be 92.26 g of carbon and 7.74 g of hydrogen. We next convert these masses into moles:

$$92.26\,\text{g C} \times \frac{1\,\text{mol C}}{12.01\,\text{g C}} = 7.682\,\text{mol C} \quad 7.74\,\text{g H} \times \frac{1\,\text{mol H}}{1.008\,\text{g H}} = 7.679\,\text{mol H}$$

We then reduce the moles of each to the simplest ratio by dividing by the smaller number of moles:

$$\frac{7.682\,\text{mol C}}{7.679\,\text{mol H}} = 1.018\,\text{C} \;\text{ to each }\; \frac{7.679\,\text{mol H}}{7.679\,\text{mol H}} = 1.000\,\text{H}$$

Therefore, the empirical formula is CH. The empirical molar mass is 12.01 + 1.008 = 13.02 g/mol. The number of empirical units in the formula is

$$78.11\,\text{g/mol formula} \times \frac{1\,\text{empirical unit}}{13.02\,\text{g/mol}} = 6.00\,\text{empirical units/formula}$$

so the molecular formula is C_6H_6.

3.9. We will again start this problem with 100 g of the compound, yielding 40.0 g C, 6.7 g H, and 53.3 g O. Converting into moles:

$$40.0\,\text{g C} \times \frac{1\,\text{mol C}}{12.01\,\text{g C}} = 3.33\,\text{mol C}$$

$$6.7\,\text{g H} \times \frac{1\,\text{mol H}}{1.008\,\text{g H}} = 6.65\,\text{mol H}$$

$$53.3\,\text{g O} \times \frac{1\,\text{mol O}}{16.00\,\text{g O}} = 3.33\,\text{mol O}$$

Creating the ratios by dividing by the smallest number of moles:

$$\frac{3.33\,\text{mol C}}{3.33\,\text{mol C}} = 1.00\,\text{C},\ \frac{6.65\,\text{mol H}}{3.33\,\text{mol C}} = 2.00\,\text{H},\ \frac{3.33\,\text{mol O}}{3.33\,\text{mol C}} = 1.00\,\text{O}$$

So the empirical formula is CH_2O, which has an empirical mass of 12.01 + 2(1.008) + 16.00 = 30.03 g/mol. Since the molecular molar mass is 60.06 g/mol, there are 2 empirical units in the formula. The molecular formula is $C_2H_4O_2$.

3.10. In looking at the equation $Ca_3(PO_4)_2 + H_2SO_4 \rightarrow CaSO_4 + H_3PO_4$, we use the most complicated formula as the starting point, $Ca_3(PO_4)_2$. It has 3 calcium atoms and 2 phosphorus atoms. To balance these atoms on the right side of the equation, we multiply $CaSO_4$ by 3 and H_3PO_4 by 2:

$$Ca_3(PO_4)_2 + H_2SO_4 \rightarrow 3CaSO_4 + 2H_3PO_4$$

When we count the number of hydrogen atoms on each side, we find that there are 6 on the right and only 2 on the left. We can multiply H_2SO_4 by 3 to balance the hydrogen atoms:

$$Ca_3(PO_4)_2 + 3H_2SO_4 \rightarrow 3CaSO_4 + H_3PO_4.$$

At this point, we have balanced Ca, P, and H explicitly; we need to check to make sure the remaining S and O are balanced. Counting oxygen atoms, we find 20 on each side, and counting sulfur atoms, we find 3 on each side. Double-checking, we find 6 hydrogen atoms on each side, 3 calcium atoms on each side, and 2 phosphorus atoms on each side. The equation is balanced.

3.11. We see from the equation for the reaction that there is 1 mole of stearic acid produced for each mole of oleic acid consumed. This ratio is the conversion factor that we need, along with the molar masses, to complete the calculation.

$$72.55\,\text{g}\,C_{18}H_{36}O_2 \times \frac{1\,\text{mol}\,C_{18}H_{36}O_2}{284.5\,\text{g}\,C_{18}H_{36}O_2} \times \frac{1\,\text{mol}\,C_{18}H_{34}O_2}{1\,\text{mol}\,C_{18}H_{36}O_2} \times \frac{282.5\,\text{g}\,C_{18}H_{34}O_2}{1\,\text{mol}\,C_{18}H_{34}O_2}$$

$$= 72.04\,\text{g}\,C_{18}H_{34}O_2$$

3.12.

$$673\,\text{kg}\,C_{18}H_{34}O_2 \times \frac{1000\,\text{g}}{1\,\text{kg}} \times \frac{1\,\text{mol}\,C_{18}H_{34}O_2}{282.5\,\text{g}\,C_{18}H_{34}O_2} \times \frac{1\,\text{mol}\,C_{18}H_{36}O_2}{1\,\text{mol}\,C_{18}H_{34}O_2}$$

$$\times \frac{284.5\,\text{g}\,C_{18}H_{36}O_2}{1\,\text{mol}\,C_{18}H_{36}O_2} \times \frac{1\,\text{kg}}{1000\,\text{g}} = 678\,\text{kg}\,C_{18}H_{36}O_2$$

3.13. We first need to balance the equation $CH_4 + O_2 \rightarrow CO_2 + H_2O$. There is already 1 carbon on each side, so no adjustment needs to be made there. We can next balance the hydrogen atoms by multiplying the water by 2, which gives 4 hydrogen atoms on each side: $CH_4 + O_2 \rightarrow CO_2 + 2H_2O$. We finish by noting that there are 4 oxygen atoms on the right. Multiplying the O_2 by 2 balances the equation: $CH_4 + 2O_2 \rightarrow CO_2 + 2H_2O$. The correct molar masses are as follows: CH_4 (16.05 g/mol), O_2 (32.00 g/mol), CO_2 (44.01 g/mol), and H_2O (18.02 g/mol). We finish the problem using the stoichiometric coefficients to create the needed conversion factors:

$$26\,\text{g}\,CH_4 \times \frac{1\,\text{mol}\,CH_4}{16.04\,\text{g}\,CH_4} \times \frac{1\,\text{mol}\,CO_2}{1\,\text{mol}\,CH_4} \times \frac{44.01\,\text{g}\,CO_2}{1\,\text{mol}\,CO_2} = 71\,\text{g}\,CO_2$$

3.14. The balanced equation is $Ca(OH)_2 + 2HCl \rightarrow CaCl_2 + 2H_2O$. We calculate the amount of $CaCl_2$ that could be formed from each reactant, assuming there is an excess of the other reactant:

$$32.15\,g\,HCl \times \frac{1\,mol\,HCl}{36.46\,g\,HCl} \times \frac{1\,mol\,CaCl_2}{2\,mol\,HCl} \times \frac{110.98\,g\,CaCl_2}{1\,mol\,CaCl_2} = 48.93\,g\,CaCl_2$$

$$12.33\,g\,Ca(OH)_2 \times \frac{1\,mol\,Ca(OH)_2}{74.10\,g\,Ca(OH)_2} \times \frac{1\,mol\,CaCl_2}{1\,mol\,Ca(OH)_2} \times \frac{110.98\,g\,CaCl_2}{1\,mol\,CaCl_2} = 18.47\,g\,CaCl_2$$

Since less $CaCl_2$ is formed by complete use of the $Ca(OH)_2$, the $Ca(OH)_2$ is the limiting reagent and only 18.47 g of $CaCl_2$ is produced.

3.15. To complete this calculation, we need to calculate the theoretical amount of sodium phosphate that would have been produced and then complete the stoichiometric calculation of the amount of sodium hydroxide needed.

$$100.\,g\,Na_3PO_4\,(actual) \times \frac{100.00\,g\,Na_3PO_4\,(theoretical)}{82.52\,g\,Na_3PO_4\,(actual)} = 121\,g\,Na_3PO_4\,(theoretical)$$

$$\text{Then } 121\,g\,Na_3PO_4 \times \frac{1\,mol\,Na_3PO_4}{163.94\,g\,Na_3PO_4} \times \frac{3\,mol\,NaOH}{1\,mol\,Na_3PO_4} \times \frac{40.00\,g\,NaOH}{1\,mol\,NaOH} = 88.6\,g\,NaOH.$$

Solutions to Student Problems

1. a. CO has a molecular mass of 1(12.01 amu) + 1(16.00 amu) = 28.01 amu.

b. SiO_2 has a molecular mass of 1(28.09 amu) + 2(16.00 amu) = 60.09 amu.

c. NH_3 has a molecular mass of 1(14.01 amu) + 3(1.008 amu) = 17.03 amu.

d. $Na_2S_2O_3$ has a formula mass of 2(22.99 amu) + 2(32.07 amu) + 3(16.00 amu) = 158.12 amu.

e. $C_{57}H_{110}O_6$ has a molecular mass of
57(12.01 amu) + 110(1.008 amu) + 6(16.00 amu) = 891.5 amu.

3. H_2O(18.02 amu) < CO(28.01 amu) < C_2H_4OH(45.06 amu) < C_6H_6(78.11 amu) < $CaCl_2$(110.98 amu)

5. a. saccharin, $C_7H_5NO_3S$:
7(12.01 amu) + 5(1.008 amu) + 1(14.01 amu) + 3(16.00 amu) + 1(32.07 amu) = 183.19 amu
aspartame, $C_{14}H_{18}N_2O_5$:
14(12.01 amu) + 18(1.008 amu) + 2(14.01 amu) + 5(16.00 amu) = 294.3 amu

b. aspartame:saccharine = 294.3 amu: 183.19 amu = 1.607 : 1.000 mass ratio

c. Since the mass ratio represents a molecule-to-molecule ratio, we can use it to convert between the two.

$$42.0\,g\,aspartame \times \frac{1.000\,g\,saccharine}{1.6065\,g\,aspartame} = 26.1\,g\,saccharine$$

7. $$\frac{\$10,000}{1\,kg} \times \frac{1\,kg}{1000\,g} \times \frac{1.6605 \times 10^{-24}\,g}{1\,amu} \times \frac{196.97\,amu}{1\,atom} = \$3 \times 10^{-21}/\text{atom}$$

9. a. $6.02\times10^{22}\text{ atoms}\times\frac{107.87\text{ amu}}{1\text{ atom}}\times\frac{1.6605\times10^{-24}\text{ g}}{1\text{ amu}}=10.8\text{ g}$

b. $10^{12}\text{ atoms}\times\frac{196.97\text{ amu}}{1\text{ atom}}\times\frac{1.6605\times10^{-24}\text{ g}}{1\text{ amu}}=3.271\times10^{-10}\text{ g}$

c. $24\text{ molecules}\times\frac{18.02\text{ amu}}{1\text{ molecule}}\times\frac{1.6605\times10^{-24}\text{ g}}{1\text{ amu}}=7.181\times10^{-22}\text{ g}$

d. $1\text{ molecule}\times\frac{44.09\text{ amu}}{1\text{ molecule}}\times\frac{1.6605\times10^{-24}\text{ g}}{1\text{ amu}}=7.321\times10^{-23}\text{ g}$

11. a. $12.01\text{ g}\times\frac{1\text{ amu}}{1.6605\times10^{-24}\text{ g}}\times\frac{1\text{ molecule}}{18.02\text{ amu}}=4.014\times10^{23}\text{ molecules}$

b. $68.3\text{ g}\times\frac{1\text{ amu}}{1.6605\times10^{-24}\text{ g}}\times\frac{1\text{ unit}}{84.01\text{ amu}}=4.896\times10^{23}\text{ molecules}$

c. $100\text{ g}\times\frac{1\text{ amu}}{1.6605\times10^{-24}\text{ g}}\times\frac{1\text{ molecule}}{16.04\text{ amu}}=4\times10^{24}\text{ molecules}$

d. $2.3\text{ mg}\times\frac{1\text{ g}}{1000\text{ mg}}\times\frac{1\text{ amu}}{1.6605\times10^{-24}\text{ g}}\times\frac{1\text{ molecule}}{180.16\text{ amu}}=7.7\times10^{18}\text{ molecules}$

13. a. Urea, CH_4ON_2, has a mass of
1(12.01 amu) + 4(1.008 amu) + 1(16.00 amu) + 2(14.01 amu) = 60.06 amu

b. $6.022\times10^{23}\text{ molecules}\times\frac{60.06\text{ amu}}{1\text{ molecule}}\times\frac{1.6605\times10^{-24}\text{ g}}{1\text{ amu}}=60.06\text{ g}$

c. $6.022\times10^{22}\text{ molecules}\times\frac{60.06\text{ amu}}{1\text{ molecule}}\times\frac{1.6605\times10^{-24}\text{ g}}{1\text{ amu}}=6.006\text{ g}$

15. Acetaminophen has a molecular mass of 151.16 amu.

$0.500\text{ g}\times\frac{1\text{ amu}}{1.6605\times10^{-24}\text{ g}}\times\frac{1\text{ molecule}}{151.16\text{ amu}}=1.99\times10^{21}\text{ molecules}$

17. a. To complete the problem, we need to convert all the choices into the same units. NaCl has a formula mass of 58.44 amu, or a molar mass of 58.44 g/mol.

$0.100\text{ mol}\times\frac{58.44\text{ g}}{\text{mol}}=5.84\text{ g}$

b. $4.2\times10^{23}\text{ units}\times\frac{58.44\text{ amu}}{1\text{ unit}}\times\frac{1.6605\times10^{-24}\text{ g}}{1\text{ amu}}=41\text{ g}$

c. 1.60 g
An amount of 4.2×10^{23} formula units has the greatest mass.

19. a. $65.0\,g \times \dfrac{1\,mol}{44.01\,g} = 1.48\,mol$

b. $1.5 \times 10^{22}\ atoms \times \dfrac{1\,mol}{6.022 \times 10^{23}\ atoms} = 0.025\,mol$

c. $25\,g \times \dfrac{1\,mol}{44.09\,g} = 0.57\,mol$

d. $4.5 \times 10^{24}\ molecules \times \dfrac{1\,mol}{6.022 \times 10^{23}\ atoms} = 7.5\,mol$

21. a. The substance with the greatest number of moles of atoms will have the greatest number of atoms. We can more simply calculate the number of moles in each. For molecules, we will need to multiply by the number of atoms per molecule.

$454\,g\,Au \times \dfrac{1\,mol}{196.97\,g\,Au} = 2.30\,mol$

b. $56.0\,g\,O_2 \times \dfrac{1\,mol\,O_2}{32.00\,g\,O_2} \times \dfrac{2\,atoms}{1\,mol\,O_2} = 3.50\,mol\,atoms$

c. $245\,g\,C \times \dfrac{1\,mol\,C}{12.01\,g\,C} = 20.4\,mol$

The sample of graphite has the greatest number of moles of atoms; therefore, it has the greatest number of atoms.

23. a. $15.0\,g\,CaCl_2 \times \dfrac{1\,mol\,CaCl_2}{110.98\,g\,CaCl_2} \times \dfrac{6.022 \times 10^{23}\ units}{1\,mol\,CaCl_2} = 8.14 \times 10^{22}\ units$

b. $15.0\,mol\,CaCl_2 \times \dfrac{6.022 \times 10^{23}\ units}{1\,mol\,CaCl_2} = 9.03 \times 10^{24}\ units$

c. $15.0\,mL\,CaCl_2 \times \dfrac{1\,L}{1000\,mL} \times \dfrac{0.42\,mol\,CaCl_2}{1\,L\,CaCl_2} \times \dfrac{6.022 \times 10^{23}\ units}{1\,mol}$

$= 3.8 \times 10^{21}\ units$

25. a. $1500\,mol\,Ne \times \dfrac{20.18\,g\,Ne}{1\,mol\,Ne} = 3.0 \times 10^{4}\,g\,Ne$

b. $12.5\,mol\,Na \times \dfrac{22.99\,g\,Na}{1\,mol\,Na} = 287\,g\,Na$

c. $0.42\,mol\,N_2 \times \dfrac{28.02\,g\,N_2}{1\,mol\,N_2} = 12\,g\,N_2$

27. $5.00\,g\,Tc \times \dfrac{1\,mol\,Tc}{98\,g\,Tc} \times \dfrac{6.022 \times 10^{23}\ atoms\,Tc}{1\,mol\,Tc} = 3.1 \times 10^{22}\ atoms\,Tc$

29. a. $1\,\text{atom Fe} \times \frac{1\,\text{mol Fe}}{6.022\times 10^{23}\,\text{atoms Fe}} \times \frac{55.85\,\text{g Fe}}{1\,\text{mol Fe}} = 9.274\times 10^{-23}\,\text{g Fe}$ (Note that this is the average mass of 1 atom of Fe based on the mass of many atoms of a mixed sample of iron.)

b. $15\,\text{mg Fe} \times \frac{1\,\text{g}}{1000\,\text{mg}} \times \frac{1\,\text{mol Fe}}{55.85\,\text{g Fe}} \times \frac{6.022\times 10^{23}\,\text{atoms}}{1\,\text{mol Fe}} = 1.6\times 10^{20}\,\text{atoms}$

c. $15\,\text{mg Fe} \times \frac{1\,\text{g}}{1000\,\text{mg}} \times \frac{1\,\text{mol Fe}}{55.85\,\text{g Fe}} = 2.7\times 10^{-4}\,\text{mol Fe}$

31. a. $1.5\,\text{mg Co} \times \frac{1\,\text{g}}{1000\,\text{mg}} \times \frac{1\,\text{mol Co}}{58.93\,\text{g Co}} = 2.5\times 10^{-5}\,\text{mol}$

b. $2.5\times 10^{-5}\,\text{mol Co} \times \frac{6.022\times 10^{23}\,\text{atoms}}{1\,\text{mol Co}} = 1.5\times 10^{19}\,\text{atoms Co}$

33. a. $10.5\,\text{mol}\,P_4 \times \frac{123.88\,\text{g}\,P_4}{1\,\text{mol}\,P_4} = 1.30\times 10^{3}\,\text{g}\,P_4$

b. $454\,\text{g}\,S_8 \times \frac{1\,\text{mol}\,S_8}{256.56\,\text{g}\,S_8} = 1.77\,\text{mol}\,S_8$

35. Pyridoxin, $C_8H_{11}NO_3$, has a molar mass of 169.18 g/mol.

$0.156\,\text{g}\,C_8H_{11}NO_3 \times \frac{1\,\text{mol}\,C_8H_{11}NO_3}{169.18\,\text{g}\,C_8H_{11}NO_3} = 9.22\times 10^{-4}\,\text{mol}\,C_8H_{11}NO_3$

$9.22\times 10^{-4}\,\text{mol}\,C_8H_{11}NO_3 \times \frac{6.022\times 10^{23}\,\text{molecules}\,C_8H_{11}NO_3}{1\,\text{mol}\,C_8H_{11}NO_3} = 5.55\times 10^{20}\,\text{molecules}$

$5.55\times 10^{20}\,\text{molecules}\,C_8H_{11}NO_3 \times \frac{8\,\text{atoms C}}{1\,\text{molecule}\,C_8H_{11}NO_3} = 4.44\times 10^{21}\,\text{atoms C}$

37. a. For each of the problems, we calculate the percentage of carbon by taking the ratio of the total mass of carbon to the total mass of the molecule (times 100%).

C_3H_6: $\frac{3(12.01\,\text{amu})}{3(12.01\,\text{amu}) + 6(1.008\,\text{amu})} \times 100\% = 85.63\%\,\text{C}$

b. CH_3OH: $\frac{1(12.01\,\text{amu})}{1(12.01\,\text{amu}) + 4(1.008\,\text{amu}) + 1(16.00\,\text{amu})} \times 100\% = 37.48\%\,\text{C}$

c. C_6H_6: $\frac{6(12.01\,\text{amu})}{6(12.01\,\text{amu}) + 6(1.008\,\text{amu})} \times 100\% = 92.26\%\,\text{C}$

d. $C_{12}H_{22}O_{11}$: $\frac{12(12.01\,\text{amu})}{12(12.01\,\text{amu}) + 22(1.008\,\text{amu}) + 11(16.00\,\text{amu})} \times 100\% = 42.10\%\,\text{C}$

39. For each of the molecules, we calculate the percentage of sulfur by taking the ratio of the total mass of sulfur to the total mass of the molecule (times 100%).

$$H_2S: \frac{1(32.07\text{ amu})}{2(1.008\text{ amu})+1(32.07\text{ amu})}\times 100\% = 94.09\%\text{S}$$

$$SO_2: \frac{1(32.07\text{ amu})}{1(32.07\text{ amu})+2(16.00\text{ amu})}\times 100\% = 50.05\%\text{S}$$

$$Na_2S_2O_4: \frac{2(32.07\text{ amu})}{2(22.99\text{ amu})+2(32.07\text{ amu})+4(16.00\text{ amu})}\times 100\% = 36.84\%\text{S}$$

$$H_2SO_3: \frac{1(32.07\text{ amu})}{2(1.008\text{ amu})+1(32.07\text{ amu})+3(16.00\text{ amu})}\times 100\% = 39.07\%\text{S}$$

Placing these in order of decreasing percentage sulfur: $H_2S > SO_2 > H_2SO_3 > Na_2S_2O_4$.

41. For each of the compounds, we calculate the percentage of boron by taking the ratio of the total mass of boron to the total mass of the compound (times 100%).

$$B_{13}C_2: \frac{13(10.81\text{ amu})}{13(10.81\text{ amu})+2(12.01\text{ amu})}\times 100\% = 85.40\%\text{B}$$

$$Ti_3B_4: \frac{4(10.81\text{ amu})}{3(47.87\text{ amu})+4(10.81\text{ amu})}\times 100\% = 23.14\%\text{B}$$

$$CaB_6: \frac{6(10.81\text{ amu})}{1(40.08\text{ amu})+6(10.81\text{ amu})}\times 100\% = 61.81\%\text{B}$$

43. a. For each of the compounds, we calculate the percentage of oxygen by taking the ratio of the total mass of oxygen to the total mass of the compound (times 100%).

$$FeSO_4: \frac{4(16.00\text{ amu})}{1(55.85\text{ amu})+1(32.07\text{ amu})+4(16.00\text{ amu})}\times 100\% = 42.13\%\text{O}$$

$$Na_2SO_4: \frac{4(16.00\text{ amu})}{2(22.99\text{ amu})+1(32.07\text{ amu})+4(16.00\text{ amu})}\times 100\% = 45.05\%\text{O}$$

Na_2SO_4 has the greater percent oxygen.

b. $$K_2FeO_4: \frac{4(16.00\text{ amu})}{2(39.10\text{ amu})+1(55.85\text{ amu})+4(16.00\text{ amu})}\times 100\% = 32.32\%\text{O}$$

$$KMnO_4: \frac{4(16.00\text{ amu})}{1(39.10\text{ amu})+1(54.94\text{ amu})+4(16.00\text{ amu})}\times 100\% = 40.50\%\text{O}$$

$KMnO_4$ has the greater percent oxygen.

c. $$Fe(NO_3)_3: \frac{9(16.00\text{ amu})}{1(55.85\text{ amu})+3(14.01\text{ amu})+9(16.00\text{ amu})}\times 100\% = 59.53\%\text{O}$$

$$HNO_3: \frac{3(16.00\text{ amu})}{1(1.008\text{amu})+1(14.01\text{ amu})+3(16.00\text{ amu})}\times 100\% = 76.17\%\text{O}$$

HNO_3 has the greater percent oxygen.

45. $$\frac{2(12.01\,\text{amu})}{1(40.08\,\text{amu})+2(12.01\,\text{amu})+4(16.00\,\text{amu})}\times 100\% = 18.75\%\,\text{C}$$

$$\frac{1(40.08\,\text{amu})}{1(40.08\,\text{amu})+2(12.01\,\text{amu})+4(16.00\,\text{amu})}\times 100\% = 31.29\%\,\text{Ca}$$

$$\frac{4(16.00\,\text{amu})}{1(40.08\,\text{amu})+2(12.01\,\text{amu})+4(16.00\,\text{amu})}\times 100\% = 49.96\%\,\text{C}$$

47. The mass percentage of carbon in the saturated hydrocarbon would decrease relative to that in the unsaturated hydrocarbon. The percentage of carbon decreases because the total mass of the compound increases while the mass of carbon is unchanged; when the mass of carbon is divided by the larger total mass, the percentage decreases.

49. The minimum number in the group is three; if we divide all values by that number, we reach the orchestral empirical formula of 8 violinists: 6 brasses: 2 cellos: 1 percussionist.

51. We start the problem by assuming 100.0 g of compound, so there will be 85.7 g of carbon and 14.3 g of hydrogen. Next, we convert these masses to moles:

$$85.7\,\text{g C}\times\frac{1\,\text{mol C}}{12.01\,\text{g C}} = 7.14\,\text{mol C} \qquad 14.3\,\text{g H}\times\frac{1\,\text{mol H}}{1.008\,\text{g H}} = 14.2\,\text{mol H}$$

Dividing by the smallest number of moles (7.1), we get ratios:

$$\frac{7.14\,\text{mol C}}{7.14\,\text{mol C}} = 1\,\text{C} \qquad \frac{14.2\,\text{mol H}}{7.14\,\text{mol C}} = 2\,\text{H}$$

So the empirical formula is CH_2.

53. We start the problem by assuming 100.0 g of compound, so there will be 57.1 g of carbon, 4.76 g of hydrogen, and 38.1 g of oxygen. Next, we convert these masses to moles:

$$57.1\,\text{g C}\times\frac{1\,\text{mol C}}{12.01\,\text{g C}} = 4.75\,\text{mol C},\ 4.76\,\text{g H}\times\frac{1\,\text{mol H}}{1.008\,\text{g H}} = 4.72\,\text{mol H},$$

$$38.1\,\text{g O}\times\frac{1\,\text{mol O}}{16.00\,\text{g O}} = 2.38\,\text{mol O}$$

Dividing by the smallest number of moles (2.38), we get ratios:

$$\frac{4.75\,\text{mol C}}{2.38\,\text{mol O}} = 2\,\text{C} \qquad \frac{4.72\,\text{mol H}}{2.38\,\text{mol O}} = 2\,\text{H} \qquad \frac{2.38\,\text{mol O}}{2.38\,\text{mol O}} = 1\,\text{O}$$

So the empirical formula is C_2H_2O, which has an empirical molar mass of 42. The only compound molar mass exactly divisible by 42 is the molar mass of 84. It is the only possible molar mass consistent with the data.

55. We start the problem by assuming 100.0 g of compound, so there will be 19.6 g of potassium, 54.3 g of silver, 12.1 g of carbon, and 14.7 g of nitrogen. Next, we convert these masses to moles:

$$19.6\,\text{g K}\times\frac{1\,\text{mol K}}{39.10\,\text{g K}} = 0.501\,\text{mol K},\ 54.3\,\text{g Ag}\times\frac{1\,\text{mol Ag}}{107.87\,\text{g Ag}} = 0.503\,\text{mol Ag},$$

$$12.1\,\text{g C}\times\frac{1\,\text{mol C}}{12.01\,\text{g C}} = 1.01\,\text{mol C},\ 14.7\,\text{g N}\times\frac{1\,\text{mol N}}{14.01\,\text{g N}} = 1.05\,\text{mol N}$$

Dividing by the smallest number of moles (0.501), we get ratios:

$$\frac{0.501\,\text{mol K}}{0.501\,\text{mol K}} = 1\,\text{K} \qquad \frac{0.503\,\text{mol Ag}}{0.501\,\text{mol K}} = 1\,\text{Ag} \qquad \frac{1.01\,\text{mol C}}{0.501\,\text{mol K}} = 2\,\text{C} \qquad \frac{1.05\,\text{mol N}}{0.501\,\text{mol K}} = 2\,\text{N}$$

So the empirical formula is $KAgC_2N_2$.

57. We start the problem by assuming 100.0 g of compound, so there will be 77.1 g of carbon, 11.4 g of hydrogen, and 11.4 g of oxygen. Next, we convert these masses to moles:

$$77.1\,\text{g C} \times \frac{1\,\text{mol C}}{12.01\,\text{g C}} = 6.42\,\text{mol C},\ 11.4\,\text{g H} \times \frac{1\,\text{mol H}}{1.008\,\text{g H}} = 11.3\,\text{mol H},$$

$$11.4\,\text{g O} \times \frac{1\,\text{mol O}}{16.00\,\text{g O}} = 0.712\,\text{mol O}$$

Dividing by the smallest number of moles (0.712), we get ratios:

$$\frac{6.42\,\text{mol C}}{0.712\,\text{mol O}} = 9\,\text{C} \qquad \frac{11.3\,\text{mol H}}{0.712\,\text{mol O}} = 16\,\text{H} \qquad \frac{0.712\,\text{mol O}}{0.712\,\text{mol O}} = 1\,\text{O}$$

So the empirical formula is $C_9H_{16}O$, which has an empirical molar mass of 140.2 g/mol. Dividing the empirical mass into the molar mass, we find that there are 280 / 140 = 2 empirical units in the molecular formula. The formula for linoleic acid is $C_{18}H_{32}O_2$.

59. a. $2\ N_2H_4 + 1\ N_2O_4 \rightarrow 3\ N_2 + 4\ H_2O$

b. $1\ Pb(C_2H_3O_2)_2 + 2\ KI \rightarrow 1\ PbI_2 + 2\ KC_2H_3O_2$

c. $1\ PCl_5 + 4\ H_2O \rightarrow 1\ H_3PO_4 + 5\ HCl$

d. $1\ Ba_3N_2 + 6\ H_2O \rightarrow 3\ Ba(OH)_2 + 2\ NH_3$

61. $3\ CaCl_2 + 2\ Na_3PO_4 \rightarrow Ca_3(PO_4)_2 + 6\ NaCl$

63. $TiCl_4 + 2\ H_2O \rightarrow TiO_2 + 4\ HCl$

65. a. $2\ C_6H_5Cl + 1\ C_2HOCl_3 \rightarrow 1\ C_{14}H_9Cl_5 + 1\ H_2O$

b. $$45.0\,\text{g}\ C_6H_5Cl \times \frac{1\,\text{mol}\ C_6H_5Cl}{112.55\,\text{g}\ C_6H_5Cl} \times \frac{1\,\text{mol}\ C_{14}H_9Cl_5}{2\,\text{mol}\ C_6H_5Cl} \times \frac{354.5\,\text{g}\ C_{14}H_9Cl_5}{1\,\text{mol}\ C_{14}H_9Cl_5}$$

$$= 70.9\,\text{g}\ C_{14}H_9Cl_5$$

67. The balanced equation is:

$$3\ NaHCO_3\ (aq) + 1\ H_3C_6H_5O_7\ (aq) \rightarrow 3\ CO_2\ (g) + Na_3C_6H_5O_7\ (aq) + 3\ H_2O$$

$$0.487\,\text{g}\ H_3C_6H_5O_7 \times \frac{1\,\text{mol}\ H_3C_6H_5O_7}{192.12\,\text{g}\ H_3C_6H_5O_7} \times \frac{3\,\text{mol}\ NaHCO_3}{1\,\text{mol}\ H_3C_6H_5O_7} \times \frac{84.01\,\text{g}\ NaHCO_3}{1\,\text{mol}\ NaHCO_3}$$

$$= 0.639\,\text{g}\ NaHCO_3$$

69. a. $4\ NH_3 + 5\ O_2 \rightarrow 4\ NO + 6\ H_2O$

b. $$34.0\,\text{g}\ NH_3 \times \frac{1\,\text{mol}\ NH_3}{17.03\,\text{g}\ NH_3} \times \frac{4\,\text{mol NO}}{4\,\text{mol}\ NH_3} \times \frac{30.01\,\text{g NO}}{1\,\text{mol NO}} = 59.9\,\text{g NO formed}$$

c. $34.0\,g\,NH_3 \times \frac{1\,mol\,NH_3}{17.03\,g\,NH_3} \times \frac{6\,mol\,H_2O}{4\,mol\,NH_3} \times \frac{18.02\,g\,H_2O}{1\,mol\,H_2O} = 54.0\,g\,H_2O$ formed

d. $59.9\,g\,NO\,(theor) \times \frac{35.0\,g\,NO\,(actual)}{100.0\,g\,NO\,(theor.)} = 21.0\,g\,NO\,(actual)$ formed

71. a. $3\,Ca(OH)_2 + 2\,H_3PO_4 \rightarrow 1\,Ca_3(PO_4)_2 + 6\,H_2O$

b. $$0.567\,g\,Ca_3(PO_4)_2 \times \frac{1\,mol\,Ca_3(PO_4)_2}{310.2\,g\,Ca_3(PO_4)_2} \times \frac{3\,mol\,Ca(OH)_2}{1\,mol\,Ca_3(PO_4)_2} \times \frac{74.10\,g\,Ca(OH)_2}{1\,mol\,Ca(OH)_2}$$
$$= 0.406\,g\,Ca(OH)_2 \text{ required}$$

c. $$0.567\,g\,Ca_3(PO_4)_2 \times \frac{1\,mol\,Ca_3(PO_4)_2}{310.2\,g\,Ca_3(PO_4)_2} \times \frac{2\,mol\,H_3PO_4}{1\,mol\,Ca_3(PO_4)_2} \times \frac{97.99\,g\,H_3PO_4}{1\,mol\,H_3PO_4}$$
$$= 0.358\,g\,H_3PO_4 \text{ required}$$

73. We can calculate the maximum number of sandwiches from each ingredient to determine the limiting ingredient:

$$200 \text{ pieces bread} \times \frac{1 \text{ sandwich}}{2 \text{ pieces bread}} = 100 \text{ sandwiches}$$

$$200 \text{ meat patties} \times \frac{1 \text{ sandwich}}{1 \text{ meat patty}} = 200 \text{ sandwiches}$$

$$200 \text{ cheese slices} \times \frac{1 \text{ sandwich}}{2 \text{ cheese slices}} = 100 \text{ sandwiches}$$

$$200 \text{ pickle slices} \times \frac{1 \text{ sandwich}}{4 \text{ pickle slices}} = 50 \text{ sandwiches}$$

The maximum number of sandwiches that can be prepared is 50, which is limited by the number of pickle slices available.

75. a. $125\,g\,fat \times \frac{1\,mol\,fat}{891.5\,g\,fat} \times \frac{3\,mol\,soap}{1\,mol\,fat} \times \frac{306.4\,g\,soap}{1\,mol\,soap} = 129\,g\,soap\,(theor.)$

b. $\frac{31\,g\,soap\,(actual)}{129\,g\,soap\,(theor.)} \times 100\% = 24\%$ yield of soap

77. This is a limiting reagent problem. To solve for the maximum yield, we can solve for the yield separately for each reagent.

$$0.155\,mol\,HCl \times \frac{1\,mol\,NaCl}{1\,mol\,HCl} \times \frac{58.44\,g\,NaCl}{1\,mol\,NaCl} = 9.06\,g\,NaCl$$

$$4.55\,g\,NaOH \times \frac{1\,mol\,NaOH}{40.00\,g\,NaOH} \times \frac{1\,mol\,NaCl}{1\,mol\,NaOH} \times \frac{58.44\,g\,NaCl}{1\,mol\,NaCl} = 6.65\,g\,NaCl$$

NaOH is the limiting reagent in this reaction; the maximum yield of NaCl is 6.65 g.

79. This is a limiting reagent problem. To solve for the maximum yield, we can solve for the yield separately for each reagent.

$$10.0\,g\,KHC_4H_4O_6 \times \frac{1\,mol\,KHC_4H_4O_6}{188.18\,g\,KHC_4H_4O_6} \times \frac{1\,mol\,NaKC_4H_4O_6}{1\,mol\,KHC_4H_4O_6} \times \frac{210.16\,g\,NaKC_4H_4O_6}{1\,mol\,NaKC_4H_4O_6}$$
$$= 11.2\,g\,NaKC_4H_4O_6$$

$$10.0\,g\,NaHCO_3 \times \frac{1\,mol\,NaHCO_3}{84.01\,g\,NaHCO_3} \times \frac{1\,mol\,NaKC_4H_4O_6}{1\,mol\,NaHCO_3} \times \frac{210.16\,g\,NaKC_4H_4O_6}{1\,mol\,NaKC_4H_4O_6}$$
$$= 25.0\,g\,NaKC_4H_4O_6$$

Since 10.0 g $KHC_4H_4O_6$ yields less $NaKC_4H_4O_6$, it is the limiting reagent and the maximum yield is 11.2 g $NaKC_4H_4O_6$.

81. We need to calculate first the mass of ethanol converted and then the percentage converted.

$$4.00\,g\,C_2H_4O_2 \times \frac{1\,mol\,C_2H_4O_2}{60.05\,g\,C_2H_4O_2} \times \frac{1\,mol\,C_2H_5OH}{1\,mol\,C_2H_4O_2} \times \frac{46.07\,g\,C_2H_5OH}{1\,mol\,C_2H_5OH}$$
$$= 3.07\,g\,C_2H_5OH\,\text{oxidized}$$

$$\frac{3.07\,g\,C_2H_5OH\,\text{oxidized}}{12.0\,g\,C_2H_5OH} \times 100\% = 25.6\%\,C_2H_5OH\,\text{oxidized}$$

83. The balanced equation is 4 Fe(s) + 3 O_2(g) → 2 Fe_2O_3(s). The reactants used are:

$$454\,g\,Fe_2O_3 \times \frac{1\,mol\,Fe_2O_3}{159.7\,g\,Fe_2O_3} \times \frac{3\,mol\,O_2}{2\,mol\,Fe_2O_3} \times \frac{32.00\,g\,O_2}{1\,mol\,O_2} = 136\,g\,O_2$$

$$454\,g\,Fe_2O_3 \times \frac{1\,mol\,Fe_2O_3}{159.7\,g\,Fe_2O_3} \times \frac{4\,mol\,Fe}{2\,mol\,Fe_2O_3} \times \frac{55.85\,g\,Fe}{1\,mol\,Fe} = 318\,g\,Fe$$

85. The balanced reaction is 1 $C_6H_{12}O_6$ → 2 C_2H_5OH + 2 CO_2. The number of moles of CO_2 is

$$25.0\,g\,C_6H_{12}O_6 \times \frac{1\,mol\,C_6H_{12}O_6}{180.16\,g\,C_6H_{12}O_6} \times \frac{2\,mol\,CO_2}{1\,mol\,C_6H_{12}O_6} = 0.278\,mol\,CO_2$$

87. The effect of a chemical on the body often depends critically on the amount administered, with one dose having a beneficial effect and another dose having a detrimental effect. In the case of vitamin C, there is strong evidence that small amounts are crucial for a healthy life; however, large doses, though promoted by some, may possibly be harmful.

89. The atomic masses given in the periodic table can be thought of in two ways: First, atomic mass can be thought of as the total mass of 1 mol (6.022×10^{23} atoms) of a substance (with all isotopes present). Second, it can be thought of as the weighted average mass of all that element's isotopes. In either case, a single atom's mass is not the same as the average mass (unless there is only one possible isotope of that element).

91. To complete this problem, the number of atoms need not be calculated; we just need to realize that having equivalent numbers of moles is the same as having equivalent numbers of atoms.

$$10.0\,\text{g Ne} \times \frac{1\,\text{mol Ne}}{20.18\,\text{g Ne}} \times \frac{1\,\text{mol Ar}}{1\,\text{mol Ne}} \times \frac{39.95\,\text{g Ar}}{1\,\text{mol Ar}} = 19.8\,\text{g Ar}$$

93. The percentage of iron in hemoglobin can be used as a conversion factor: 0.373 g Fe : 100 g hemoglobin.

$$\frac{55.85\,\text{g Fe}}{1\,\text{mol Fe}} \times \frac{100.0\,\text{g hemoglobin}}{0.373\,\text{g Fe}} \times \frac{4\,\text{mol Fe}}{1\,\text{mol hemoglobin}} = 5.99 \times 10^4\,\text{g / mol hemoglobin}$$

95. When a hydrocarbon is combusted in oxygen, all of the carbon ends up as CO_2 and all of the hydrogen as H_2O. We can calculate the moles of carbon and hydrogen in these samples and then use those data to calculate the empirical formula. We can then use the number of moles to get the mass of each to calculate the mass percentages.

$$3.05\,\text{g CO}_2 \times \frac{1\,\text{mol CO}_2}{44.01\,\text{g CO}_2} \times \frac{1\,\text{mol C}}{1\,\text{mol CO}_2} = 0.0693\,\text{mol C}$$

$$1.50\,\text{g H}_2\text{O} \times \frac{1\,\text{mol H}_2\text{O}}{18.02\,\text{g H}_2\text{O}} \times \frac{2\,\text{mol H}}{1\,\text{mol H}_2\text{O}} = 0.166_5\,\text{mol H}$$

Taking the ratios of the moles to the lesser of the two:

$$\frac{0.166\,\text{mol H}}{0.0693\,\text{mol C}} = 2.40\,\text{H} \qquad \frac{0.0693\,\text{mol C}}{0.0693\,\text{mol C}} = 1\,\text{C}$$

We multiply each of these by 5 to get whole numbers, and the empirical formula is C_5H_{12}. We can now multiply the moles of each by its molar mass:

$$0.0693\,\text{mol C} \times \frac{12.01\,\text{g C}}{1\,\text{mol C}} = 0.832\,\text{g C} \qquad 0.166_5\,\text{mol H} \times \frac{1.008\,\text{g H}}{1\,\text{mol H}} = 0.168\,\text{g H}$$

Out of 1.00 g of sample, these masses represent 83.2% C and 16.8% H.

97. Since different compounds will contain differing numbers of atoms of each element, there is no reason why the coefficients should add up to equal numbers on both sides of a reaction. It is important that there be the same number of atoms of each element on both sides of the equation.

98. a. We start the problem by assuming 100.0 g of compound, so there will be 90.51 g of carbon and 9.49 g of hydrogen. We next convert these masses into moles:

$$90.51\,\text{g C} \times \frac{1\,\text{mol C}}{12.01\,\text{g C}} = 7.536\,\text{mol C} \qquad 9.49\,\text{g H} \times \frac{1\,\text{mol H}}{1.008\,\text{g H}} = 9.415\,\text{mol H}$$

We then reduce the moles of each to the simplest ratio by dividing by the smaller number of moles:

$$\frac{7.536\,\text{mol C}}{7.536\,\text{mol C}} = 1.000\,\text{C} \text{ to each } \frac{9.415\,\text{mol H}}{7.536\,\text{mol C}} = 1.249\,\text{H}$$

To get whole numbers, we multiply each by 4 and get an empirical formula of C_4H_5.

b. The empirical molar mass is 4(12.01) + 5(1.008) = 53.08 g/mol. The number of empirical units in the formula is

$$106.17\,\text{g / mol formula} \times \frac{1\,\text{empirical unit}}{53.08\,\text{g / mol}} = 2.00\,\text{empirical units / formula}$$

Therefore, the molecular formula is C_8H_{10}.

c. $2\ C_8H_{10} + 13\ O_2 \rightarrow 16\ CO + 10\ H_2O$

d. $$12.5\,g\,C_8H_{10} \times \frac{1\,mol\,C_8H_{10}}{106.17\,g\,C_8H_{10}} \times \frac{13\,mol\,O_2}{2\,mol\,C_8H_{10}} \times \frac{32.00\,g\,O_2}{1\,mol\,O_2} = 24.5\,g\,O_2$$

e. To solve this limiting reagent problem, we can calculate the maximum mass of CO that would be produced by each reagent:

$$45.8\,mL\,C_8H_{10} \times \frac{0.8787\,g\,C_8H_{10}}{1\,mL\,C_8H_{10}} \times \frac{1\,mol\,C_8H_{10}}{106.17\,g\,C_8H_{10}} \times \frac{16\,mol\,CO}{2\,mol\,C_8H_{10}} \times \frac{28.01\,g\,CO}{1\,mol\,CO}$$
$$= 84.9\,g\,CO$$

$$31.0\,g\,O_2 \times \frac{1\,mol\,O_2}{32.00\,g\,O_2} \times \frac{16\,mol\,CO}{13\,mol\,O_2} \times \frac{28.01\,g\,CO}{1\,mol\,CO} = 33.4\,g\,CO$$

So O_2 is the limiting reagent, and the maximum CO produced is 33.4 g.

f. The percentage yield is

$$\frac{1.40\,g\,CO\,(actual)}{33.4\,g\,CO\,(theor.)} \times 100\% = 4.19\%\ \text{of the theoretical yield of CO}$$

Chapter 4: Solution Stoichiometry and Types of Reactions

The Bottom Line

- Water is an extremely versatile solvent, partly thanks to its polarity, which is due to the uneven distribution of electrons in the molecule. (Section 4.1)
- When ionic compounds dissolve in water, the ions dissociate and become surrounded by water molecules—a process known as hydration. (Section 4.1)
- The concentration of a solution can be expressed in units known as molarity. This term indicates how many moles of the chemical concerned would be present if we had one liter of the solution. (Section 4.2)

$$\text{Molarity} = \frac{\text{moles of solute}}{\text{liter of solution}} = M$$

- Chemicals present at very low levels are often measured in terms of parts per million (ppm), parts per billion (ppb), or parts per trillion (ppt). (Section 4.2)
- The quantitative analysis of the chemicals present in solutions is of great importance in medicine, industry, and environmental science. (Sections 4.3, 4.8)
- Precipitation reactions involve an insoluble precipitate forming when soluble chemical species combine. (Section 4.5)
- Acid–base reactions involve acids and bases reacting in ways that can neutralize the acidic and basic character of each. (Section 4.6)
- Oxidation–reduction reactions are electron transfer processes in which some reactants lose electrons (are oxidized) while others gain electrons (are reduced). (Section 4.7)

Solutions to Practice Problems

4.1. Molarity is expressed in units of mol solute / liter solution.

1. We can solve for the molarity by converting the units of g K^+ / L solution into mol K^+ / L solution using dimensional analysis.

$$\frac{6.5\,\text{g K}^+}{1\,\text{L}} \times \frac{1\,\text{mol}}{39.10\,\text{g K}^+} = 0.17\,\text{mol/L K}^+$$

2. We can solve for the molarity here by converting the mg Na^+ first into grams, then into moles and converting the ounces into milliliters, then into liters.

$$\frac{110.\,\text{mg Na}^+}{8\,\text{oz}} \times \frac{1\,\text{g}}{1000\,\text{mg}} \times \frac{1\,\text{mol Na}^+}{22.99\,\text{g Na}^+} \times \frac{1\,\text{oz}}{29.6\,\text{mL}} \times \frac{1000\,\text{mL}}{1\,\text{L}} = 0.0202\,\text{mol/L Na}^+$$

3. To solve for the mass of ethanol, we use both the volume and the concentration to get the moles of ethanol, and then we convert to grams of ethanol using the molar mass.

$$600.0\,\text{mL soln} \times \frac{1\,\text{L soln}}{1000\,\text{mL soln}} \times \frac{1.200\,\text{mol C}_2\text{H}_5\text{OH}}{1\,\text{L soln}} \times \frac{46.07\,\text{g C}_2\text{H}_5\text{OH}}{1\,\text{mol C}_2\text{H}_5\text{OH}} = 33.17\,\text{g C}_2\text{H}_5\text{OH}$$

4.2. We use our molarity as a conversion constant in the calculation to find the number of moles in a solution:

$$5.00\,\text{L soln} \times \frac{0.77\,\text{mol Na}_3\text{PO}_4}{1\,\text{L soln}} = 3.85\,\text{mol Na}_3\text{PO}_4 = 3.8\,\text{mol Na}_3\text{PO}_4$$

We now can use a stoichiometric ratio to obtain the moles of Na^+ from the moles of Na_3PO_4:

$$3.85\,\text{mol}\,Na_3PO_4 \times \frac{3\,\text{mol}\,Na^+}{1\,\text{mol}\,Na_3PO_4} = 11.55\,\text{mol}\,Na^+ = 12\,\text{mol}\,Na^+$$

4.3. To convert from moles of a solute to a volume of a solution, we again use the molarity as a conversion constant; however, in this case, the ratio is inverted in the calculation (the volume will be in the numerator and the moles in the denominator.)

1. $4.76\,\text{mol}\,CuSO_4 \times \frac{1\,\text{L soln}}{3.40\,\text{mol}\,CuSO_4} = 1.40\,\text{L soln}$

2. $5.5\,\text{mol}\,Ca(NO_3)_2 \times \frac{1\,\text{L soln}}{2.25\,\text{mol}\,Ca(NO_3)_2} = 2.4\,\text{L soln}$

To complete the second question, we need to include a stoichiometric ratio (of nitrate to calcium nitrate.)

$$5.5\,\text{mol}\,NO_3^- \times \frac{1\,\text{mol}\,Ca(NO_3)_2}{2\,\text{mol}\,NO_3^-} \times \frac{1\,\text{L soln}}{2.25\,\text{mol}\,Ca(NO_3)_2} = 1.2\,\text{L soln}$$

4.4. Parts per million is equivalent to mg solute / L solution. We can then convert the mg solute to moles to get the concentration.

$$\frac{0.200\,\text{mg}\,CN^-}{1\,\text{L soln}} \times \frac{1\,\text{g}}{1000\,\text{mg}} \times \frac{1\,\text{mol}\,CN^-}{26.02\,\text{g}\,CN^-} = 7.69 \times 10^{-6}\,\text{mol/L}\,CN^-$$

4.5. When using molarities, we can use the equation $C_{init} \times V_{init} = C_{final} \times V_{final}$ to complete the calculation:

$$V_{init} = \frac{C_{final} \times V_{final}}{C_{init}} = \frac{2.00\,\text{L} \times 0.15\,\text{mol/L}}{16\,\text{mol/L}} = 0.019\,\text{L of con}\,HNO_3$$

$$0.019\,\text{L of con}\,HNO_3 \times \frac{1000\,\text{mL}}{1\,\text{L}} = 19\,\text{mL of con}\,HNO_3$$

4.6. To calculate the concentration of the sulfuric acid, we will use the molarity and volume of the sodium hydroxide and the stoichiometric ratio from the balanced equation to find the moles of sulfuric acid. We can then use the moles and volume of the sulfuric acid to find its concentration.

$$22.25\,\text{mL NaOH soln} \times \frac{1\,\text{L}}{1000\,\text{mL}} \times \frac{0.100\,\text{mol NaOH}}{1\,\text{L NaOH soln}} \times \frac{1\,\text{mol}\,H_2SO_4}{2\,\text{mol NaOH}} = 0.00111\,\text{mol}\,H_2SO_4$$

$$\frac{0.00111\,\text{mol}\,H_2SO_4}{25.00\,\text{mL}\,H_2SO_4\,\text{solution}} \times \frac{1000\,\text{mL}\,H_2SO_4\,\text{solution}}{1\,\text{L}\,H_2SO_4\,\text{solution}} = 0.0445\,\text{mol/L}\,H_2SO_4 = 0.0445\,\text{M}\,H_2SO_4$$

4.7. The molecular equation is

$$K_2C_2O_{4(aq)} + 2\,HNO_{3(aq)} \rightarrow H_2C_2O_{4(aq)} + 2\,KNO_{3(aq)}$$

For the complete ionic equation, we split all the ionic compounds ($H_2C_2O_{4(aq)}$ is a nonelectrolyte and doesn't split):

$$2K^+_{(aq)} + C_2O_4^{2-}{}_{(aq)} + 2\,H^+_{(aq)} + 2\,NO_3^-{}_{(aq)} \rightarrow H_2C_2O_{4(aq)} + 2\,K^+_{(aq)} + 2\,NO_3^-{}_{(aq)}$$

To get the net ionic equation, all the spectator ions (those that appear unchanged on both sides of the equation) are removed:

$$C_2O_4^{2-}{}_{(aq)} + 2\,H^+_{(aq)} \rightarrow H_2C_2O_{4(aq)}$$

4.8. Splitting the ions from the salts in the given solutions, we have $Ag^{+}_{(aq)}$, $NO_3^{-}{}_{(aq)}$, $Na^{+}_{(aq)}$, $Cl^{-}_{(aq)}$, $S^{2-}_{(aq)}$, $Zn^{2+}_{(aq)}$, $SO_4^{2-}{}_{(aq)}$. Consulting the solubility rules, we find that the following combinations are the only ones that are insoluble: AgCl, Ag_2S, Ag_2SO_4, and ZnS. Therefore, the following combinations of solutions will produce a precipitate:

$AgNO_3 + NaCl$ forms $AgCl_{(s)}$; $AgNO_3 + Na_2S$ forms $Ag_2S_{(s)}$
$AgNO_3 + ZnSO_4$ forms $Ag_2SO_{4(aq)}$ $Na_2S + ZnSO_4$ forms $ZnS_{(s)}$

4.9. To tackle the problem, we first will write the balanced equation and then determine the limiting reagent to find the maximum amount of precipitate formed. The balanced equation is

$$Na_2CO_{3(aq)} + BaCl_{2(aq)} \rightarrow BaCO_{3(s)} + 2\ NaCl_{(aq)}$$

To determine the limiting reagent, we can calculate how much barium carbonate is formed from each reagent, assuming the second reagent is in excess:

$$0.35\,\text{L Na}_2\text{CO}_3\text{ soln} \times \frac{0.25\,\text{mol Na}_2\text{CO}_3}{1\,\text{L Na}_2\text{CO}_3\text{ soln}} \times \frac{1\,\text{mol BaCO}_3}{1\,\text{mol Na}_2\text{CO}_3} \times \frac{197.34\,\text{g BaCO}_3}{1\,\text{mol BaCO}_3}$$
$$= 17\,\text{g BaCO}_3$$

$$0.55\,\text{L BaCl}_2\text{ soln} \times \frac{0.35\,\text{mol BaCl}_2}{1\,\text{L BaCl}_2\text{ soln}} \times \frac{1\,\text{mol BaCO}_3}{1\,\text{mol BaCl}_2} \times \frac{197.34\,\text{g BaCO}_3}{1\,\text{mol BaCO}_3}$$
$$= 38\,\text{g BaCO}_3$$

The sodium carbonate is the limiting reagent; therefore, the amount of $BaCO_3$ formed by mixing the solutions is 17 g.

4.10. To finish the problem, we need to write the balanced equation and the complete a calculation that will use the molarities of both solutions along with a stoichiometric ratio of the reagents in the dimensional analysis. The balanced reaction is

$$H_3PO_4 + 3\ NaOH \rightarrow 3\ H_2O + Na_3PO_4$$

$$50.0\,\text{mL NaOH soln} \times \frac{1\,\text{L}}{1000\,\text{mL}} \times \frac{0.250\,\text{mol NaOH}}{1\,\text{L NaOH soln}} \times \frac{1\,\text{mol H}_3\text{PO}_4}{3\,\text{mol NaOH}} \times \frac{1\,\text{L H}_3\text{PO}_4\text{ soln}}{0.550\,\text{mol H}_3\text{PO}_4}$$

$$= 0.00758\,\text{L H}_3\text{PO}_4\text{ soln} \times \frac{1000\,\text{mL}}{1\,\text{L}} = 7.58\,\text{mL H}_3\text{PO}_4\text{ soln}$$

4.11.

- KCl is a neutral compound. K has an oxidation number of +1, so Cl has an oxidation number of –1.
- Fe_2O_3 is a neutral compound. O has an oxidation number of –2 per atom, so each Fe has an oxidation number of +3 and that all the oxidation numbers sum to zero.
- P_4 is an element, so the oxidation number is 0.
- CH_2Cl_2 is a neutral compound. H will have an oxidation number of +1 per atom, and Cl will have an oxidation number of –1 per atom, so C has an oxidation number of 0.
- Al is an element, so the oxidation number is 0.
- PBr_3 is a neutral compound. Br will have an oxidation number of –1 per atom, so P must have an oxidation number of +3 so that all the oxidation numbers sum to 0.
- HCN is a neutral compound. H will have an oxidation number of +1, and N will have an oxidation number of –3, so carbon has an oxidation number of +2 so that all the oxidation numbers sum to 0.

4.12. This is not a redox equation because none of the oxidation numbers change.

Solutions to Student Problems

1. Because the water molecule contains both partial positive charges (on the hydrogens) and partial negative charges (on the oxygen), it can interact favorably with both cations and anions.

3. The hydration sphere is the cage of water molecules that surrounds a charged particle as it dissolves in water.

5. Water tends to dissolve those compounds that have some type of charge on the molecule or ion; however, oil molecules have very little, if any, charges on the molecule that water can attract. Oil, then, doesn't dissolve because it cannot interact favorably with water.

7. When some compounds dissolve, they form anions and cations in the water. Even though the particles formed a neutral compound before dissolving, these ions exist separately in the water and are free to move. Because the current requires freely moving charges, the new ions in the water can carry the current.

9. A conductivity apparatus like that shown in the chapter will have the brightness of the bulb somewhat related to the amount of ions present in the solution. A certain minimum current will be needed in order for us to see the bulb get brighter. Even though a solution of a very weak electrolyte will be conducting some current, that current may not be enough to visibly brighten the bulb. In this case, that apparatus would not be able to distinguish between nonelectrolytes and very weak electrolytes. The bulb will light brightly for strong electrolyte solutions, allowing for ready identification of those solutions.

11. When $MgCl_2$ dissolves, there is a single cation (Mg^{2+}) that would attract the negative end of water (mostly O) and two anions (Cl^-) that would attract the positive end of water (mostly H). Therefore, (c) is the correct answer.

13. $MgCl_2$ dissociates into 3 ions per formula unit, so

$$0.100\text{ mol MgCl}_2 \times \frac{3\text{ mol ions}}{1\text{ mol MgCl}_2} = 0.300\text{ mol ions}$$

15. a. The molar mass of $C_6H_8O_6$ is 176.12 g/mol.

$$\text{Molarity} = \frac{\text{moles solute}}{\text{liters solution}} = \frac{0.150\text{ g Vit C} \times \dfrac{1\text{ mol Vit C}}{176.12\text{ g Vit C}}}{1.50\text{ L}} = 5.68\times10^{-4}\ M\text{ Vit C}$$

b. $$\text{Molarity} = \frac{0.250\text{ g Vit C} \times \dfrac{1\text{ mol Vit C}}{176.12\text{ g Vit C}}}{0.500\text{ L}} = 2.84\times10^{-3}\ M\text{ Vit C}$$

c. $$\text{Molarity} = \frac{3.50\text{ g Vit C} \times \dfrac{1\text{ mol Vit C}}{176.12\text{ g Vit C}}}{2.0\text{ L}} = 9.9\times10^{-3}\ M\text{ Vit C}$$

17. a. From the volume and concentration, we get the moles of glycine. We can then convert moles into mass to finish the calculation.

$$100.0\,\text{mL glycine soln} \times \frac{1\,\text{L}}{1000\,\text{mL}} \times \frac{0.015\,\text{mol glycine}}{1\,\text{L glycine soln}} \times \frac{75.07\,\text{g glycine}}{1\,\text{mol glycine}} = 0.11\,\text{g glycine}$$

b. $$125.0\,\text{mL glycine soln} \times \frac{1\,\text{L}}{1000\,\text{mL}} \times \frac{0.0145\,\text{mol glycine}}{1\,\text{L glycine soln}} \times \frac{75.07\,\text{g glycine}}{1\,\text{mol glycine}}$$
$$= 0.136\,\text{g glycine}$$

c. $$74.6\,\text{mL glycine soln} \times \frac{1\,\text{L}}{1000\,\text{mL}} \times \frac{1.44\,\text{mol glycine}}{1\,\text{L glycine soln}} \times \frac{75.07\,\text{g glycine}}{1\,\text{mol glycine}} = 8.06\,\text{g glycine}$$

19. In parts a and b, we need to convert the mass into moles, then use a stoichiometric ratio to get the appropriate moles, and finish by dividing by the volume to get the molarity. In part c, we start with the assumption of 1 kg water per L water.

a. $$\frac{24.55\,\text{g CaCl}_2 \times \dfrac{1\,\text{mol CaCl}_2}{110.98\,\text{g CaCl}_2} \times \dfrac{1\,\text{mol Ca}^+}{1\,\text{mol CaCl}_2}}{1.00\,\text{L soln}} = 0.221\,M\ \text{Ca}^+$$

b. $$\frac{24.55\,\text{g CaCl}_2 \times \dfrac{1\,\text{mol CaCl}_2}{110.98\,\text{g CaCl}_2} \times \dfrac{2\,\text{mol Cl}^-}{1\,\text{mol CaCl}_2}}{1.00\,\text{L soln}} = 0.442\,M\ \text{Cl}^-$$

c. $$\frac{1.00\,\text{kg H}_2\text{O}}{1.00\,\text{L H}_2\text{O}} \times \frac{1000\,\text{g}}{1\,\text{kg}} \times \frac{1\,\text{mol H}_2\text{O}}{18.02\,\text{g H}_2\text{O}} = 55.5\,M\ \text{H}_2\text{O}$$

21. We use the molarity to create a conversion constant for the calculation: 1 L seawater = 0.00081 mol Br^-.

$$1\,\text{mol Br}^- \times \frac{1\,\text{L seawater}}{0.00081\,\text{mol Br}^-} = 1.2 \times 10^3\ \text{L seawater}$$

23. In each case, we need to get the moles of $MgCl_2$ using stoichiometric ratios (if needed); then we can use the molarity to convert to a volume.

a. $$0.10\,\text{mol MgCl}_2 \times \frac{1\,\text{L soln}}{0.125\,\text{mol MgCl}_2} = 0.80\,\text{L soln}$$

b. $$0.10\,\text{mol Mg}^{2+} \times \frac{1\,\text{mol MgCl}_2}{1\,\text{mol Mg}^{2+}} \times \frac{1\,\text{L soln}}{0.125\,\text{mol MgCl}_2} = 0.80\,\text{L soln}$$

c. $$2.33\,\text{mol Cl}^- \times \frac{1\,\text{mol MgCl}_2}{2\,\text{mol Cl}^-} \times \frac{1\,\text{L soln}}{0.125\,\text{mol MgCl}_2} = 9.32\,\text{L soln}$$

25. A concentration of 1 mg/L is equivalent to ppm. If we solve for the concentration in mg/L, then we have the value for ppm concentration.

a. 1000.0 mL = 1.0000 L, so the concentration is 2.5 mg/L or 2.5 ppm

b. 500.0 mL = 0.5000 L, so the concentration is $\dfrac{5.25\,\text{mg}}{0.500\,\text{L}} = 10.5\,\text{mg/L} = 10.5\,\text{ppm}$

c. 300.0 mL = 0.3000 L, so the concentration is $\frac{12.5\,\text{mg}}{0.300\,\text{L}} = 41.7\,\text{mg/L} = 41.7\,\text{ppm}$

27. a. Because ppm = mg/L, we need to convert moles NaCl into mg NaCl (The liters are fine!)

$$\frac{0.00012\,\text{mol NaCl}}{1\,\text{L soln}} \times \frac{58.44\,\text{g NaCl}}{1\,\text{mol NaCl}} \times \frac{1000\,\text{mg}}{1\,\text{g}} = \frac{7.01\,\text{mg}}{1\,\text{L soln}} = 7.01\,\text{ppm}$$

b. 1000 ppb = 1 ppm, so 5.33 ppm = 5.33×10^3 pbb.

c. Here we need to realize that

$$\text{parts per million} = \frac{\text{mass solute}}{\text{mass solution}} \times \left(1\times10^6\right)\text{ppm}$$

$$\text{and mass percent} = \frac{\text{mass solute}}{\text{mass solution}} \times 100\%$$

Therefore, the ratio of the solute to solution masses is

$$\frac{\text{mass solute}}{\text{mass solution}} = \frac{(\text{ppm})}{1\times10^6\,\text{ppm}}$$

$$\frac{\text{mass NaOH}}{\text{mass solution}} = \frac{170\,\text{ppm}}{1\times10^6\,\text{ppm}} = 1.70\times10^{-4}$$

We can use this ratio of masses to find the mass percent of NaOH:

$$\text{Mass percent NaOH} = \frac{\text{mass NaOH}}{\text{mass solution}} \times 100\% = 1.70\times10^{-4} \times 100\% = 1.70\times10^{-2}\,\%$$

29. It is possible to answer this question without doing any calculations. The solution in part b has half the mass of KH_2PO_4, but also half the volume. Both concentrations will be the same.

31. We can determine the number of moles of ethylene glycol by multiplying the volume by the concentration. We can then convert the moles into the mass in kilograms.

$$10.0\,\text{L soln} \times \frac{16.0\,\text{mol}\,C_2H_6O_2}{1\,\text{L soln}} \times \frac{62.07\,\text{g}\,C_2H_6O_2}{1\,\text{mol}\,C_2H_6O_2} \times \frac{1\,\text{kg}}{1000\,\text{g}} = 9.93\,\text{kg}\,C_2H_6O_2$$

33. The volume of the cup does not enter explicitly into this calculation. We need only calculate the molarity from the maximum concentration given:

$$\frac{455\,\text{g caffeine}}{1\,\text{L soln}} \times \frac{1\,\text{mol caffeine}}{194.20\,\text{g caffeine}} = 2.34\,M\,\text{caffeine}$$

35. We can use the equivalence of 1 ppb = $1\,\mu\text{g/L}$ to start the problem.

a. $$3.0\,\text{ppb atrazine} = \frac{3.0\,\mu\text{g atrazine}}{1\,\text{L soln}} \times \frac{1\,\text{g}}{10^6\,\mu\text{g}} = \frac{3.0\times10^{-6}\,\text{g atrazine}}{1\,\text{L soln}}$$

b. $$3.0\,\text{ppb atrazine} = \frac{3.0\,\mu\text{g atrazine}}{1\,\text{L soln}} \times \frac{1\,\text{g}}{10^6\,\mu\text{g}} \times \frac{1\,\text{mol atrazine}}{215.69\,\text{g atrazine}} = \frac{1.4\times10^{-8}\,\text{mol atrazine}}{1\,\text{L soln}}$$

$$= 1.4\times10^{-8}\,M\,\text{atrazine}$$

37. The volume of the container is irrelevant to the calculation; you need only the actual volume consumed and the concentration of the solution:

$$50.0\,\text{mL liquid} \times \frac{1\,\text{L}}{1000\,\text{mL}} \times \frac{0.20\,\text{mol NaCl}}{1\,\text{L liquid}} \times \frac{1\,\text{mol Na}^+}{1\,\text{mol NaCl}} = 0.010\,\text{mol Na}^+$$

39. To solve this problem, we start with the volume and concentration of the neutralizing solution to get moles and use a stoichiometric factor to convert between moles of each reagent. Dividing by the volume of the desired solution yields the concentration.

a. $$\frac{35.0\,\text{mL NaOH} \times \frac{1\,\text{L}}{1000\,\text{mL}} \times \frac{0.155\,\text{mol NaOH}}{1\,\text{L NaOH}} \times \frac{1\,\text{mol HCl}}{1\,\text{mol NaOH}}}{25.0\,\text{mL HCl} \times \frac{1\,\text{L}}{1000\,\text{mL}}} = 0.217\,M\,\text{HCl}$$

b. $$\frac{35.0\,\text{mL H}_2\text{SO}_4 \times \frac{1\,\text{L}}{1000\,\text{mL}} \times \frac{0.0200\,\text{mol H}_2\text{SO}_4}{1\,\text{L H}_2\text{SO}_4} \times \frac{1\,\text{mol Sr(OH)}_2}{1\,\text{mol H}_2\text{SO}_4}}{50.0\,\text{mL Sr(OH)}_2 \times \frac{1\,\text{L}}{1000\,\text{mL}}} = 0.0140\,M\,\text{Sr(OH)}_2$$

c. $$\frac{25.0\,\text{mL NaOH} \times \frac{1\,\text{L}}{1000\,\text{mL}} \times \frac{0.35\,\text{mol KOH}}{1\,\text{L KOH}} \times \frac{1\,\text{mol HNO}_3}{1\,\text{mol KOH}}}{40.5\,\text{mL HNO}_3 \times \frac{1\,\text{L}}{1000\,\text{mL}}} = 0.22\,M\,\text{HNO}_3$$

41. To answer the question, we need to determine the number of moles in the sample from the volume and concentration. Using the resultant number of moles and the given mass, we can find the molar mass of the substance and decide its identity.

$$25.0\,\text{mL} \times \frac{1\,\text{L}}{1000\,\text{mL}} \times \frac{0.10\,\text{mol}}{1\,\text{L}} = 0.0025\,\text{mol}$$

Therefore, the molecular weight is 0.336 g / 0.0025 mol = 134 g/mol. CuCl has a molar mass of 99 g/mol, but $CuCl_2$ has a molar mass of 134 g/mol. The label should read 0.10 *M* $CuCl_2$.

43. In words: potassium hydroxide + hydrochloric acid → potassium chloride + water

Molecular equation: $KOH_{(aq)} + HCl_{(aq)} \rightarrow KCl_{(aq)} + H_2O_{(l)}$

Ionic equation: $K^+{}_{(aq)} + OH^-{}_{(aq)} + H^+{}_{(aq)} + Cl^-{}_{(aq)} \rightarrow K^+{}_{(aq)} + Cl^-{}_{(aq)} + H_2O_{(l)}$

Net ionic equation: $OH^-{}_{(aq)} + H^+{}_{(aq)} \rightarrow H_2O_{(l)}$

45. Molecular equation: $2\,NaCl_{(aq)} + Ca(NO_3)_{2(aq)} \rightarrow 2\,NaNO_{3(aq)} + CaCl_{2(aq)}$

Ionic equation:

$$2\,Na^+{}_{(aq)} + 2\,Cl^-{}_{(aq)} + Ca^{2+}{}_{(aq)} + 2\,NO_3^-{}_{(aq)} \rightarrow 2\,Na^+{}_{(aq)} + 2\,NO_3^-{}_{(aq)} + Ca^{2+}{}_{(aq)} + 2\,Cl^-{}_{(aq)}$$

Net ionic equation: NONE (all ions remain in solution)

47. To solve this problem, we start with volume and concentration of the iodine solution to get moles and use a stoichiometric factor to obtain moles of vitamin C. Dividing by the volume of the vitamin C solution yields the concentration.

a. $$\frac{10.5\,\text{mL}\,I_2 \times \frac{1\,\text{L}}{1000\,\text{mL}} \times \frac{0.0855\,\text{mol}\,I_2}{1\,\text{L}\,I_2} \times \frac{1\,\text{mol VitC}}{1\,\text{mol}\,I_2}}{250.0\,\text{mL VitC} \times \frac{1\,\text{L}}{1000\,\text{mL}}} = 3.59\times10^{-3}\,M\text{ VitC}$$

b. $$\frac{12.0\,\text{mL}\,I_2 \times \frac{1\,\text{L}}{1000\,\text{mL}} \times \frac{0.0855\,\text{mol}\,I_2}{1\,\text{L}\,I_2} \times \frac{1\,\text{mol VitC}}{1\,\text{mol}\,I_2}}{300.0\,\text{mL VitC} \times \frac{1\,\text{L}}{1000\,\text{mL}}} = 3.42\times10^{-3}\,M\text{ VitC}$$

Brand A has the higher concentration of vitamin C. Because Brand B required the greater amount of I_2 in the reaction, it has the greater amount of vitamin C.

49. We can calculate the moles of NaOH from the mass of KHP, its molar mass, and the stoichiometric ratio of NaOH to KHP. Dividing the moles by the volume of NaOH gives the concentration.

Trial A: $$\frac{0.5467\,\text{g KHP} \times \frac{1\,\text{mol KHP}}{204.22\,\text{g KHP}} \times \frac{1\,\text{mol NaOH}}{1\,\text{mol KHP}}}{45.12\,\text{mL NaOH} \times \frac{1\,\text{L}}{1000\,\text{mL}}} = 0.05933\,M\text{ NaOH}$$

Trial B: $$\frac{0.5475\,\text{g KHP} \times \frac{1\,\text{mol KHP}}{204.22\,\text{g KHP}} \times \frac{1\,\text{mol NaOH}}{1\,\text{mol KHP}}}{44.89\,\text{mL NaOH} \times \frac{1\,\text{L}}{1000\,\text{mL}}} = 0.05972\,M\text{ NaOH}$$

Trial C: $$\frac{0.5501\,\text{g KHP} \times \frac{1\,\text{mol KHP}}{204.22\,\text{g KHP}} \times \frac{1\,\text{mol NaOH}}{1\,\text{mol KHP}}}{46.50\,\text{mL NaOH} \times \frac{1\,\text{L}}{1000\,\text{mL}}} = 0.05793\,M\text{ NaOH}$$

The average molarity: $$\frac{0.05933\,M + 0.05972\,M + 0.05793\,M}{3} = 0.05899\,M\text{ NaOH}$$

51. The reaction of $HCl + NaOH \rightarrow NaCl + H_2O$ has a net ionic equation of $H^+ + OH^- \rightarrow H_2O$.

$$\frac{45.55\,\text{mL NaOH} \times \frac{1\,\text{L}}{1000\,\text{mL}} \times \frac{0.9876\,\text{mol NaOH}}{1\,\text{L NaOH}} \times \frac{1\,\text{mol HCl}}{1\,\text{mol NaOH}}}{10.00\,\text{mL HCl} \times \frac{1\,\text{L}}{1000\,\text{mL}}} = 4.499\,M\text{ HCl}$$

53. a. The balanced reaction is $CaCO_3 + 2\,HCl \rightarrow CaCl_2 + H_2O + CO_2$.

b. $$5.00\,\text{mL HCl} \times \frac{1\,\text{L}}{1000\,\text{mL}} \times \frac{0.500\,\text{mol HCl}}{1\,\text{L HCl}} \times \frac{1\,\text{mol}\,CaCO_3}{2\,\text{mol HCl}} \times \frac{100.09\,\text{g}\,CaCO_3}{1\,\text{mol}\,CaCO_3}$$
$$= 0.125\,\text{g}\,CaCO_3$$

55. a. $2\ C_4H_{10}(g) + 13\ O_2(g) \rightarrow 8\ CO_2(g) + 10\ H_2O(l)$, redox (combustion)

b. $Ca(OH)_2\ (aq) + 2\ HNO_3(aq) \rightarrow Ca(NO_3)_2(aq) + 2\ H_2O(l)$, acid–base

Net ionic equation: $H^+(aq) + OH^-(aq) \rightarrow H_2O(l)$

c. $Pb(NO_3)_2(aq) + 2\ NaCl(aq) \rightarrow PbCl_2(s) + 2\ NaNO_3(aq)$, precipitation

Net ionic equation: $Pb^{2+}(aq) + 2\ Cl^-(aq) \rightarrow PbCl_2(s)$

57. a. $CuCO_3$ is not soluble. (The only soluble carbonates are ammonium and Group 1A carbonates.)

b. NiS is not soluble. (The only soluble sulfides are Group 1A and ammonium sulfides.)

c. $(NH_4)_2CO_3$ is soluble.

d. KOH is soluble. (It is a Group 1A hydroxide.)

e. Lead acetate is soluble.

59. a. $BaCl_2(aq) + 2\ NaNO_3(aq) \rightarrow Ba(NO_3)_2(aq) + 2\ NaCl(aq)$

Net ionic equation: None, all compounds are soluble.

b. $2\ Fe(NO_3)_3(aq) + 3\ (NH_4)_2SO_4(aq) \rightarrow Fe_2(SO_4)_3(aq) + 6\ NH_4NO_3$

Net ionic equation: None, all compounds are soluble.

c. $CaCl_2(aq) + K_2SO_4(aq) \rightarrow CaSO_4(s) + 2\ KCl(aq)$

Net ionic equation: $Ca^{2+}(aq) + SO_4^{2-}(aq) \rightarrow CaSO_4(s)$

61. a. $Cu(NO_3)_2(aq) + 2\ KOH(aq) \rightarrow Cu(OH)_2(s) + 2\ KNO_3(aq)$

Net ionic equation: $Cu^{2+}(aq) + 2\ OH^-(aq) \rightarrow Cu(OH)_2(s)$

b. $3\ Na_2CO_3(aq) + 2\ AlCl_3(aq) \rightarrow 6\ NaCl(aq) + Al_2(CO_3)_3(s)$

Net ionic equation: $3\ CO_3^{2-}(aq) + 2\ Al^{3+}(aq) \rightarrow Al_2(CO_3)_3(s)$

c. $2\ (NH_4)_3PO_4(aq) + 3\ ZnCl_2(aq) \rightarrow 6\ NH_4Cl(aq) + Zn_3(PO_4)_2(s)$

Net ionic equation: $2\ PO_4^{3-} + 3\ Zn^{2+}(aq) \rightarrow Zn_3(PO_4)_2(s)$

63. Many possible answers exist. Generally, the anion in the insoluble salt would be paired with a Group 1A ion or NH_4^+, while the cation in the insoluble salt would be paired with nitrate, acetate, perchlorate, chlorate, or halides (unless the cation is Pb^{2+}, Ag^+, or Hg_2^{2+}).

a. Barium nitrate plus sodium sulfide would give barium sulfide.

b. Copper(II) nitrate plus sodium hydroxide would give copper(II) hydroxide.

c. Lead(II) nitrate plus sodium sulfate would give lead(II) sulfate.

65. Adding sodium chloride to the sample would precipitate silver chloride, leaving Ba^{2+} and Cu^{2+} in solution. Adding sodium sulfate to the supernatant (the liquid leftover after the precipitation occurred) would precipitate barium sulfate, leaving Cu^{2+} in solution. Last, adding sodium sulfide to the supernatant would precipitate copper(II) sulfide.

67. a. The balanced equation is $AgNO_3(aq) + NaCl(aq) \rightarrow NaNO_3(aq) + AgCl(s)$. The net ionic equation is $Ag^+(aq) + Cl^-(aq) \rightarrow AgCl(s)$.

b.
$$25.0\,\text{mL NaCl soln} \times \frac{1\,\text{L}}{1000\,\text{mL}} \times \frac{0.242\,\text{mol NaCl}}{1\,\text{L NaCl soln}} \times \frac{1\,\text{mol AgCl}}{1\,\text{mol NaCl}} \times \frac{143.32\,\text{g AgCl}}{1\,\text{mol AgCl}} = 0.867\,\text{g AgCl}$$

$$25.0\,\text{mL NaCl soln} \times \frac{1\,\text{L}}{1000\,\text{mL}} \times \frac{0.242\,\text{mol NaCl}}{1\,\text{L NaCl soln}} \times \frac{1\,\text{mol Ag}^+}{1\,\text{mol NaCl}} \times \frac{107.87\,\text{g Ag}^+}{1\,\text{mol Ag}^+} = 0.653\,\text{g Ag}^+$$

69. The balanced molecular equation is $2\,HBr(aq) + Mg(OH)_2(aq) \rightarrow 2\,H_2O(l) + MgBr_2(aq)$. The net ionic equation is $H^+(aq) + OH^-(aq) \rightarrow H_2O(l)$.

71. a. Because molarity = moles solute / volume solution, we find the moles of the base by using the volume and molarity of the acid along with the stoichiometric ratio of the base to the acid. Dividing the moles of the base by the volume of the base gives the concentration.

The reaction is $KOH + HCl \rightarrow H_2O + KCl$.

$$\frac{34.5\,\text{mL HCl soln} \times \frac{1\,\text{L}}{1000\,\text{mL}} \times \frac{0.50\,\text{mol HCl}}{1\,\text{L HCl soln}} \times \frac{1\,\text{mol KOH}}{1\,\text{mol HCl}}}{22.4\,\text{mL KOH solution} \times \frac{1\,\text{L}}{1000\,\text{mL}}} = 0.770\,M\,\text{KOH}$$

b. Here we use the molarity and volume of the KOH solution along with the stoichiometric ratio of acid to base to find the moles of H_2SO_4. We then use the molarity of the acid solution to convert the moles into a volume. The reaction is $2\,KOH + H_2SO_4 \rightarrow 2\,H_2O + K_2SO_4$.

$$22.4\,\text{mL KOH soln} \times \frac{1\,\text{L}}{1000\,\text{mL}} \times \frac{0.770\,\text{mol KOH}}{1\,\text{L KOH soln}} \times \frac{1\,\text{mol H}_2\text{SO}_4}{2\,\text{mol KOH}} \times \frac{1\,\text{L H}_2\text{SO}_4}{0.50\,\text{mol H}_2\text{SO}_4}$$

$$= 0.0172\,\text{L H}_2\text{SO}_4 \times \frac{1000\,\text{mL}}{1\,\text{L}} = 17.2\,\text{mL H}_2\text{SO}_4$$

73. The balanced reaction here is $2\,NaOH + H_2C_2O_4 \rightarrow 2\,H_2O + Na_2C_2O_4$.

$$\frac{0.255\,\text{g H}_2\text{C}_2\text{O}_4 \times \frac{1\,\text{mol H}_2\text{C}_2\text{O}_4}{90.04\,\text{g H}_2\text{C}_2\text{O}_4} \times \frac{2\,\text{mol NaOH}}{1\,\text{mol H}_2\text{C}_2\text{O}_4}}{25.7\,\text{mL NaOH} \times \frac{1\,\text{L}}{1000\,\text{mL}}} = 0.220\,M\,\text{NaOH}$$

75. a. The balanced molecular equation is $H_2CO_3 + 2\,NaOH \rightarrow 2\,H_2O + Na_2CO_3$. The net ionic equation is $H^+ + OH^- \rightarrow H_2O$.

b.
$$\frac{38.98\,\text{mL NaOH} \times \frac{1\,\text{L}}{1000\,\text{mL}} \times \frac{0.1445\,\text{mol NaOH}}{1\,\text{L NaOH}} \times \frac{1\,\text{mol H}_2\text{CO}_3}{2\,\text{mol NaOH}}}{50.00\,\text{mL H}_2\text{CO}_3 \times \frac{1\,\text{L}}{1000\,\text{mL}}} = 0.05633\,M\,\text{H}_2\text{CO}_3$$

77. a. The balanced reaction with $Al(OH)_3$ is $Al(OH)_3 + 3\ HCl \rightarrow AlCl_3 + 3\ H_2O$.

$$50.0\ \text{mL HCl soln} \times \frac{1\ \text{L}}{1000\ \text{mL}} \times \frac{0.0100\ \text{mol HCl}}{1\ \text{L HCl soln}} \times \frac{1\ \text{mol Al(OH)}_3}{3\ \text{mol HCl}} \times \frac{78.00\ \text{g Al(OH)}_3}{1\ \text{mol Al(OH)}_3}$$
$$= 0.0130\ \text{g Al(OH)}_3$$

b. The balanced reaction with $Mg(OH)_2$ is $Mg(OH)_2 + 2\ HCl \rightarrow MgCl_2 + 2\ H_2O$.

$$50.0\ \text{mL HCl soln} \times \frac{1\ \text{L}}{1000\ \text{mL}} \times \frac{0.0100\ \text{mol HCl}}{1\ \text{L HCl soln}} \times \frac{1\ \text{mol Mg(OH)}_2}{2\ \text{mol HCl}} \times \frac{58.33\ \text{g Mg(OH)}_2}{1\ \text{mol Mg(OH)}_2}$$
$$= 0.0146\ \text{g Mg(OH)}_2$$

c. The balanced reaction with $CaCO_3$ is $CaCO_3 + 2\ HCl \rightarrow CaCl_2 + H_2O + CO_2$.

$$50.0\ \text{mL HCl soln} \times \frac{1\ \text{L}}{1000\ \text{mL}} \times \frac{0.0100\ \text{mol HCl}}{1\ \text{L HCl soln}} \times \frac{1\ \text{mol CaCO}_3}{2\ \text{mol HCl}} \times \frac{100.09\ \text{g CaCO}_3}{1\ \text{mol CaCO}_3}$$
$$= 0.0250\ \text{g CaCO}_3$$

79. To answer this question, simply follow the rules for assigning oxidation numbers found in the text. The atom with the negative oxidation number is more likely to have a full or partial negative charge.

a. C b. F c. O d. P e. O

81. a. In N_2O_5, O has an oxidation number of –2 each, so N must be +5 each to make the molecule neutral.

b. In PO_4^{3-}, O has an oxidation number of –2 each, so P must be +5 to make the ion –3 overall.

c. $CuCO_3$, Cu must have an oxidation number of +2, because carbonate has an overall charge of –2. Because O has an oxidation number of –2 each, C has an oxidation number of +4 to give the carbonate its –2 overall charge.

d. N_2 is in its elemental form and has an oxidation number of 0.

e. In H_2SO_3, H has an oxidation number of +1 each, and O has an oxidation number of –2 each; therefore, S must have an oxidation number of +4 to make the compound neutral.

83. This reaction is a redox reaction. On the reactant side the oxidation numbers are Fe(+2); H(+1); O(–2); and Cr(+6). On the product side, the oxidation numbers are Fe(+3); Cr(+3); H(+1); and O(–2). Both iron and chromium change their oxidation state.

85. Water is called the universal solvent because it dissolves so many molecules of varying sizes, from small ionic compounds to very large proteins and DNA.

87.

$$100.0\ \text{mL NaOH} \times \frac{1\ \text{L}}{1000\ \text{mL}} \times \frac{0.230\ \text{mol NaOH}}{1\ \text{L NaOH}} \times \frac{1\ \text{mol HCl}}{1\ \text{mol NaOH}} \times \frac{1\ \text{L HCl}}{0.530\ \text{mol HCl}}$$
$$= 0.0434\ \text{L HCl} \times \frac{1000\ \text{mL}}{1\ \text{L}} = 43.4\ \text{mL HCl}$$

89. Nitrate and most chloride compound are soluble, so those ions will not be involved. Sodium and ammonium compounds are soluble, so those ions will not be involved. The only soluble hydroxides are Group 1A and some Group 2A (including Ba), so $Ba(OH)_2$ will not precipitate, but there will be CuOH as a precipitate. The only soluble carbonates are Group 1A and ammonium, so both $BaCO_3$ and Cu_2CO_3 will be precipitated. Overall, three precipitates are likely to form: CuOH, $BaCO_3$, and Cu_2CO_3.

91. The balanced molecular reaction is

$$Ba(NO_3)_2(aq) + Na_2SO_4(aq) \rightarrow BaSO_4(s) + 2\ NaNO_3(aq).$$

The balanced ionic equation is

$$Ba^{2+}(aq) + 2\ NO_3^-(aq) + 2\ Na^+(aq) + SO_4^{2-}(aq) \rightarrow BaSO_4(s) + 2\ Na^+(aq) + 2\ NO_3^-(aq)$$

To find the maximum barium sulfate that could be formed, we must calculate which reagent limits the reaction:

$$125.0\ \text{mL Na}_2\text{SO}_4 \times \frac{1\ \text{L}}{1000\ \text{mL}} \times \frac{0.567\ \text{mol Na}_2\text{SO}_4}{1\ \text{L Na}_2\text{SO}_4} \times \frac{1\ \text{mol BaSO}_4}{1\ \text{mol Na}_2\text{SO}_4} \times \frac{233.4\ \text{g BaSO}_4}{1\ \text{mol BaSO}_4} = 16.5\ \text{g BaSO}_4$$

$$75.0\ \text{mL Ba(NO}_3)_2 \times \frac{1\ \text{L}}{1000\ \text{mL}} \times \frac{0.786\ \text{mol Ba(NO}_3)_2}{1\ \text{L Ba(NO}_3)_2} \times \frac{1\ \text{mol BaSO}_4}{1\ \text{mol Ba(NO}_3)_2} \times \frac{233.4\ \text{g BaSO}_4}{1\ \text{mol BaSO}_4} = 13.8\ \text{g BaSO}_4$$

The $Ba(NO_3)_3$ solution limits the reaction, and 13.8 g of $BaSO_4$ is the maximum that could be formed by mixing the solutions.

92. a. Oxalic acid has –OH structural units along with addition O atoms that would interact very strongly with water. Oxalic acid is expected to be soluble in water.

b. $H_2C_2O_4(s) \rightarrow H_2C_2O_4(aq) \rightarrow 2\ H^+(aq) + C_2O_4^{2-}(aq)$

c. Carbon has an oxidation number of +3 in $C_2O_4^{2-}$ and of +4 in CO_2; therefore, CO_2 is the more oxidized species.

d. Manganese has an oxidation number of +7 in MnO_4^- and of +2 in Mn^{2+}; therefore, MnO_4^- is the more oxidized species.

e. The 2 Mn atoms in 2 MnO_4^- will gain a total of 10 electrons (5 each) to become 2 Mn^{2+}. The 10 carbon atoms in 5 $C_2O_4^{2-}$ will lose a total of 10 electrons (1 each) to become 10 CO_2.

f. Because MnO_4^- gains electrons, it is reduced. Because $C_2O_4^{2-}$ loses electrons, it is oxidized.

g. $$33.5\ \text{mL KMnO}_4 \times \frac{1\ \text{L}}{1000\ \text{mL}} \times \frac{0.00976\ \text{mol KMnO}_4}{1\ \text{L KMnO}_4} \times \frac{1\ \text{mol MnO}_4^-}{1\ \text{mol KMnO}_4} \times \frac{5\ \text{mol C}_2\text{O}_4^{2-}}{2\ \text{mol MnO}_4^-} \times \frac{1\ \text{mol H}_2\text{C}_2\text{O}_4}{1\ \text{mol C}_2\text{O}_4^{2-}} \times \frac{90.04\ \text{g H}_2\text{C}_2\text{O}_4}{1\ \text{mol H}_2\text{C}_2\text{O}_4} = 0.0736\ \text{g H}_2\text{C}_2\text{O}_4$$

Chapter 5: Energy

The Bottom Line

- Chemical changes are accompanied by the gain or release of energy. (Section 5.1)
- Chemicals can store huge amounts of energy and then release it as soon as a chemical reaction begins. (Section 5.1)
- Thermodynamics is the study of energy changes and exchanges. (Section 5.1)
- Energy comes in two basic forms, which we call kinetic energy (the energy of motion) and potential energy (positional energy). (Section 5.1)
- Energy is never created or destroyed; it is only transferred from place to place and converted from one form into another. This is the law of conservation of energy. (Section 5.1)
- A chemical reaction that releases energy from the chemicals involved is known as an exothermic reaction, because energy is flowing out of the system and into the surroundings. A reaction that absorbs energy into the chemicals involved is known as an endothermic reaction. (Section 5.1)
- All chemical reactions begin with an input of energy, needed to "jolt" the chemicals into reacting. The energy required to make this happen is known as the activation energy of the reaction. (Section 5.1)
- The total change in the energy of a chemical system, as it undergoes a chemical reaction, is equal to the heat flow (q), known as the heat of reaction, and the work done (w): $\_U = q + w$. (Section 5.1)
- The SI unit of energy is the joule, which we can relate to the more familiar units of calories and (in the context of food) Calories. (Section 5.2)
- Each substance has a particular specific heat capacity (c), often simply called its specific heat, which is the amount of heat needed to raise the temperature of 1 gramof the substance by 1 degree Celsius (or 1 kelvin) when the pressure is constant. (Section 5.3)
- Hess's law states that the enthalpy change of a chemical reaction is independent of the chemical path or mechanism involved in the reaction. (Section 5.5)
- The standard enthalpy change for a reaction can be calculated by subtracting the sum of the enthalpies of formation of the reactants of the reaction from the sum of the enthalpies of formation of the products of the reaction. (Section 5.5)
- Making appropriate choices about our energy sources will be an important part of building a successful and sustainable future for humanity. (Section 5.6)

Solutions to Practice Problems

5.1. Plants use the energy in the environment (from the sun, in the chemical bonds of water, carbon dioxide, and minerals) to create sugars and starches. These sugars and starches, along with everything else that makes a plant, are storage systems for chemical energy, is the total of the kinetic and potential energies due to the motion and position of the atoms of the chemicals. In this way, plants serve as storage of chemical energy for animals that eat them, and even as future fossil fuels as the plant material is converted to coal or oil.

5.2. Work due to the expansion of a gas can be calculated according to the equation $w = -P\Delta V$. The volume change in the system is $\Delta V = 300\text{ L} - 30\text{ L} = 270\text{ L}$, so $w = -(1\text{ atm})(270\text{ L}) = -270$ L•atm. Because the sign on the work is negative, work is being done *by the system*. Converting to joules yields -270 L•atm × (101.3 J / L•atm) = -2.7×10^4 J = -27 kJ.

5.3. Because 84.7 J of work is done *on the system*, this is a positive change: w = 84.7 J. If the system *loses* 39.9 J of heat, this change is negative: q = −39.9 J.

Then $\Delta U = q + w = (-39.9\text{ J}) + 84.7\text{ J} = 44.8\text{ J}$.

5.4. First we need to find the energy content of one apple; then we can find the energy equivalent of one bag of peanuts. Using the conversion factors 4.184 kJ = 1 Cal and 200 g = 1 apple:

$$\text{in kilojoules}: \frac{30\,\text{Cal}}{100\,\text{g}} \times \frac{4.184\,\text{kJ}}{1\,\text{Cal}} \times \frac{200\,\text{g}}{1\,\text{apple}} = 250\,\text{kJ / apple}$$

Then:
$$\frac{726\,\text{kJ}}{\text{bag}} \times \frac{1\,\text{apple}}{250\,\text{kJ}} = 2.9\,\text{apple / bag peanuts}$$

Therefore, 2.9 apples have the energy equivalent of the small bag of peanuts.

5.5. This problem has two parts: First, we calculate the total heat released in the reaction from the heat capacity and the temperature change of the bomb calorimeter; second, we use the number of moles of the benzoic acid and the heat released to calculate the energy of combustion for benzoic acid. The heat released is the product of the heat capacity and the temperature change:

$$q = C \times \Delta T = (5.02\,\text{kJ/°C}) \times (39.20\text{°C} - 26.34\text{°C})$$
$$= -64.4\,\text{kJ}$$

The molar mass of benzoic acid is 122.12 g/mol, so there is 2.442 g / (122.12 g/mol) = 0.02000 mol of benzoic acid. Therefore, the energy of combustion is

$$\frac{-64.4\,\text{kJ}}{0.02000\,\text{mol}} = -3220\,\text{kJ / mol}$$

Because the sign of the energy of combustion is negative, heat is released in the process.

5.6. In the appendix, the heat of formation of water at 25°C is −285.8 kJ/mol. We can use this figure and the molar mass as conversion constants to complete the calculation:

$$38\,\text{g H}_2\text{O} \times \frac{1\,\text{mol H}_2\text{O}}{18.02\,\text{g H}_2\text{O}} \times \frac{-285.8\,\text{kJ}}{1\,\text{mol H}_2\text{O}} = -6.03 \times 10^2\,\text{kJ}$$

5.7. Realizing that the heat of reaction refers to "1 mol of reaction," we can use the reaction stoichiometry to assist in the conversions. We will use the absolute value of the enthalpy of reaction to find the amount of moles involved. We can then consider the effect of the sign on the direction of reaction required.

a. $$250\,\text{kJ} \times \frac{1\,\text{mol reaction}}{1012\,\text{kJ}} \times \frac{2\,\text{mol NH}_3}{1\,\text{mol reaction}} = 0.494\,\text{mol NH}_3.$$

If the energy is released, the reaction proceeds as written, to give a (–) enthalpy, resulting in the consumption of 0.494 mol NH_3. If the enthalpy change is (+), the reaction must proceed in the opposite direction, resulting in an increase of 0.494 mol NH_3.

b. $$250\,\text{kJ} \times \frac{1\,\text{mol reaction}}{164\,\text{kJ}} \times \frac{1\,\text{mol O}_2}{1\,\text{mol reaction}} = 1.52\,\text{mol O}_2.$$

If the energy is released, the reaction proceeds as written, to give a (–) enthalpy, resulting in the consumption of 1.52 mol O_2. If the enthalpy change is (+), the reaction must proceed in the opposite direction, resulting in an increase of 1.52 mol O_2.

5.8. To solve the problem, we need to find a way to combine the given equations to yield the reaction we are seeking. Given:

$$6\ Fe(s) + 4\ O_2(g) \rightarrow 2\ Fe_3O_4(s) \qquad \Delta H = -1787\ kJ$$
$$2\ Fe_3O_4(s) + \tfrac{1}{2}\ O_2(g) \rightarrow 3\ Fe_2O_3(s) \qquad \Delta H = -186\ kJ$$

Note that the first equation has the iron on the reactant side, but we want to finish with Fe(s) as a product, so we will reverse the reaction and the sign of the enthalpy change. However, the reaction would then contain $Fe_3O_4(s)$ as a reactant, which we want to eliminate. If we reverse the second equation (and the sign of the enthalpy change), we get

$$2\ Fe_3O_4(s) \rightarrow 6\ Fe(s) + 4\ O_2(g) \qquad \Delta H = +1787\ kJ$$
$$3\ Fe_2O_3(s) \rightarrow 2\ Fe_3O_4(s) + \tfrac{1}{2}\ O_2(g) \qquad \Delta H = +186\ kJ$$

Adding these two equations yields

$$3\ Fe_2O_3\ (s) \rightarrow 6\ Fe\ (s) + 9/2\ O_2(g)$$

This is the equation we wanted! (Note that 2 $Fe_3O_4(s)$ molecules also canceled from each side.) To get the new enthalpy change, we also add the enthalpies according to Hess's Law:

$$\Delta H = (+1787\ kJ) + (+186\ kJ) = +1973\ kJ$$

5.9. To solve the problem, we need to find a way to combine the given equations to yield the reaction we are seeking. Given:

$$2\ NH_3(g) + 3\ N_2O(g) \rightarrow 4\ N_2(g) + 3\ H_2O(l) \qquad \Delta H = -1012\ kJ$$
$$2\ N_2O(g) \rightarrow O_2(g) + 2\ N_2(g) \qquad \Delta H = -164\ kJ$$

Note that the first equation has the ammonia on the reactant side, matching what we want; however the reaction contains N_2O as a reactant, which we want to eliminate. If we reverse the second equation (and the sign of the enthalpy change) we get:

$$2\ NH_3(g) + 3\ N_2O(g) \rightarrow 4\ N_2(g) + 3\ H_2O(l) \qquad \Delta H = -1012\ kJ$$
$$O_2(g) + 2\ N_2(g) \rightarrow 2\ N_2O(g) \qquad \Delta H = +164\ kJ$$

If we multiply the first equation (and its enthalpy change) by 2 and the second equation (and its enthalpy change) by 3, we will be able to cancel out the N_2O, because it will appear on opposite sides of the equations in equal numbers:

$$4\ NH_3(g) + 6\ N_2O(g) \rightarrow 8\ N_2(g) + 6\ H_2O(l) \qquad \Delta H = -2024\ kJ$$
$$3\ O_2(g) + 6\ N_2(g) \rightarrow 6\ N_2O(g) \qquad \Delta H = +492\ kJ$$

Adding these two equations, we get

$$4\ NH_3(g) + 3\ O_2(g) \rightarrow 2\ N_2(g) + 6\ H_2O(l)$$

This is the equation we wanted! (Note that 6 N_2 molecules also canceled from each side.) To get the new enthalpy change, we also add the enthalpies according to Hess's Law:

$$\Delta H = -2024\ kJ + 492\ kJ = -1532\ kJ$$

Solutions to Student Problems

1. At the top of one side, the skateboarder has only potential energy. As the skater begins down one side of the half-pipe, the potential energy begins to decrease as it is converted into the kinetic energy of the skater. As the skater hits the bottom of the half-pipe, the kinetic energy is at a maximum and the potential energy is at a minimum. All of the available potential energy due to the height difference from the bottom to the top has been converted into kinetic energy—the skater is moving the fastest here. As the skater begins to climb the opposite wall of the half-pipe, the kinetic energy decreases and some of it is converted into potential energy. At the top of the wall, the kinetic energy reaches zero and the potential energy is at a maximum. (This analysis assumes that no energy is lost to friction and that the skater does not increase the kinetic energy by pushing off at any point.)

3. In all cases, we can use kinetic energy = $KE = \frac{1}{2} mv^2$, where m is the mass in kilograms and v is the velocity in meters per second. Using these units for the mass and velocity will give the kinetic energy units of joules.

a. $$KE = \frac{1}{2} \times \left(185\,\text{lb} \times \frac{1\,\text{kg}}{2.205\,\text{lb}}\right) \times \left(\frac{8.0\,\text{mi}}{1\,\text{hr}} \times \frac{1\,\text{hr}}{60\,\text{min}} \times \frac{1\,\text{min}}{60\,\text{s}} \times \frac{1.609\,\text{km}}{\text{mi}} \times \frac{1000\,\text{m}}{1\,\text{km}}\right)^2$$
$$= 5.4 \times 10^2\ \text{J}$$

b. $$KE = \frac{1}{2} \times \left(42\,\text{g} \times \frac{1\,\text{kg}}{1000\,\text{g}}\right) \times \left(\frac{0.78\,\text{m}}{1\,\text{s}}\right)^2 = 1.3 \times 10^{-2}\ \text{J}$$

c. $$KE = \frac{1}{2} \times \left(\frac{44.01\,\text{g}}{1\,\text{mol}} \times \frac{1\,\text{mol}}{6.022 \times 10^{23}} \times \frac{1\,\text{kg}}{1000\,\text{g}}\right) \times \left(\frac{560\,\text{m}}{1\,\text{s}}\right)^2 = 1.1 \times 10^{-20}\ \text{J}$$

5. a. The attractive force between the negative electron and the positive nucleus is a function of position (or distance between the two) and therefore is a form of potential energy.
b. In this case, the average potential energy and the average kinetic energy of the vibrating molecule are exactly the same. There is a force holding the atoms together (it is based on their distance apart), which is potential energy, and the atoms are moving within the molecule, which is kinetic energy.
c. The *movement* of the hydrogen molecule is kinetic-energy-dominated.

7. Some of the kinetic energy in the particles is what is transferred between the system and the surroundings. The transfer is completed as the more energetic (higher-temperature) particles collide and transfer energy to the less energetic (lower-temperature) particles.

9. The system is the chemicals involved in the combustion, both reactant and product: the gasoline and oxygen that react and the carbon dioxide and water that are produced. The surroundings are the engine, including the piston where the combustion takes place, along with the rest of the universe. The system is losing energy both as heat is lost to the surroundings and as it does work on the piston. The sign on the work is (–) because the volume change is positive and $w = -P\Delta V$.

11. If $q = +24$ J and $w = +12$ J, then $\Delta U = q + w = +36$ J, representing a gain of energy in the system. Because energy is conserved, this also represents a loss of energy by the surroundings.

13. a. $\Delta U = q + w = (+45\ \text{J}) + (+45\ \text{J}) = 90$ J. Surroundings lose energy.
b. $\Delta U = q + w = (-266\ \text{J}) + (+1200\ \text{J}) = 930$ J. Surroundings lose energy. Note the change in units for the work: 1.2 kJ = 1200 J.
c. $\Delta U = q + w = (23.4\ \text{kJ}) + (-14\ \text{kJ}) = +9$ kJ. Surroundings lose energy.
d. The gas inside the cylinder is the system. Work is done on the system to compress the gas, so $w = +67$ kJ. Heat is lost to the surroundings, so $q = -23$ kJ. $\Delta U = q + w = (-23\ \text{kJ}) + (+67\ \text{kJ}) = +44$ kJ. Surroundings lose energy.

15. The formation of one mole of CO_2 from its elements releases much heat ($\Delta_f H^\circ = -393.5$ kJ/mol), so we can conclude that the formation of carbon–oxygen bonds is a heat-releasing process. Molecules already containing C—O bonds will not be able to release as much energy as those that do not, and CO_2 is fully oxidized and will react no further; therefore, CO_2 will release the least amount of heat in either a combustion or a digestion process.

17. Because the force due to gravity is less on the surface of the moon than on the surface of the earth, a ball must not be struck with as much force on the moon for it to cover the same distance. An additional factor lowering the force needed on the moon is the lack of atmosphere on the moon. The drag of the atmosphere on the earthbound ball requires additional force to be applied so that the ball can cover the same distance.

19. An endothermic process is defined by the absorption of heat *by the system*, so q is (+). Work *on the surroundings* would involve a loss of energy by the system, so the work for that process would be negative (–).

21. The kinetic energy of the apple is

$$KE = \frac{1}{2}mv^2 = \frac{1}{2}\times\left(275\,\text{g}\times\frac{1\,\text{kg}}{1000\,\text{g}}\right)\times\left(\frac{15\,\text{m}}{1\,\text{s}}\right)^2 = 30.94\,\text{J}.$$

Setting the kinetic energy of the orange to the same value, we can solve for the velocity:

$$KE = \frac{1}{2}mv^2$$

$$v^2 = \frac{2KE}{m}$$

$$v = \sqrt{\frac{2KE}{m}} = \sqrt{\frac{2\times 30.94\text{J}}{\left(175\,\text{g}\times\frac{1\text{kg}}{1000\,\text{g}}\right)}} = 19\,\text{m/s}$$

23. To solve for the kinetic energy, it is necessary to find the mass of the N_2 molecule in kilograms:

$$KE = \frac{1}{2}\times\left(\frac{28.02\,\text{g}}{1\,\text{mol}}\times\frac{1\,\text{mol}}{6.022\text{x}10^{23}}\times\frac{1\text{kg}}{1000\,\text{g}}\right)\times\left(\frac{420\,\text{m}}{1\text{s}}\right)^2 = 4.1\times10^{-21}\,\text{J}$$

25. In Calories per gram: 90 Cal / 34 g = 2.647 Cal/g = 2.6 Cal/g
There are 1000 cal / 1 Cal, so in cal/g: 2.647 Cal/g × 1000 cal / 1 Cal.= 2647 cal/g = 2.6 kcal/g
There are 4.184 J / cal, so in J/g: 2647 cal/g × 4.184 J/cal =11,075 J/g = 11 kJ/g

27. To calculate the heat energy (q) required to change the temperature of a sample, we can use the specific heat (c), the mass (m), and the temperature change (ΔT):

$$q = mc\Delta T$$

$$q = 50.0\,\text{g}\times 4.184\,\text{J/g}\cdot°\text{C}\times(37.0°\text{C} - 23.0°\text{C})$$

$$q = 2929\,\text{J} = 2.93\,\text{kJ}$$

29. We can use the same equation, $q = mc\Delta T$, to calculate the heat, but we need to convert all the values into the appropriate units. (Note that the conversion from Fahrenheit to Celsius works only for temperature changes!)

$$1\,\text{lb} \times \frac{1\,\text{kg}}{2.2051\,\text{lb}} \times \frac{1000\,\text{g}}{1\,\text{kg}} = 454\,\text{g}$$

$$1°\text{F} \times \frac{1°\text{C}}{1.8°\text{F}} = 0.556°\text{C}$$

$$\text{so } q = mc\Delta T$$

$$q = 454\,\text{g} \times 4.184\,\text{J/g}\cdot°\text{C} \times (0.556°\text{C})$$

$$q = 1056\,\text{J} = 1.06\,\text{kJ}$$

31. a. We start this problem using $q = mc\Delta T$, but we solve the equation for ΔT.

$$q = mc\Delta T$$

$$\Delta T = \frac{q}{mc} = \frac{1\,\text{mol} \times \frac{410.0\,\text{kJ}}{1\,\text{mol}} \times \frac{1000\,\text{J}}{1\,\text{kJ}}}{2000.0\,\text{g} \times 4.184\,\text{J/g}\cdot°\text{C}}$$

$$\Delta T = +49.00\,°\text{C}$$

$$\text{so } T = 22.0 + 49.00 = 71.0°\text{C}$$

b. A quick look at the problem reveals that 1/10 of the heat will be generated, but only 1/10 of the water is heated, so we should find that the temperature change is the same.

$$\Delta T = \frac{q}{mc} = \frac{0.10\,\text{mol} \times \frac{410.0\,\text{kJ}}{1\,\text{mol}} \times \frac{1000\,\text{J}}{1\,\text{kJ}}}{200.0\,\text{g} \times 4.184\,\text{J/g}\cdot°\text{C}}$$

$$\Delta T = +49.00\,°\text{C}$$

$$\text{so } T = 22.0 + 49.00 = 71.0°\text{C}$$

33. We can set up a ratio of heat released to mass consumed to describe the process; we need only to convert the mass to moles to answer the question. The result is the enthalpy of combustion, $\Delta_c H$, for ethanol.

$$\Delta_c H = \frac{268\,\text{kJ}}{10.0\,\text{g}} \times \frac{46.07\,\text{g}\,C_2H_5OH}{1\,\text{mol}\,C_2H_5OH} = 1.23 \times 10^3\,\text{kJ/mol}$$

35. Specific heat involves the heat required relative to the *temperature change* for a certain amount of the substance. Even though the actual Celsius and Kelvin temperatures are always different, the size of a single degree on each scale is exactly the same, so it makes no difference, in terms of specific units, whether the temperature is given in Kelvin or Celsius.

37. Because $q = mc\Delta T$ and both samples are water, a quick answer can be obtained by multiplying the mass by the temperature change. For part a, this product is 10.0 kg × 35 °C = 350 kg•°C; for part b, using 1 L = 1 kg, the product is 8.0 kg × 77 °C = 616 kg•°C. The sample in part b requires more heat. The exact calculations are

$$q = mc\Delta T$$
$$\text{Part a: } q = 10000\,\text{g} \times 4.184\,\text{J/g}\cdot°\text{C} \times (10.0°\text{C} - 45.0°\text{C})$$
$$q = -1.46 \times 10^6\,\text{J} = -1.46\,\text{MJ}$$

$$q = mc\Delta T$$
$$\text{Part b: } q = 8000\,\text{g} \times 4.184\,\text{J/g}\cdot°\text{C} \times (99.0°\text{C} - 22.0°\text{C})$$
$$q = 2.6 \times 10^6\,\text{J} = 2.6\,\text{MJ}$$

39. The full calculations are shown below the table.

Heat, q	Specific Heat (J/g·°C)	Mass (g)	Δt (°C)
10.0 joules	4.184	**0.239**	10.0
5.8 joules	0.115	10.0	5.0
15.5 joules	**0.0243**	42.5	15.0

$$m = \frac{q}{c\Delta T} = \frac{10.0\,\text{J}}{4.184\,\text{J/g}\cdot°\text{C} \times (10.0°\text{C})} = 0.239\,\text{g}$$
$$q = mc\Delta T = 10.0\,\text{g} \times 0.115\,\text{J/g}\cdot°\text{C} \times (5.0°\text{C}) = 5.8\,\text{J}$$
$$c = \frac{q}{m\Delta T} = \frac{15.5\,\text{J}}{42.5\,\text{g} \times (15.0°\text{C})} = 0.0243\,\text{J/g}\cdot°\text{C}$$

41. $$c = \frac{q}{m\Delta T} = \frac{311\,\text{J}}{35.0\,\text{g} \times (20.0°\text{C})} = 0.444\,\text{J/g}\cdot°\text{C}$$

43. We can break this problem into two parts: (1) Find the energy released by the combustion of benzoic acid. (2) Use this energy to find the heat capacity of the calorimeter.

$$q_c = 2.000\,\text{g}\,C_6H_5COOH \times \frac{1\,\text{mol}\,C_6H_5COOH}{122.12\,\text{g}\,C_6H_5COOH} \times \frac{-3227\,\text{kJ}}{1\,\text{mol}\,C_6H_5COOH} = -52.85\,\text{kJ}$$

The heat released by the benzoic acid combustion (q_c) is absorbed by the calorimeter (q_{cal}). So

$$q_{cal} = -q_c = +52.85\,\text{kJ}$$
$$q_{cal} = C\Delta T$$
$$C = \frac{q_{cal}}{\Delta T} = \frac{52.85\,\text{kJ}}{1.978\,°\text{C}} = 26.72\,\text{kJ/°C}$$

45. This problem is the reverse of Problems 43 and 44 in that we will use the calorimeter constant and temperature change to find the amount of heat transferred and then use that information to find the energy of combustion for the sugar.

a. From the calorimeter data: $q = C\Delta T = 28.9\,\text{kJ}/°\text{C} \times (2.56\,°\text{C}) = 74.0\,\text{kJ}$

$$\text{Heat of combustion per gram: } \Delta_c H = \frac{74.0\,\text{kJ}}{1.500\,\text{g}} = 49.3\,\text{kJ/g}$$

b. In order to determine the molar heat of combustion for the sugar, we would need to convert the grams to moles using the *molar mass* of the sugar.

47. We first must find the total heat required to warm the water and then use the heat of combustion of propane (C_3H_8) to find the amount of propane used. (Note that the heat used by water was released by the propane and the sign on the heat changes!)

Heat used: $q = mc\Delta T = 1000\,\text{g} \times 4.184\,\text{J/g}\cdot°\text{C} \times (100.0°\text{C} - 22.0°\text{C}) = 326352\,\text{J} = 326\,\text{kJ}$

$$\text{Propane used: } -326.3\,\text{kJ} \times \frac{1\,\text{mol}\,C_3H_8}{-2.2\times10^3\,\text{kJ}} \times \frac{44.09\,\text{g}\,C_3H_8}{1\,\text{mol}\,C_3H_8} = 6.5\,\text{g}\,C_3H_8$$

49. In each case, we can divide the molar heat of combustion by the molar mass to get the heat of combustion per gram:

$$\text{Acetylene: } \Delta_c H = \frac{-1300\,\text{kJ}}{1\,\text{mol}} \times \frac{1\,\text{mol}\,C_2H_2}{26.04\,\text{g}\,C_2H_2} = -5.0\times10^1\,\text{kJ/g}$$

$$\text{Methane: } \Delta_c H = \frac{-890\,\text{kJ}}{1\,\text{mol}} \times \frac{1\,\text{mol}\,CH_4}{16.04\,\text{g}\,CH_4} = -55\,\text{kJ/g}$$

Therefore, methane provides more heat per gram.

51. To solve the problem we equate the two heat transfers, including a change of sign to account for the fact that the gold is losing heat while the calorimeter is gaining heat.

$$q_{\text{Au}} = -q_{\text{Cal}}$$

$$m_{\text{Au}} c_{\text{Au}} \Delta T_{\text{Au}} = -C_{\text{Cal}} \Delta T_{\text{Cal}}$$

$$(15.0\,\text{g}) \times (0.13\,\text{J/g}\cdot°\text{C}) \times (T_{\text{final}} - 99.0°\text{C}) = -25.0\,\text{J}/°\text{C} \times (T_{\text{final}} - 25.0°\text{C})$$

$$1.95 \times (T_{\text{final}} - 99.0°\text{C}) = -25.0\,\text{J}/°\text{C} \times (T_{\text{final}} - 25.0°\text{C})$$

$$1.95T_{\text{final}} - 193.05°\text{C} = -25.0T_{\text{final}} + 625°\text{C}$$

$$26.95T_{\text{final}} = 818.05°\text{C}$$

$$T_{\text{final}} = 30.3°\text{C}$$

Therefore, an observed temperature change of 5.3°C in the calorimeter might indicate gold (or any metal with a similar specific heat).

53. Any gases that are generated as a result of the chemical process will expand (or contract) until the pressure of the gas matches the atmospheric pressure. Because the entire process begins and ends at the same pressure, we can consider the process to be a constant-pressure process.

55. Enthalpy is equivalent to the amount of heat energy transferred in a constant-pressure process.

57. An exothermic reaction involves a release of heat by the system, so the surroundings are gaining the heat.

59. Standard conditions imply a pressure of 1 bar (also commonly referred to as 1 atmosphere) and concentrations of species in solution at 1 mol/L. Standard values are often tabulated at 25°C, but this is not a standard temperature.

61. A standard formation reaction is written with elemental reactants in their standard states and 1 mol of the product in its standard state.

a. This is a proper formation reaction.

b. This is not a proper formation reaction because it results in the formation of 2 mol of ammonia.

c. This is not a formation reaction; it actually describes only a phase change.

d. This is not a proper formation reaction because hydrogen is $H_2(g)$ in its standard state, not atomic hydrogen.

e. This is not a formation reaction (more than a single product, not elemental reactants).

f. This is not a formation reaction (more than a single product, not elemental reactants).

63. In a constant-pressure process, the enthalpy change and energy change differ by the term $P\Delta V$. Because there is a change of 29 mol of gas from reactant to product, that term is large; therefore, the difference between the enthalpy change and the energy change is large.

65. a. $C(s) + \frac{1}{2} O_2(g) \rightarrow CO(g)$

b. The negative value for the heat of formation indicates a loss of heat from the system—an exothermic process.

c. For the reverse reaction, $\Delta H = +110.5$ kJ/mol. (Reversing a reaction reverses the sign of the enthalpy change.)

67. The standard heat of combustion reaction is written for the combustion of a single mole of reactant plus oxygen, regardless of whether that will require fractional coefficients:

$$C_4H_{10}(g) + 13/2\ O_2(g) \rightarrow 4\ CO_2(g) + 5\ H_2O(l)$$

69. Standard heat of formation reactions always form 1 mol of product from the elements in their standard states:

$$10\ C(s, \text{graphite}) + 15/2\ H_2(g) + 5/2\ N_2(g) + 13/2\ O_2(g) + 3P(s, \alpha\ \text{white}) \rightarrow C_{10}H_{15}N_5O_{13}P_3(s)$$

71. a. Reversing the reaction changes the sign, but not the value: $\Delta H = +1012$ kJ.

b. We need to realize that the enthalpy change refers to 1 mol of "reaction," so we can use the stoichiometry of the reaction to see what the enthalpy changes would be for specific amounts of reactants or products:

$$1\,\text{mol}\,NH_3 \times \frac{1\,\text{mol reaction}}{2\,\text{mol}\,NH_3} \times \frac{-1012\,\text{kJ}}{1\,\text{mol reaction}} = -506\,\text{kJ}$$

It makes sense that consuming 1 mol of NH_3 would release half as much energy as consuming 2 mol.

c. $$4\,\text{mol}\,NH_3 \times \frac{1\,\text{mol reaction}}{2\,\text{mol}\,NH_3} \times \frac{-1012\,\text{kJ}}{1\,\text{mol reaction}} = -2024\,\text{kJ}$$

73. The combustion reaction is $C_3H_8(g) + 5\ O_2(g) \rightarrow 3\ CO_2(g) + 4\ H_2O(l)$ with ΔH_c(propane). The additional combustion reactions we need are

$$H_2(g) + \tfrac{1}{2} O_2(g) \rightarrow H_2O(l) \qquad \Delta_c H(\text{hydrogen})$$
$$C(s) + O_2(g) \rightarrow CO_2(g) \qquad \Delta_c H(\text{carbon})$$

We want to combine these three reactions in such a way to end up with the formation reaction:

$$3\ C(s) + 4\ H_2(g) \rightarrow C_3H_8(g)$$

To have propane as a product, we reverse the first reaction and then use the remaining two to ensure that the water and carbon dioxide cancel out:

$$3\ CO_2(g) + 4\ H_2O(l) \rightarrow C_3H_8(g) + 5\ O_2(g) \quad \Delta H = -\Delta_c H(\text{propane})$$
$$4\{\ H_2(g) + \tfrac{1}{2} O_2(g) \rightarrow H_2O(l)\ \} \quad \Delta H = 4\ \Delta_c H(\text{hydrogen})$$
$$3\{\ C(s) + O_2(g) \rightarrow CO_2(g)\ \} \quad \Delta H = 3\ \Delta_c H(\text{carbon})$$

$$3\ C(s) + 4\ H_2(g) \rightarrow C_3H_8(g)$$

and $\Delta H(\text{total}) = \Delta_f H(\text{propane}) = -\Delta_c H(\text{propane}) + 4\ \Delta_c H(\text{hydrogen}) + 3\ \Delta_c H(\text{carbon})$

75. We are given the following reactions:

$$B_2O_3\ (s) + 3H_2O\ (g) \rightarrow B_2H_6\ (g) + 3\ O_2\ (g) \qquad \Delta H = +2035\ \text{kJ}$$
$$2\ H_2O\ (l) \rightarrow 2\ H_2O\ (g) \qquad \Delta H = +88\ \text{kJ}$$
$$H_2\ (g) + 1/2\ O_2\ (g) \rightarrow H_2O\ (l) \qquad \Delta H = -286\ \text{kJ}$$
$$2\ B(s) + 3\ H_2\ (g) \rightarrow B_2H_6\ (g) \qquad \Delta H = +36\ \text{kJ}$$

To calculate ΔH for $2\ B\ (s) + 3/2\ O_2\ (g) \rightarrow B_2O_3\ (s)$, we need to reverse the first equation to get B_2O_3 as a product and then use the remaining reactions to cancel the unwanted reactants and products:

$$B_2H_6\ (g) + 3\ O_2\ (g) \rightarrow B_2O_3\ (s) + 3H_2O\ (g) \qquad \Delta H = -2035\ \text{kJ}$$
$$3/2\ \{\ 2\ H_2O\ (g) \rightarrow 2\ H_2O\ (l)\ \} \qquad \Delta H = 3/2\ (-88\ \text{kJ}) = -132\ \text{kJ}$$
$$3\ \{\ H_2O\ (l) \rightarrow H_2\ (g) + 1/2\ O_2\ (g)\ \} \qquad \Delta H = 3(+286\ \text{kJ}) = +858\ \text{kJ}$$
$$2\ B(s) + 3\ H_2\ (g) \rightarrow B_2H_6\ (g) \qquad \Delta H = +36\ \text{kJ}$$

$$2\ B\ (s) + 3/2\ O_2\ (g) \rightarrow B_2O_3\ (s) \qquad \Delta H = (-2035) + (-132) + (+858) + (+36) = \mathbf{-1273\ kJ}$$

77. a. We are given the heat of formation of $C_2H_5OH(l)$ (–278 kJ), and we look up the values for $H_2O(l)$ (–286 kJ) and $CO_2(g)$ (–393.5 kJ).
For the reaction $C_2H_5OH(l) + 3\ O_2(g) \rightarrow 2\ CO_2(g) + 3\ H_2O(l)$:

$$\Sigma n_p \Delta_f H°(\text{products}) = 2\times(-393.5\ \text{kJ}) + 3\times(-286\ \text{kJ}) = -1645\ \text{kJ}$$
$$\Sigma n_r \Delta_f H°(\text{reactants}) = 1\times(-278\ \text{kJ}) + 3\times(0\ \text{kJ}) = -278\ \text{kJ}$$
$$\text{so } \Delta_{rxn} H° = \Delta_c H° = \Sigma n_p \Delta_f H°(\text{products}) - \Sigma n_r \Delta_f H°(\text{reactants})$$
$$= -1645\ \text{kJ} - (-278\ \text{kJ}) = -1367\ \text{kJ/mol}$$

b. $$100.0\ \text{g}\ C_2H_5OH \times \frac{1\ \text{mol}\ C_2H_5OH}{46.07\ \text{g}\ C_2H_5OH} \times \frac{1\ \text{mol reaction}}{1\ \text{mol}\ C_2H_5OH} \times \frac{-1367\ \text{kJ}}{1\ \text{mol reaction}} = -2967\ \text{kJ}$$

79. The heats of formation of octane, carbon dioxide, and water(l) are –249.9 kJ, –393.5 kJ, and –286 kJ, respectively. The combustion reaction is $C_8H_{18}(l) + 25/2\ O_2(g) \rightarrow 8\ CO_2(g) + 9\ H_2O(l)$.

$$\Sigma n_p \Delta_f H°(\text{products}) = 8\times(-393.5\ \text{kJ}) + 9\times(-286\ \text{kJ}) = -5722\ \text{kJ}$$
$$\Sigma n_r \Delta_f H°(\text{reactants}) = 1\times(-249.9\ \text{kJ}) + 25/2\times(0\ \text{kJ}) = -249.9\ \text{kJ}$$
$$\text{so } \Delta_{rxn} H° = \Delta_c H° = \Sigma n_p \Delta_f H°(\text{products}) - \Sigma n_r \Delta_f H°(\text{reactants})$$
$$= -5722\ \text{kJ} - (-249.9\ \text{kJ}) = -5472\ \text{kJ/mol}$$

81. To get the reaction Ca (s) + H_2O (l) → $Ca(OH)_2$ (s) + H_2 (g), we can use the given reactions, reversing only reaction (a):

$$\text{Ca (s)} + \tfrac{1}{2}\,O_2\text{ (g)} \rightarrow \text{CaO (s)} \qquad \Delta H^\circ = -635\text{ kJ}$$
$$\text{CaO (s)} + H_2O\text{ (l)} \rightarrow Ca(OH)_2\text{ (s)} \qquad \Delta H^\circ = -64\text{ kJ}$$
$$H_2O\text{ (l)} \rightarrow H_2\text{ (g)} + \tfrac{1}{2}\,O_2\text{ (g)} \qquad \Delta H^\circ = 286\text{ kJ}$$
$$\text{Ca (s)} + 2\,H_2O\text{ (l)} \rightarrow Ca(OH)_2\text{ (s)} + H_2\text{ (g)} \qquad \Delta H^\circ = (-635) + (-64) + (286) = \mathbf{-413\text{ kJ}}$$

83. There are many possible answers for each type, such as solar cells, solar thermal systems, biomass conversion, hydroelectric systems, wind power, and geothermal systems. In addition to being renewable, several of these can generate power on site (solar, solar thermal, geothermal, wind, and biomass); biomass conversion can use unwanted by-products of agriculture or even boost prices of commodities that are used; and all would probably reduce overall production of pollution due to greenhouse gases or combustion byproducts.

85. Solving for the kinetic energy:

$$KE = \frac{1}{2}mv^2 = \frac{1}{2}\left(\frac{32.0\,\text{g}\,O_2}{1\,\text{mol}\,O_2} \times \frac{1\,\text{mol}\,O_2}{6.022\text{x}10^{23}\ \text{molecules}\,O_2} \times \frac{1\,\text{kg}}{1000\,\text{g}}\right)(460\,\text{m/s})^2$$
$$= 5.62 \times 10^{-21}\,\text{J}$$

Because the mass of the nitrogen molecule is lower, it will have less kinetic energy if it is traveling at the same velocity as the oxygen molecule.

87. The activation energy is the energy that is needed "to get the reaction started." Most reactions have an activation energy that is greater than the overall change in energy for the reactions (the exceptions are some very exothermic reactions). In chemical reactions, a lot of energy is usually required to break the bonds between atoms before new bonds can be formed. This bond-breaking energy is one part of the activation energy. When new bonds form, some of that energy is regained; if there is a deficit or surplus, then the reaction is either endothermic or exothermic in nature.

89. a. If you feel warmth, the system is releasing energy, an exothermic process.

b. The beaker is part of the surroundings Because it wasn't explicitly named as part of the system.

c. Energy is flowing out of the system.

d. The temperature of the water has also increased; therefore, the kinetic energy of the water molecules has increased.

e. With the exception of some minor expansion (water, beaker, etc) due to the warming, there is no work done in the process.

91. In Calories per gram: $\dfrac{247\,\text{Cal}}{2\,\text{oz}} \times \dfrac{1\,\text{oz}}{28.35\,\text{g}} = 4.36\,\text{Cal/g}$

In joules per gram: $\dfrac{4.36\,\text{Cal}}{1\,\text{g}} \times \dfrac{1000\,\text{cal}}{1\,\text{Cal}} \times \dfrac{4.184\,\text{J}}{1\,\text{cal}} = 1.82 \times 10^4\ \text{J/g}$

In kilojoules per gram: $1.82 \times 10^4\,\text{J/g} \times \dfrac{1\,\text{kJ}}{1000\,\text{J}} = 18.2\,\text{kJ/g}$

We can use the formula for kinetic energy, $KE = 1/2\ mv^2$, to solve for the velocity:

$$v = \sqrt{\frac{2KE}{m}} = \sqrt{\frac{2(1.82\times10^4\ \text{J})}{1.5\,\text{lb}\times\frac{1\,\text{kg}}{2.205\,\text{lb}}}} = 231\,\text{m/s}$$

(For perspective, 231 m/s = 517 mph!)

93. a. We use $q = mc\Delta T$ to calculate the heat absorbed by the water:

$$q = mc\Delta T = (225\,\text{g})(4.184\,\text{J/g}\cdot{}^\circ\text{C})(98^\circ\text{C} - 15^\circ\text{C}) = 7.81\times10^4\,\text{J} = 78.1\,\text{kJ}$$

b. To determine the heat absorbed by the mug, either you would need to know its heat capacity, which would allow the calculation to proceed by multiplying by the temperature change, or you would need the total heat absorbed by the water and mug, which would allow the calculation to proceed by taking the difference between the total heat and the heat absorbed by the water.

95. The heat released by the carbon is $0.200\,\text{g C}\times\frac{1\,\text{mol C}}{12.01\,\text{g C}}\times\frac{392\,\text{kJ}}{1\,\text{mol C}} = 6.53\,\text{kJ}$.

The calorimeter's heat capacity is $C_{\text{cal}} = \frac{q}{\Delta T} = \frac{6.53\,\text{kJ}}{(25.5^\circ\text{C} - 24.0^\circ\text{C})} = 4.35\,\text{kJ}/^\circ\text{C}$.

97. The heat absorbed by the calorimeter (equal to that released by the oil) is

$$q = C_{\text{cal}}\Delta T = 7.5\,\text{J/K}\times2.5\,\text{K} = 18.75\,\text{kJ}$$

Solving for the heat of combustion of the oil yields

$$\Delta_c H = \frac{q}{m} = \frac{18.75\,\text{kJ}}{(0.500\,\text{g})} = 37.5\,\text{kJ/g} = 38\,\text{kJ/g}$$

99. a. Reversing the direction of a reaction only changes the sign on the enthapy change, so $\Delta H =$ +453 kJ.

b. $$10.0\,\text{g H}_3\text{PO}_4\times\frac{1\,\text{mol H}_3\text{PO}_4}{97.99\,\text{g H}_3\text{PO}_4}\times\frac{1\,\text{mol reaction}}{4\,\text{mol H}_3\text{PO}_4}\times\frac{-453\,\text{kJ}}{1\,\text{mol reaction}} = -11.6\,\text{kJ}$$

c. As written, this reaction releases energy, making it exothermic.

d. This is a limiting reagent problem; we solve it by calculating the amount of acid produced by each, assuming there is excess of the other reagent:

From P_4O_{10}: $$1.5\,\text{g P}_4\text{O}_{10}\times\frac{1\,\text{mol P}_4\text{O}_{10}}{283.88\,\text{g P}_4\text{O}_{10}}\times\frac{4\,\text{mol H}_3\text{PO}_4}{1\,\text{mol P}_4\text{O}_{10}}\times\frac{97.99\,\text{g H}_3\text{PO}_4}{1\,\text{mol H}_3\text{PO}_4} = 2.07\,\text{g H}_3\text{PO}_4$$

From H_2O: $$2.50\,\text{mL H}_2\text{O}\times\frac{1\,\text{g H}_2\text{O}}{1\,\text{mL H}_2\text{O}}\times\frac{1\,\text{mol H}_2\text{O}}{18.02\,\text{g H}_2\text{O}}\times\frac{4\,\text{mol H}_3\text{PO}_4}{6\,\text{mol H}_2\text{O}}\times\frac{97.99\,\text{g H}_3\text{PO}_4}{1\,\text{mol H}_3\text{PO}_4} = 9.06\,\text{g H}_3\text{PO}_4$$

Therefore, P_4O_{10} is the limiting reagent, and only 2.07 g of H_3PO_4 is produced.

e. $$2.07\,\text{g H}_3\text{PO}_4\times\frac{1\,\text{mol H}_3\text{PO}_4}{97.99\,\text{g H}_3\text{PO}_4}\times\frac{1\,\text{mol reaction}}{4\,\text{mol H}_3\text{PO}_4}\times\frac{-453\,\text{kJ}}{1\,\text{mol reaction}} = -2.39\,\text{kJ}$$

f. P_4O_{10} is the limiting reagent and will produce heat equal to

$$10.0\,\text{g}\,P_4O_{10} \times \frac{1\,\text{mol}\,P_4O_{10}}{283.88\,\text{g}\,P_4O_{10}} \times \frac{1\,\text{mol reaction}}{1\,\text{mol}\,P_4O_{10}} \times \frac{-453\,\text{kJ}}{1\,\text{mol reaction}}$$

$$= -15.9\,\text{kJ}$$

Assuming all the heat went into the water, the temperature change will be

$$\Delta T = \frac{q}{mc} = \frac{+15.9\,\text{kJ} \times \dfrac{1000\,\text{J}}{1\,\text{kJ}}}{(1000\,\text{g})(4.184\,\text{J/g}\cdot{}^\circ\text{C})} = 3.80^\circ\text{C}$$

Therefore, the final temperature will be $T = 25.0 + 3.80 =$ **28.8 °C.**

(Note that we neglected the amount of water that was consumed in the reaction, but we also neglected the heat that would be absorbed by the phosphoric acid that is now in the water. These effects are offsetting and will not change the final answer measurably.)

Chapter 6: Quantum Chemistry: The Strange World of Atoms

The Bottom Line

- We can interconvert among the wavelength, frequency, and energy of electromagnetic radiation. (Section 6.2)
- Atoms and molecules absorb and emit electromagnetic radiation to gain and lose energy, but only certain wavelengths of radiation can be absorbed and emitted. (Section 6.3)
- The wavelengths of electromagnetic radiation absorbed and emitted by an atom are characteristic of that atom and can be used to probe the atomic structure. (Section 6.3)
- One of the first models of atomic structure, the Bohr model, quantized the energies and spatial locations of the electrons to explain the hydrogen emission spectrum. Although the model was not correct in other details, quantum energy and orbital locations were breakthrough concepts and are key concepts in the modern picture of the atom. (Section 6.4)
- Electronic transitions between the quantized energy levels are responsible for atomic absorption and emission spectra. The energies are give by $\Delta E = E_f - E_i$, where the subscripts i and f stand for initial and final states. (Section 6.4)
- The energy of electromagnetic radiation absorbed or emitted in an electronic transition must exactly match the energy of the transition. (Section 6.4)
- At the atomic scale, all things show both wave and particle behavior. This concept is known as wave– particle duality. (Section 6.5)
- The wave and particle natures of quantum objects, such as electrons and photons, are linked in the de Broglie equation: $\lambda = h/p$. (Section 6.6)
- The Heisenberg uncertainty principle, $\Delta x \Delta p \geq h/4\pi$, places limits on how precisely we can simultaneously measure position and momentum. (Section 6.7)
- All quantum systems, like the hydrogen atom, can be completely described by the Schrödinger equation: $\hat{H}\Psi_n = E_n\Psi_n$ (Section 6.9)
- The probability distribution function $\Psi_n^*\Psi_n$ tells us where to find electrons in the atom, and the shape described by this function is called the atomic orbital. (Section 6.9)
- The orbitals are associated with the probability of finding the electron within a certain region of space. (Section 6.9)
- Each electron wave function has a set of four quantum numbers associated with it. They are n, the principal quantum number; l, the angular momentum quantum number; m_l, the magnetic quantum number; and m_s, the electron spin quantum number. (Section 6.10)
- The principal quantum number n indicates the electron shell that an electron is in, and n, l, and m_l determine the shape and orientation of the orbital in three-dimensional space. (Section 6.10)
- Only two possible electron spin states exist: $m_s = +1/2$ and $m_s = -1/2$. (Section 6.10)
- The Pauli exclusion principle states that in a given atom, no two electrons can have the same set of the four quantum numbers – n, l, m_l, and m_s. (Section 6.11)
- We can envisage multielectron atoms being assembled by placing electrons into the orbitals one by one, starting with the lowest energy level, according to the Aufbau principle. The complete listing of occupied orbitals is called the ground-state electron configuration. (Section 6.13)

- Hund's rule for ground-state electron configurations states that when orbitals of equal energy are available, the lowest energy configuration for an atom is the one with the maximum number of unpaired electrons with parallel spins. (Section 6.13)

Solutions to Practice Problems

6.1. To find the wavelength of the radiation from its frequency, we can use the relation $\nu = \frac{c}{\lambda}$ and solve for the wavelength: $\lambda = \frac{c}{\nu} = \frac{3.00\times10^{8}\text{ m/s}}{7.45\times10^{14}\text{ /s}} = 4.03\times10^{-7}\text{ m} = 403\text{ nm}$. Because the frequency of the light is higher, the wavelength should be shorter than in the exercise, as it is.

6.2. To find the energy of the radiation from its wavelength, we can use the relation

$$E = \frac{hc}{\lambda} = \frac{(6.626\times10^{-34}\text{ J}\cdot\text{s})\times(3.00\times10^{8}\text{ m/s})}{8.0\times10^{-3}\text{ m}} = 2.48\times10^{-23}\text{ J}.$$

6.3. For each possibility, we can use the Rydberg formula and the appropriate values for n_f and n_i:

$$\frac{1}{\lambda} = 1.0968\times10^{-2}\left[\frac{1}{n_f^2} - \frac{1}{n_i^2}\right]\text{ nm}^{-1}$$

from $n = 4$ to $n = 3$: $\frac{1}{\lambda} = 1.0968\times10^{-2}\left[\frac{1}{3^2} - \frac{1}{4^2}\right]\text{ nm}^{-1}$

$= 5.3317\times10^{-4}\text{ nm}^{-1}$

$\lambda = (5.3317\times10^{-4}\text{ nm}^{-1})^{-1} = 1875.6\text{ nm}$

from $n = 6$ to $n = 4$: $\frac{1}{\lambda} = 1.0968\times10^{-2}\left[\frac{1}{4^2} - \frac{1}{6^2}\right]\text{ nm}^{-1}$

$= 3.8083\times10^{-4}\text{ nm}^{-1}$

$\lambda = (3.8083\times10^{-4}\text{ nm}^{-1})^{-1} = 2625.8\text{ nm}$

from $n = 6$ to $n = 5$: $\frac{1}{\lambda} = 1.0968\times10^{-2}\left[\frac{1}{5^2} - \frac{1}{6^2}\right]\text{ nm}^{-1}$

$= 1.3405\times10^{-4}\text{ nm}^{-1}$

$\lambda = (1.3405\times10^{-4}\text{ nm}^{-1})^{-1} = 7459.7\text{ nm}$

6.4. As seen in Exercise 6.4, the ground-state energy of hydrogen is -2.1786×10^{-18} J. The energy difference in the transition is $\Delta E = E_f - E_i = (-2.1786\times10^{-18}\text{ J}) - (-1.5129\times10^{-20}\text{ J}) = -2.16347\times10^{-18}$ J. The wavelength of the photon is

$$\lambda = \frac{hc}{-\Delta E} = \frac{(6.626\times10^{-34}\text{ J}\cdot\text{s})\times(3.00\times10^{8}\text{ m/s})}{-(-2.16347\times10^{-18}\text{ J})} = 9.19\times10^{-8}\text{m} = 91.9\text{ nm}$$

6.5. To complete the problem, we first need to convert the mass from pounds to kilograms and the speed from miles per hour to meters per second:

$$1200\,\text{lb} \times \frac{1\,\text{kg}}{2.205\,\text{lb}} = 544.2\,\text{kg}$$

$$\frac{75\,\text{mi}}{1\,\text{h}} \times \frac{1.609\,\text{km}}{1\,\text{mi}} \times \frac{1000\,\text{m}}{1\,\text{km}} \times \frac{1\,\text{h}}{60\,\text{min}} \times \frac{1\,\text{min}}{60\,\text{s}} = 33.52\,\text{m/s}$$

The de Broglie wavelength is $\lambda = \dfrac{h}{mv} = \dfrac{6.626\times10^{-34}\ \text{J}\cdot\text{s}}{(544.2\,\text{kg})\times(33.52\,\text{m/s})} = 3.6\times10^{-38}\ \text{m}$.

This wavelength is so small as to be meaningless for everyday objects moving at normal speeds.

6.6. To find the momentum of a mole of photons, we first find the momentum of a single photon using the de Broglie relation, and then multiply that value by Avogadro's number.

For 200 nm: $p = \dfrac{h}{\lambda} = \dfrac{6.626\times10^{-34}\ \text{J}\cdot\text{s}}{200\times\ 10^{-9}\ \text{m}} = \left(3.313\times10^{-27}\,\text{kg}\cdot\text{m/s}\right)\times\left(6.022\times10^{23}\ /\text{mol}\right)$

$$= 2.00\times10^{-3}\text{kg}\cdot\text{m}\cdot\text{s}^{-1}\cdot\text{mol}^{-1}$$

For 700 nm: $p = \dfrac{h}{\lambda} = \dfrac{6.626\times10^{-34}\ \text{J}\cdot\text{s}}{700\times\ 10^{-9}\ \text{m}} = \left(9.466\times10^{-28}\,\text{kg}\cdot\text{m/s}\right)\times\left(6.022\times10^{23}\ /\text{mol}\right)$

$$= 5.70\times10^{-4}\text{kg}\cdot\text{m}\cdot\text{s}^{-1}\cdot\text{mol}^{-1}$$

For 1000 nm: $p = \dfrac{h}{\lambda} = \dfrac{6.626\times10^{-34}\ \text{J}\cdot\text{s}}{1000\times\ 10^{-9}\ \text{m}} = \left(6.626\times10^{-28}\,\text{kg}\cdot\text{m/s}\right)\times\left(6.022\times10^{23}\ /\text{mol}\right)$

$$= 3.99\times10^{-4}\text{kg}\cdot\text{m}\cdot\text{s}^{-1}\cdot\text{mol}^{-1}$$

Because the momentum is inversely proportional to the wavelength, there is a marked difference.

6.7. With $n = 2$, we know the orbital is in the second level; with $l = 1$ and $m_l = 0$, we know the orbital is one of the p orbitals. We can identify this orbital with $2p$.

6.8. Gallium in its ground state has 31 electrons and will have the configuration $1s^2 2s^2 2p^6 3s^2 3p^6 4s^2 3d^{10} 4p^1$, which could also be written $[\text{Ar}]4s^2 3d^{10} 4p^1$

Strontium in its ground state has 38 electrons and will have the configuration $1s^2 2s^2 2p^6 3s^2 3p^6 4s^2 3d^{10} 4p^6 5s^2$, which could also be written $[\text{Kr}]5s^2$.

Solutions to Student Problems

1. The answers to this question will vary widely. One "macro" example is batting averages for baseball players. While in a single "at bat," the batter will either get a hit (1.000) or not (0.000), the long-term average will state that the batter might hit safely in only 30% of the times at bat (0.300).

3. Across the visible spectrum, the colors in increasing frequency go from red to violet, so the produce would be found in this order: red tomato, yellow squash, green squash, and finally purple eggplant.

5. a. To find the frequency, we use $\nu = \dfrac{c}{\lambda} = \dfrac{3.00\times10^{8}\ \text{m/s}}{0.015\ \text{nm}\times\dfrac{10^{-9}\ \text{m}}{1\ \text{nm}}} = 2.0\times10^{19}\ /\text{s}$.

b. The energy is given by $E = h\nu = \left(6.626\times10^{-34}\ \text{J}\cdot\text{s}\right)\left(2.0\times10^{19}\ /\text{s}\right) = 1.3\times10^{-14}\ \text{J}$.

c. The energy in one mole is $E = \left(1.3\times10^{-14}\ \text{J}\right)\times\left(6.022\times10^{23}\ /\text{mol}\right) = 8.0\times10^{9}\ \text{J/mol}$

7. 2.5×10^{4} nm = 2.5×10^{-5} m, which falls in the infrared region of the electromagnetic spectrum.

The frequency is $\nu = \dfrac{c}{\lambda} = \dfrac{3.00\times10^{8}\ \text{m/s}}{2.5\times10^{4}\ \text{nm}\times\dfrac{10^{-9}\ \text{m}}{1\ \text{nm}}} = 1.2\times10^{13}\ /\text{s}$.

The energy is given by $E = h\nu = \left(6.626\times10^{-34}\ \text{J}\cdot\text{s}\right)\left(1.2\times10^{13}\ /\text{s}\right) = 8.0\times10^{-21}\ \text{J}$.

9. Your answer will depend on your actual height. The following answers are calculated for a height of 5 ft 6 in.

$$66.0\ \text{in}\times\frac{2.54\ \text{cm}}{1\ \text{in}}\times\frac{1\ \text{m}}{100\ \text{cm}} = 1.68\ \text{m}$$

$$1.68\ \text{m}\times\frac{10^{9}\ \text{nm}}{1\ \text{m}} = 1.68\times10^{9}\ \text{nm}$$

$$1.68\text{m}\times\left(\frac{1\ \text{light}-\text{s}}{3.00\times10^{8}\ \text{m}}\times\frac{1\ \text{min}}{60\ \text{s}}\times\frac{1\ \text{h}}{60\ \text{min}}\times\frac{1\ \text{day}}{24\ \text{h}}\times\frac{1\ \text{yr}}{365.25\ \text{day}}\right) = 1.77\times10^{-16}\ \text{light}-\text{years}$$

Meters are the most convenient units.

11. One megahertz = 10^{6} Hz. Converting these to wavelengths:

$$\lambda = \frac{c}{\nu} = \frac{3.00\times10^{8}\ \text{m/s}}{824\times10^{6}\ /\text{s}} = 0.364\ \text{m}$$

$$\lambda = \frac{c}{\nu} = \frac{3.00\times10^{8}\ \text{m/s}}{894\times10^{6}\ /\text{s}} = 0.336\ \text{m}$$

These wavelengths fall in the radio portion of the electromagnetic spectrum.

13. The wavelength equivalent of 98.3 MHz is $\lambda = \dfrac{c}{\nu} = \dfrac{3.00\times10^{8}\ \text{m/s}}{98.3\times10^{6}\ /\text{s}} = 3.05\ \text{m}$.

15. To convert directly from wavelength to energy, we use the relation $E = \dfrac{hc}{\lambda}$:

For 675 nm: $E = \dfrac{hc}{\lambda} = \dfrac{\left(6.626\times10^{-34}\ \text{J}\cdot\text{s}\right)\left(3.00\times10^{8}\ \text{m/s}\right)}{675\ \text{nm}\times\dfrac{10^{-9}\ \text{m}}{1\ \text{nm}}} = 2.94\times10^{-19}\ \text{J}$

For 440 nm: $E = \frac{hc}{\lambda} = \frac{(6.626\times10^{-34}\,\text{J}\cdot\text{s})(3.00\times10^{8}\,\text{m/s})}{440\,\text{nm}\times\frac{10^{-9}\,\text{m}}{1\,\text{nm}}} = 4.52\times10^{-19}\,\text{J}$

17. We can multiply the energies found in Problem 15 by Avogadro's number to answer the question:

For 675 nm: $E = (2.94\times10^{-19}\,\text{J})\times(6.022\times10^{23}\,/\text{mol}) = 1.77\times10^{5}\,\text{J/mol} = 177\,\text{kJ/mol}$

For 440 nm: $E = (4.52\times10^{-19}\,\text{J})\times(6.022\times10^{23}\,/\text{mol}) = 2.72\times10^{5}\,\text{J/mol} = 272\,\text{kJ/mol}$

19. a. To convert directly from wavelength to energy, we use the relation $E = \frac{hc}{\lambda}$.

For 565 nm: $E = \frac{hc}{\lambda} = \frac{(6.626\times10^{-34}\,\text{J}\cdot\text{s})(3.00\times10^{8}\,\text{m/s})}{565\,\text{nm}\times\frac{10^{-9}\,\text{m}}{1\,\text{nm}}} = 3.52\times10^{-19}\,\text{J}$

b. The energy in 8 photons (producing one molecule of O_2) =

$$E = \frac{3.52\times10^{-19}\,\text{J}}{\text{photon}}\times 8\,\text{photons} = 2.81\times10^{-18}\,\text{J}$$

21. The energy of 290-nm light:

$$E = \frac{hc}{\lambda} = \frac{(6.626\times10^{-34}\,\text{J}\cdot\text{s})(3.00\times10^{8}\,\text{m/s})}{290\,\text{nm}\times\frac{10^{-9}\,\text{m}}{1\,\text{nm}}} = 6.85\times10^{-19}\,\text{J/photon}$$

The frequency: $\nu = \frac{c}{\lambda} = \frac{3.00\times10^{8}\,\text{m/s}}{290\,\text{nm}\times\frac{10^{-9}\,\text{m}}{1\,\text{nm}}} = 1.03\times10^{15}\,/\text{s}$

23. The relationship of energy to wavelength is $E = \frac{hc}{\lambda}$, so

$$\lambda = \frac{hc}{E} = \frac{(6.626\times10^{-34}\,\text{J}\cdot\text{s})(3.00\times10^{8}\,\text{m/s})}{1.6\times10^{-8}\,\text{J}} = 1.2\times10^{-17}\,\text{m}$$

The frequency is $\nu = \frac{c}{\lambda} = \frac{3.00\times10^{8}\,\text{m/s}}{1.2\times10^{-17}\,\text{m}} = 2.5\times10^{25}\,/\text{s}$.

25. One angstrom (Å) is 1×10^{-10} m, so $\nu = \frac{c}{\lambda} = \frac{3.00\times10^{8}\,\text{m/s}}{1.54\times10^{-10}\,\text{m}} = 1.95\times10^{18}\,/\text{s}$. A wavelength of 1.54×10^{-10} m falls in the X-ray portion of the electromagnetic spectrum.

27. In meters: $500\,\text{nm} \times \dfrac{1\,\text{m}}{10^9\,\text{nm}} = 5.00 \times 10^{-7}\,\text{m}$

In angstroms: $500\,\text{nm} \times \dfrac{10\,\text{Å}}{1\,\text{nm}} = 5.00 \times 10^3\,\text{Å}$

In centimeters: $500\,\text{nm} \times \dfrac{1\,\text{m}}{10^9\,\text{nm}} \times \dfrac{100\,\text{cm}}{1\,\text{m}} = 5.00 \times 10^{-5}\,\text{cm}$

In inches: $500\,\text{nm} \times \dfrac{1\,\text{m}}{10^9\,\text{nm}} \times \dfrac{100\,\text{cm}}{1\,\text{m}} \times \dfrac{1\,\text{in}}{2.54\,\text{cm}} = 1.97 \times 10^{-5}\,\text{in}$.

29. Although the whole-number difference between the level is only one in each case, the emitted wavelengths are proportional to the difference in the inverse squares of the numbers. When we use the inverse squares, the values come out much different. In other words, both 5 − 4 = 1 and 2 − 1 = 1, but the differences between the inverse squares are much different:

$$\left[\frac{1}{4^2} - \frac{1}{5^2}\right] = 0.0225 \qquad \left[\frac{1}{1^2} - \frac{1}{2^2}\right] = 0.75$$

31. The first two wavelengths would correspond with $n_i = 3$ and $n_i = 4$, and the next three lines would have $n_i = 5$, 6, and 7 in the Balmer equation:

$$\frac{1}{\lambda} = 1.0968 \times 10^{-2} \left[\frac{1}{2^2} - \frac{1}{n^2}\right] \text{nm}^{-1}$$

$$\lambda = 91.174\,\text{nm} \times \frac{1}{\left[\frac{1}{2^2} - \frac{1}{n^2}\right]}$$

For $n_i = 5$: $\lambda = 91.174\,\text{nm} \times \dfrac{1}{\left[\frac{1}{2^2} - \frac{1}{5^2}\right]} = 434.16\,\text{nm} = 4341.6\,\text{Å}$

$$\nu = \frac{c}{\lambda} = \frac{3.00 \times 10^8\,\text{m/s}}{434.16\text{nm} \times \dfrac{10^{-9}\,\text{m}}{1\,\text{nm}}} = 6.91 \times 10^{14}\,/\text{s}$$

$$E = \frac{hc}{\lambda} = \frac{(6.626 \times 10^{-34}\,\text{J} \cdot \text{s})(3.00 \times 10^8\,\text{m/s})}{434.16\,\text{nm} \times \dfrac{10^{-9}\,\text{m}}{1\,\text{nm}}} = 4.58 \times 10^{-19}\,\text{J/photon}$$

For $n_i = 6$: $\lambda = 91.174\,\text{nm} \times \dfrac{1}{\left[\frac{1}{2^2} - \frac{1}{6^2}\right]} = 410.28\,\text{nm} = 4102.8\,\text{Å}$

$$\nu = \frac{c}{\lambda} = \frac{3.00 \times 10^8\,\text{m/s}}{410.28\text{nm} \times \dfrac{10^{-9}\,\text{m}}{1\,\text{nm}}} = 7.31 \times 10^{14}\,/\text{s}$$

$$E = \frac{hc}{\lambda} = \frac{(6.626\times10^{-34}\,\mathrm{J\cdot s})(3.00\times10^{8}\,\mathrm{m/s})}{410.28\,\mathrm{nm}\times\dfrac{10^{-9}\,\mathrm{m}}{1\,\mathrm{nm}}} = 4.84\times10^{-19}\,\mathrm{J/photon}$$

For $n_i = 7$: $\lambda = 91.174\,\mathrm{nm}\times\dfrac{1}{\left[\dfrac{1}{2^2} - \dfrac{1}{7^2}\right]} = 397.11\,\mathrm{nm} = 3971.1\,\text{Å}$

$$\nu = \frac{c}{\lambda} = \frac{3.00\times10^{8}\,\mathrm{m/s}}{397.11\,\mathrm{nm}\times\dfrac{10^{-9}\,\mathrm{m}}{1\,\mathrm{nm}}} = 7.55\times10^{14}\,/\mathrm{s}\,.$$

$$E = \frac{hc}{\lambda} = \frac{(6.626\times10^{-34}\,\mathrm{J\cdot s})(3.00\times10^{8}\,\mathrm{m/s})}{397.11\,\mathrm{nm}\times\dfrac{10^{-9}\,\mathrm{m}}{1\,\mathrm{nm}}} = 5.01\times10^{-19}\,\mathrm{J/photon}$$

33. a. $\nu = \dfrac{c}{\lambda} = \dfrac{3.00\times10^{8}\,\mathrm{m/s}}{214.439\,\mathrm{nm}\times\dfrac{10^{-9}\,\mathrm{m}}{1\,\mathrm{nm}}} = 1.40\times10^{15}\,/\mathrm{s}$

b. The wavelength of 214.439 nm falls in the ultraviolet region of the electromagnetic spectrum.

c. Zinc, which emits light at 213.857 nm, may be difficult to distinguish from cadmium.

35. a. The Bohr energy levels are given by $E_n = -\dfrac{2.1786\times10^{-18}\,\mathrm{J}}{n^2}$, so

for $n=1$: $E_1 = -\dfrac{2.1786\times10^{-18}\,\mathrm{J}}{1^2} = -2.1786\times10^{-18}\,\mathrm{J}$.

b. For $n = 3$: $E_3 = -\dfrac{2.1786\times10^{-18}\,\mathrm{J}}{3^2} = -2.4207\times10^{-19}\,\mathrm{J}$

c. For $n = 5$: $E_5 = -\dfrac{2.1786\times10^{-18}\,\mathrm{J}}{5^2} = -8.7144\times10^{-20}\,\mathrm{J}$

d. For $n = 7$: $E_7 = -\dfrac{2.1786\times10^{-18}\,\mathrm{J}}{7^2} = -4.4461\times10^{-20}\,\mathrm{J}$

37. a. We can use the formula

$$E_{h\nu} = -\Delta E = -\left(-2.1786\times10^{-18}\,\mathrm{J}\left[\frac{1}{n_f^2} - \frac{1}{n_i^2}\right]\right) = +2.1786\times10^{-18}\,\mathrm{J}\left[\frac{1}{n_f^2} - \frac{1}{n_i^2}\right]$$

directly to calculate the energy changes and photon energy. Note that for an emitted photon (n decreases), ΔE is negative but $E_{h\nu}$ is always positive.

For $n = 4$ to $n = 1$: $E_{h\nu} = 2.1786\times10^{-18}\,\mathrm{J}\left[\dfrac{1}{1^2} - \dfrac{1}{4^2}\right] = 2.0424\times10^{-18}\,\mathrm{J}$

b. For n=3 to n=1: $E_{hv} = 2.1786 \times 10^{-18}\text{ J}\left[\frac{1}{1^2} - \frac{1}{3^2}\right] = 1.9365 \times 10^{-18}\text{ J}$

c. For n=5 to n=4: $E_{hv} = 2.1786 \times 10^{-18}\text{ J}\left[\frac{1}{4^2} - \frac{1}{5^2}\right] = 4.9018 \times 10^{-20}\text{ J}$

d. For n = 7 to n = 2: $E_{hv} = 2.1786 \times 10^{-18}\text{ J}\left[\frac{1}{2^2} - \frac{1}{7^2}\right] = 5.0019 \times 10^{-19}\text{ J}$

39. a. We can use the formula

$$\frac{1}{\lambda} = -1.0968 \times 10^{-2}\left[\frac{1}{n_f^{\,2}} - \frac{1}{n_i^{\,2}}\right]\text{nm}^{-1}$$

$$\lambda = -91.174\,\text{nm} \times \frac{1}{\left[\frac{1}{n_f^{\,2}} - \frac{1}{n_i^{\,2}}\right]}$$

directly to calculate the photon wavelength.

For n = 1 to n = 2: $\lambda = -91.174\text{nm} \times \frac{1}{\left[\frac{1}{2^2} - \frac{1}{1^2}\right]} = 121.57\,\text{nm}$

b. For n = 3 to n = 5: $\lambda = -91.174\text{nm} \times \frac{1}{\left[\frac{1}{5^2} - \frac{1}{3^2}\right]} = 1282.1\,\text{nm}$

c. For n = 2 to n = 4: $\lambda = -91.174\text{nm} \times \frac{1}{\left[\frac{1}{4^2} - \frac{1}{2^2}\right]} = 486.26\,\text{nm}$

d. For n = 1 to n = 6: $\lambda = -91.174\text{nm} \times \frac{1}{\left[\frac{1}{6^2} - \frac{1}{1^2}\right]} = 93.779\,\text{nm}$

e. For n = 1 to n = 5: $\lambda = -91.174\text{nm} \times \frac{1}{\left[\frac{1}{5^2} - \frac{1}{1^2}\right]} = 94.973\,\text{nm}$

f. For n = 3 to n = 4: $\lambda = -91.174\text{nm} \times \frac{1}{\left[\frac{1}{4^2} - \frac{1}{3^2}\right]} = 1875.6\,\text{nm}$

41. The Brackett series has $n_f = 4$; the shortest wavelength (highest energy) occurs with $n_i = \infty$.

$$\frac{1}{\lambda} = 1.0968 \times 10^{-2}\left[\frac{1}{4^2} - \frac{1}{n^2}\right]\text{nm}^{-1}$$

$$\lambda = 91.174\text{nm} \times \frac{1}{\left[\frac{1}{4^2} - \frac{1}{n^2}\right]} = 91.174\text{nm} \times \frac{1}{\left[\frac{1}{4^2} - \frac{1}{\infty^2}\right]} = 1458.8\,\text{nm}$$

The longest wavelength (lowest energy) in the series occurs with $n_i = 5$.

$$\lambda = 91.174\text{nm} \times \frac{1}{\left[\frac{1}{4^2} - \frac{1}{5^2}\right]} = 4052.2\,\text{nm}$$

43. An energy of 4.84 × 10^{-19} J corresponds very closely with the emission of a photon from the transition $n = 6$ to $n = 2$:

$$\Delta E = -2.1786 \times 10^{-18}\ \text{J}\left[\frac{1}{n_f^2} - \frac{1}{n_i^2}\right] = -2.1786 \times 10^{-18}\ \text{J}\left[\frac{1}{2^2} - \frac{1}{6^2}\right] = -4.8413 \times 10^{-19}\ \text{J}$$

This emission is possibly from hydrogen.

45. $\Delta E = -2.1786 \times 10^{-18}\ \text{J}\left[\frac{1}{n_f^2} - \frac{1}{n_i^2}\right] = -2.1786 \times 10^{-18}\ \text{J}\left[\frac{1}{\infty^2} - \frac{1}{1^2}\right] = 2.1786 \times 10^{-18}\ \text{J}$

Note that the positive value indicates that energy is required.

47. Classically, matter has mass in discrete particles and can therefore have momentum; further, any possible value of energy (and momentum) or position is allowed. Both position and momentum can be measured (at the same time) to infinite precision. Waves can have specific, but infinitely variable, energies and amplitudes, but they do not have a specific location because they are spread out over space and time.

49. Momentum, p, is given by the relation $p = m \times v$, so for this electron,

$$p = \left(9.1\times10^{-31}\ \text{kg}\right)\times\left(0.68\times3.0\times10^{8}\ \text{m/s}\right) = 1.9\times10^{-22}\ \text{kg}\cdot\text{m/s}$$

51. To complete the problem, we first need to convert the mass from pounds to kilograms and the speed from miles per hour to meters per second:

$$415000\,\text{lb} \times \frac{1\,\text{kg}}{2.205\,\text{lb}} = 1.88\times10^{5}\ \text{kg}$$

$$\frac{100.0\,\text{mi}}{1\,\text{h}} \times \frac{1.609\,\text{km}}{1\,\text{mi}} \times \frac{1000\,\text{m}}{1\,\text{km}} \times \frac{1\,\text{h}}{60\,\text{min}} \times \frac{1\,\text{min}}{60\,\text{s}} = 44.7\,\text{m/s}$$

The momentum is $p = \left(1.88\times10^{5}\ \text{kg}\right)\times\left(44.7\,\text{m/s}\right) = 8.4\times10^{6}\text{kg}\cdot\text{m/s}$.

53. For photons, it is necessary to use the de Broglie relation:

$$p = \frac{h}{\lambda} = \frac{6.626\times10^{-34}\ \text{J}\cdot\text{s}}{540\text{nm}\times\frac{10^{-9}\ \text{m}}{1\ \text{nm}}} = 1.23\times10^{-27}\ \text{kg}\cdot\text{m/s}$$

55. Two particle-like properties are that electrons have mass and can have discrete position.

57. Using the relation, we can solve for the wavelength:

$$\lambda = \frac{2\pi r}{n} = \frac{2\pi\left(2.116\times10^{-10}\,\text{m}\right)}{2} = 6.648\times10^{-10}\ \text{m} = 0.6648\,\text{nm}$$

59. To calculate the wavelength, we must first covert the speed in mph to m/s.

$$\frac{78\,\text{mi}}{1\,\text{h}}\times\frac{1.609\,\text{km}}{1\,\text{mi}}\times\frac{1000\,\text{m}}{1\,\text{km}}\times\frac{1\,\text{h}}{60\,\text{min}}\times\frac{1\,\text{min}}{60\,\text{s}}=34.8\,\text{m/s}=35\,\text{m/s}$$

The wavelength according to de Broglie's relation is

$$\lambda=\frac{h}{p}=\frac{h}{mv}=\frac{6.626\times10^{-34}\ \text{J}\cdot\text{s}}{5.00\,\text{g}\times\frac{1\,\text{kg}}{1000\,\text{g}}\times34_{.8}\ \text{m/s}}=3.8\times10^{-33}\,\text{m}$$

A softball moving at the same speed, with its higher mass, would have a smaller wavelength. (Note in the equation that the mass and wavelength are inversely proportional.)

61. Here p represents the momentum of the particle, and x represents its position. Because the uncertainties are multiplied, the more certain the measurement of one is, the more *uncertain* the value for the other must be. It is not possible to measure both properties to infinite precision.

63. For 5.0%: The uncertainty in the velocity is (0.050)(2.1 × 10^6 m/s) = 1.0$_5$ × 10^5 m/s. The uncertainty in the momentum, according to the Heisenberg Uncertainty Principle, would be

$$\Delta x\Delta p_x \ \geq \frac{h}{4\pi},\ \Delta p_x = m\,\Delta v_x$$

$$\Delta x \geq \frac{h}{4\pi(m\,\Delta v_x)}=\frac{6.626\text{x}10^{-34}\,\text{J}\cdot\text{s}}{4\pi(9.11\times10^{-31}\,\text{kg})\times(1.0_5\times10^5\,\text{m/s})}$$

$$\Delta x \geq 5.5\times10^{-10}\,\text{m}$$

For 10.0%: The uncertainty in the velocity is (0.100)(2.1 × 10^6 m/s) = 2.1 × 10^5 m/s. The uncertainty in the momentum, according to the Heisenberg Uncertainty Principle, would be

$$\Delta x\Delta p_x \ \geq \frac{h}{4\pi},\ \Delta p_x = m\,\Delta v_x$$

$$\Delta x \geq \frac{h}{4\pi(m\,\Delta v_x)}=\frac{6.626\times10^{-34}\,\text{J}\cdot\text{s}}{4\pi(9.11\times10^{-31}\,\text{kg})\times(2.1\times10^5\,\text{m/s})}$$

$$\Delta x \geq 2.8\times10^{-10}\,\text{m}$$

Both values are larger than the size of the hydrogen radius (3.7 × 10^{-11} m)!

65. Using the de Broglie relation: $p=\frac{h}{\lambda}=\frac{6.626\times10^{-34}\ \text{J}\cdot\text{s}}{6.5\times10^{-10}\text{m}}=1.0\times10^{-24}\,\text{kg}\cdot\text{m/s}$

67. Simply multiply the value from Problem 65 by Avogadro's number:

$$p=\left(1.0\times10^{-24}\,\text{kg}\cdot\text{m/s}\right)\times\left(6.022\times10^{23}\ /\ \text{mol}\right)=0.60\,\text{kg}\cdot\text{m}\cdot\text{s}^{-1}\cdot\text{mol}^{-1}$$

69. Converting the mass to kilograms and the speed to meters per second:

$$190\,\text{lb}\times\frac{1\,\text{kg}}{2.205\,\text{lb}}=86.2\,\text{kg}$$

$$\frac{3.0\,\text{mi}}{1\,\text{h}}\times\frac{1.609\,\text{km}}{1\,\text{mi}}\times\frac{1000\,\text{m}}{1\,\text{km}}\times\frac{1\,\text{h}}{60\,\text{min}}\times\frac{1\,\text{min}}{60\,\text{s}}=1.3\,\text{m/s}$$

The momentum is $p = m\times v = (82.6\,\text{kg})\times(1.3\,\text{m/s}) = 1.1\times10^{2}\,\text{kg}\cdot\text{m/s}$.

71. The electron orbital describes the energy state and relative position in space of the electron within the atom. Because the position is described in terms of a probability, we cannot specify exactly where the electron is, but we can in some cases say where it is not and where it is most likely (but not required) to be.

73. It is easier to start thinking about this problem in terms of the Bohr model. The Bohr orbits are stable because the wavelengths perfectly overlap with themselves after completing the orbit. In other words, they constructively interfere and become stable standing waves. At any other radius, the electron wavelengths will not overlap correctly after completing the orbit and will destructively interfere. No stable orbit is possible. Because there are no stable orbits possible between two adjacent states, the transition from one state to another is abrupt and not gradual. The same thinking applies to the more complicated Schrödinger equation.

75. Only sequence (e) is valid. Sequences (a), (b), and (d) are not possible because l can have only *positive* integer values that *must be less than n*. Sequence (c) is also not allowed because m_l must have a magnitude less than or equal to l.

77. In the fifth energy level ($n = 5$), there can be five sublevels associated with $l = 0, 1, 2, 3, 4$. For each l value there are a set number of orbitals because m_l can have values from $-l$ to $+l$ totaling $2l + 1$. There are nine orbitals for $l = 4$, seven orbitals for $l = 3$, five orbitals for $l = 2$, three orbitals for $l = 1$, and one orbital for $l = 0$. Total orbitals = 9 + 7 + 5 + 3 + 1 = 25. In general, there are n^2 orbitals in the nth level.

79. For each l value there are a set number of orbitals because m_l can have values from $-l$ to $+l$ totaling $2l + 1$. $l = 3$ has 2(3) +1 = 7 degenerate orbitals.

81. For n=5 there are n^2 distinct orbitals = 25. (See question 77.) Each orbital can accept two electrons with distinct quantum spins giving 2(25) = 50 possible sets of quantum numbers.

83. Usually, the "shape" of an orbital refers to the surface within which 90% of the total probability of finding an electron is found. Very little of the electron probability occurs exactly on the surface, and the electron probability is spread throughout much of that space. By specifying these "shapes," we can better visualize where the most probable places for finding the electron are. There is still a 10% chance of finding the electron outside the usual surface.

85. A radial node exists at all angles at a given distance from the nuclear center. For example, the 2*s* orbital has a probability of 0 at a given distance from the nucleus. A planar node exists for a particular angle at all radial distances and usually passes through the nucleus. For example, the 2*p* orbitals (*x*, *y*, and *z*) have a node passing through the nucleus, separating the two parts of the orbital.

87. (a) Nothing actually touches in the conventional sense. When the atomic-sized needle tip comes close enough to the sample material, the wavefunction of the atom at the very tip of the needle overlaps the wavefunction of the nearest atoms in the material. With sufficient overlap, the electrons from the tip can move to the sample. This movement of electrons produces a current that is detected by the STM. (b) Tunneling is the movement of a particle between two allowed spaces (or orbitals) through a space where it could not be classically. Classically, this current could not flow until the two physically touched. The closer the tip to the surface of an atom, the greater the overlap of the orbitals and the greater the current that arises as the electrons move through space from one allowed orbital to the other.

89. We use arrows to represent the electron spin quantum number, m_s. An upwards arrow represents the +½ possibility.

91. C and O have unpaired electrons. The remaining atoms completely fill all sets of orbitals, leaving no electrons unpaired.

93. In the extreme example, we could place all electrons in the 1*s* orbital, choosing half with spin up and half with spin down. (However, we could also choose to have every electron spin up and all unpaired if the Pauli Exclusion Principle did not apply.) There are many other possibilities.

95. In multielectron atoms, the shapes of the orbitals are still very similar, the quantum numbers for the orbitals and the rules for using them remain the same, and the behavior of the nodes within the orbitals remains the same. The energies of the orbitals are changed!

97. It will be easier to remove the electron from He. Removing an electron from He leaves behind a +1 ion, whose attraction for the electron must be overcome to remove it. But removing an electron from He^+ leaves behind a +2 ion, whose attraction for the electron will be at least twice as strong, making it more difficult to remove the electron.

99. a. $N = 1s^22s^22p_x^{\ 1}2p_y^{\ 1}2p_z^{\ 1} = 1s^22s^22p^3$
b. $N^{3-} = 1s^22s^22p^6$

101. a. Si b. Cl c. K d. Sr

103. Element 21 (Sc): $1s^22s^22p^63s^23p^64s^23d^1$. Because of the effects of shielding on the orbital energies, the 4*s* orbitals lie at lower energy than the 3*d* orbitals and fill first.

105. K and Fe have unpaired electrons in the ground state. All the remaining atoms have fully filled shells or subshells.

107. In order to fill the *d* orbitals in the ground state of Cu, one 4*s* electron must be moved to fill the last 3*d* orbital. The configuration would be $1s^22s^22p^63s^23p^64s^13d^{10}$.

109. a. The speed of sound is 345 m/s in dry air at sea level. The ratio of the speed of light to the speed of sound is $\dfrac{3.00\times10^8\ \text{m/s}}{345\ \text{m/s}} = 8.70\times10^5$.

b. The speed of light in mph is

$$\frac{3.00\times10^8\ \text{m}}{1\text{s}}\times\frac{1\,\text{km}}{1000\,\text{m}}\times\frac{1\,\text{mi}}{1.609\,\text{km}}\times\frac{60\,\text{s}}{1\,\text{min}}\times\frac{60\,\text{min}}{1\,\text{h}} = 6.71\times10^8\ \text{mph}$$

111. The Lyman series has $n_f = 1$. The longest wavelength (smallest energy) occurs in the transition from $n_i = 2$.

$$\frac{1}{\lambda} = 1.0968 \times 10^{-2}\left[\frac{1}{1^2} - \frac{1}{n^2}\right]\text{nm}^{-1}$$

$$\lambda = 91.174\text{nm} \times \frac{1}{\left[\frac{1}{1^2} - \frac{1}{n^2}\right]} = 91.174\text{nm} \times \frac{1}{\left[\frac{1}{1^2} - \frac{1}{2^2}\right]} = 121.56\,\text{nm}$$

113. Using the equation $r = 0.052917$ nm $\times n^2$, the first six radii are

$r_1 = 0.052917\text{nm} \times 1^2 = 0.052917$ nm
$r_2 = 0.052917\text{nm} \times 2^2 = 0.21167$ nm $\quad \Delta r_{1,2} = 0.15875$ nm
$r_3 = 0.052917\text{nm} \times 3^2 = 0.47625$ nm $\quad \Delta r_{2,3} = 0.26458$ nm
$r_4 = 0.052917\text{nm} \times 4^2 = 0.84667$ nm $\quad \Delta r_{3,4} = 0.37042$ nm
$r_5 = 0.052917\text{nm} \times 5^2 = 1.3229$ nm $\quad \Delta r_{4,5} = 0.47623$ nm
$r_6 = 0.052917\text{nm} \times 6^2 = 1.9050$ nm $\quad \Delta r_{5,6} = 0.5821$ nm

The distance between successive levels continues to increase; the shells are becoming increasingly further apart.

115. We must think about the wavelength of the electron inside the atom as the electron orbits the nucleus. In order to be stable, the electron must create a standing wave. The only way to create a standing wave is for the wave to perfectly overlap itself after a single circumference. Perfect overlap occurs only if an integer number of waves lie on the circumference, so n must be a positive integer.

117. a. (1, 0, 0) The first level with the s orbital is n = 1, all s orbitals have $l = 0$ and $m_l = 0$.
b. (2, 1, 0) The first level with the p orbital is $n = 2$, all p orbitals have $l = 1$ and $m_l = -1, 0, +1$. (Any of the three values for m_l are allowed, depending on the orientation.)
c. (3, 2, 1) The first level with the d orbital is $n = 3$, all d orbitals have $l = 2$ and $m_l = -2, -1, 0, +1, +2$. (Any of the three values for m_l are allowed, depending on the orientation, although $m_l = 0$ is usually taken to be the d_{z^2} orbital.)

119. a. For one photon:

$$E = \frac{hc}{\lambda} = \frac{(6.626\times10^{-34}\,\text{J}\cdot\text{s})(3.00\times10^{8}\,\text{m/s})}{2165\,\text{nm}\times\frac{10^{-9}\,\text{m}}{1\,\text{nm}}} = 9.18\times10^{-20}\,\text{J}$$

For one mole: $E = (9.18\times10^{-20}\,\text{J})\times(6.022\times10^{23}/\text{mol}) = 5.53\times10^{4}$ J/mol

b. $$\nu = \frac{c}{\lambda} = \frac{3.00\times10^{8}\,\text{m/s}}{2165\,\text{nm}\times\frac{10^{-9}\,\text{m}}{1\,\text{nm}}} = 1.39\times10^{14}\,\text{s}^{-1}$$

c. 2165 nm is in the infrared region of the electromagnetic spectrum.

d. For an absorbed photon, we will need to add a negative sign before the constant in the expression we have used to calculate wavelengths of emitted light.

$$\frac{1}{\lambda} = -1.0968 \times 10^{-2} \left[\frac{1}{n^2} - \frac{1}{4^2}\right] \text{nm}^{-1}$$

$$\frac{1}{2165\,\text{nm}} = -1.0968 \times 10^{-2} \left[\frac{1}{n^2} - \frac{1}{4^2}\right] \text{nm}^{-1}$$

$$-0.04211 = \frac{1}{n^2} - \frac{1}{4^2}$$

$$0.02039 = \frac{1}{n^2}$$

$$n^2 = 49.0$$

$$n = \sqrt{49.0} = 7.0$$

e. Starting with the expression for emitted wavelength from the Bohr model:

$$\frac{1}{\lambda} = 1.0968 \times 10^{-2} \left[\frac{1}{{n_f}^2} - \frac{1}{{n_i}^2}\right] \text{nm}^{-1}$$

$$\lambda = 91.174\,\text{nm} \times \frac{1}{\left[\frac{1}{{n_f}^2} - \frac{1}{{n_i}^2}\right]} = 91.174\,\text{nm} \times \frac{1}{\left[\frac{1}{2^2} - \frac{1}{7^2}\right]} = 397.11\,\text{nm}$$

This wavelength is right on the dividing line between ultraviolet and visible light.

f. The configuration would be either $2s^1$ or $2p^1$, with a single electron in the $n = 2$ level.

Chapter 7: Periodic Properties of the Elements

The Bottom Line

- The structure of the periodic table was initially developed by scientists trying to make sense of the differing reactivities of all the elements found in the natural environment. (Section 7.1)
- The structure of the periodic table includes "blocks" defined in terms of which type of orbital is being filled as we imagine filling up the available orbitals using the Aufbau principle. This gives us the *s*-block, *p*-block, *d*-block, and *f*-block. (Section 7.1)
- The horizontal "period" of the periodic table to which an element belongs indicates how many energy levels are either fully or partially occupied by electrons in atoms of that element. (Section 7.1)
- Elements in any one main group (Groups IA through VIIIA) have the same number of electrons in their highest energy level. (Section 7.1)
- The modern form of the table first began to take shape through the work of the German Julius Lothar Meyer and the Russian Dmitri Mendeleev. (Section 7.1)
- The elements in the periodic table are arranged into three main sections: the metals, nonmetals, and metalloids (or semimetals). (Section 7.2)
- Metals and nonmetals exhibit specific types of physical properties. (Section 7.2)
- The periodic table is divided into groups, and elements in each group are similar in electronic structure, physical properties, and chemical reactivity. (Section 7.3)
- Periodicity is apparent in the way important physical and chemical characteristics recur in a periodic manner as we move through the periodic table. (Section 7.4)
- Examples of characteristics that show clear periodicity are atomic size (Section 7.5), ionization energy (Section 7.6), electron affinity (Section 7.7), electronegativity (Section 7.8), and reactivity (Section 7.9).
- The distribution of the elements on planet Earth is a result of their chemical reactivities. (Section 7.10)

Solutions to Practice Problems

7.1. a. Tin (Sn) is the *p*-block element in Period 5, Group IVA.
b. Tantalum (Ta) is the *d*-block metal in Period 6, Group VB.
c. Hydrogen (H) is the *s*-block element in Period 1, Group IA.

7.2. Li, Ni, Ce, Al, Po, Rb, and Cu are metals. Ce, Po, Rb, and Cu are also heavy metals by the definition given in the exercise (mass greater than 63.546 u). Si and Ge are metalloids, and Se is a nonmetal.

7.3. a. To place the valence electrons in the 6*s* orbital, the atom must be in Period 6. Ba (barium) has two valence electrons in the 6*s* orbital.
b. Na (sodium) has one valence electron in the 3*s* orbital.
c. With two electrons in the 2*s* orbital, the atom is in Period 2, Group VIA (sixth column): oxygen (O).
d. The only nonmetal with four valence electrons is carbon (C).

7.4. The hypothetical metal more reactive than Fr would be a Period 8, Group IA atom with atomic number 119. To get to the noble gas in Period 8, it would be necessary to fill the following sets of orbitals: 8*s* (2), 5*g* (18), 6*f* (14), 7*d* (10), and 8*p* (6) taking 50 electrons beyond the Period 7 noble gas at atomic number 118. The atomic number (and total number of electrons) would be 168.

7.5. The first ionization energies of the Group IIA metals decrease with increasing atomic number (they decrease as you move down the group); because lower ionization energies are expected to enhance the reactivity, radium and then barium should be the most reactive with water.

7.6. Based on its position in the periodic table, you would expect the energies to be high to start with and to get higher with each electron. When atoms have either a filled orbital or a half-filled orbital, it takes more energy to ionize the electron than would otherwise be expected. When the electron is ionized from a noble gas configuration (10 electrons and 2 electrons), there is a significant jump in the ionization energy, especially when electrons are removed from the 1*s* orbital.

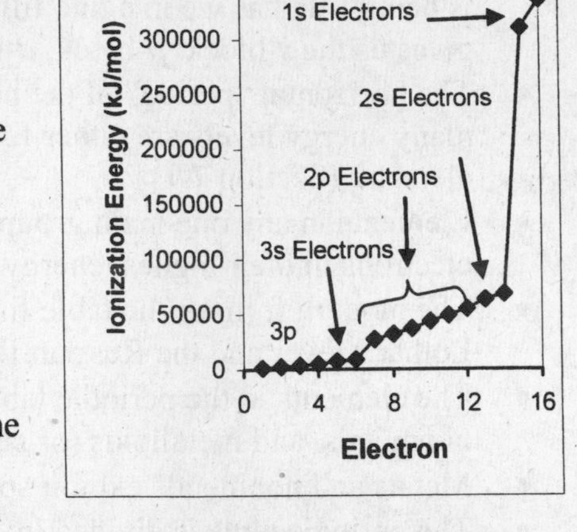

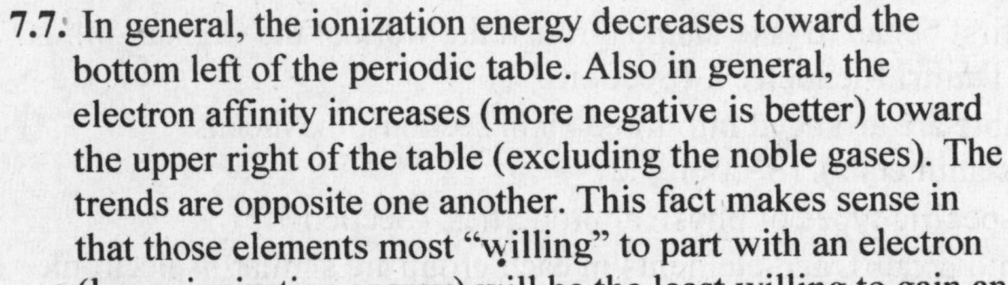

7.7. In general, the ionization energy decreases toward the bottom left of the periodic table. Also in general, the electron affinity increases (more negative is better) toward the upper right of the table (excluding the noble gases). The trends are opposite one another. This fact makes sense in that those elements most "willing" to part with an electron (lower ionization energy) will be the least willing to gain an electron (smaller electron affinity), and vice versa.

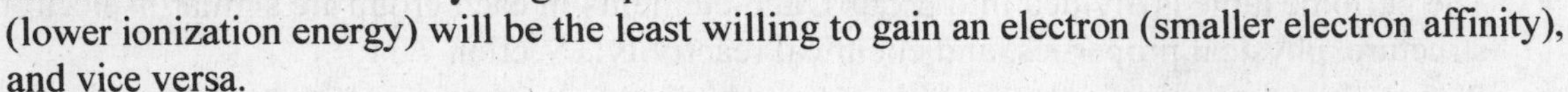

7.8. Electronegativity generally increases as an element is closer to fluorine (upper left), so P is more electronegative than Na, Cl is more electronegative than Ne (noble gases generally don't form bonds and therefore have very low electronegativities), and N is more electronegative than C.

Solutions to Student Problems

1. $_{28}$Ni (58.69 g/mol) and $_{27}$Co (58.93 g/mol) would be reversed. $_{52}$Te (127.6 g/mol) and $_{53}$I (126.9 g/mol) would be reversed. Other possibilities: $_{18}$Ar (39.95 g/mol) and $_{19}$K (39.1 g/mol); $_{90}$Th (232.04 g/mol) and $_{91}$Pa (231.04 g/mol); $_{92}$U (238.03 g/mol) and $_{93}$Np (237 g/mol).

3. a. Sr (*s*) b. Sc (*d*) c. S (*p*) d. Sn (*p*) e. Se (*p*) f. Sm (*f*) g. Sb (*p*)

5. a. C (*p*) b. Ca (*s*) c. Cu (*d*) d. Cr (*d*) e. Cs (*s*) f. Cl (*p*) g. Ce (*f*)

7. The melting point is likely to be somewhere between its neighbors to the left and right, approximately 3090°C.

9. Two trends to note: Sizes decrease from left to right in a period and increase down a group.

a. Ca is below Mg and is larger.
b. P is to the left of S and is larger.
c. Br is below Cl and is larger.
d. Cs is to the left of Ba and is larger.

11. a. Zn is a metal.
b. Ga is a metal.
c. Ge is a metalloid.
d. Sn is a metal
e. Si is a metalloid.
f. P is a nonmetal.
g. Se is a nonmetal.
h. As is a metalloid.

13. Br and Hg are liquids at room temperature.

15. a. Nickel (8.76%) and chromium (16.7%) would make up 8.76 g and 16.7 g, respectively, of each 100 g of stainless steel.
b. Converting the masses in part a to molecules:

$$8.76\,\text{g Ni} \times \frac{1\,\text{mol Ni}}{58.69\,\text{g Ni}} \times \frac{6.022\text{x}10^{23}\ \text{atoms}}{1\,\text{mol Ni}} = 8.99\text{x}10^{22}\ \text{atoms}$$

$$16.7\,\text{g Cr} \times \frac{1\,\text{mol Cr}}{52.00\,\text{g Cr}} \times \frac{6.022\text{x}10^{23}\ \text{atoms}}{1\,\text{mol Cr}} = 1.93\text{x}10^{23}\ \text{atoms}$$

17. Antimony (Sb, element 51) is classified as a metalloid. The balanced equation is

$$Sb_2S_3 + 3\ Fe \rightarrow 3\ FeS + 2\ Sb.$$

The number of moles of Sb in 60000 tons is

$$6.00\times10^4\ \text{tons} \times \frac{2000\,\text{lb}}{1\,\text{ton}} \times \frac{1\,\text{kg}}{2.205\,\text{lb}} \times \frac{1000\,\text{g}}{1\,\text{kg}} \times \frac{1\,\text{mol}}{121.76\,\text{g}} = 4.47\text{x}10^8\ \text{mol Sb}$$

19. The compounds would be Li_2O, BeO, B_2O_3, CO_2, N_2O_5, O_2, and OF_2.

21. With the exception of Po at the bottom of Group VIA, these elements would form ionic compounds with metals. These atoms will gain electrons to reach a stable noble gas configuration; this makes them anions, which will form compounds with metallic cations. (Po as a metal would form an ionic compound with nonmetals. It will lose electrons to form a cation and would form ionic compounds with nonmetallic anions.)

23. In Group VIIA, the elements have seven valence electrons in the *s* (2) and *p* (5) orbitals. One more electron could fit into the *p* orbital, leaving one electron unpaired. The ions of the halogens have 8 valence electrons, which totally fill the *s* and *p* orbitals, leaving none unpaired.

25. Based on the percentages in the table, for each 100 g, there are 61.0 g O, 23.8 g C, and 10.0 g H. Because the number of molecules is directly related to the number of moles, we can calculate the number of moles of each in those masses:

$$61.0\,\text{g O} \times \frac{1\,\text{mol O}}{16.00\,\text{g O}} = 3.81\,\text{mol O}$$

$$23.8\,\text{g C} \times \frac{1\,\text{mol C}}{12.01\,\text{g C}} = 1.98\,\text{mol C}$$

$$10.0\,\text{g H} \times \frac{1\,\text{mol H}}{1.008\,\text{g H}} = 9.92\,\text{mol H}$$

The rankings based on number of moles or molecules would move hydrogen to the top of the list: hydrogen, then oxygen, then carbon.

27. Many transition elements play a critical role as the "active site" within enzymes that assist in the metabolic reactions of the human cell.

29. Each noble gas fills its valence shell of electrons that usually include two *s* electrons and six *p* electrons. Because there are no *p* electrons in the first atomic shell, the valence shell is full upon filling the 1*s* orbital with two electrons. Once that shell is full, as it is with He, there is little reactivity, and He behaves just like the rest of the noble gases.

31. Calcium's configuration includes a filled 4*s* orbital, and potassium's configuration includes a single 4*s* electron. In general, removing an electron from a filled orbital is much more difficult, requiring more energy, than removing an electron from a partially filled orbital. Because calcium has a filled orbital and potassium does not, potassium will more easily lose its electron in a reaction, which makes it the more reactive element.

33. Tellurium has a valence shell that includes two *s* electrons and four *p* electrons, while iodine has two *s* electrons and five *p* electrons. All of the other elements in Group VIIA have two *s* electrons and five *p* electrons. Iodine has the same configuration and belongs with the other halogens, while tellurium has the same configuration as the rest of Group VIA (two *s* electrons and four *p* electrons) and belongs there.

35. The periodic trend for atomic size indicates that within a period, neutral atoms become smaller with increasing atomic number. Therefore, sodium should be largest, magnesium in the middle, and sulfur the smallest. (From left to right: sodium, sulfur, magnesium)

37. Assuming that adjacent atoms were "touching," there should be a distance equal to two radii as the bond length, so the carbon radius should be 154 pm / 2 = 77 pm.

39. For bonds between two different atoms, the bond length should be equal to the sum of the atomic radii. Subtracting the carbon radius (77 pm) from the bond length yields the chlorine radius: 171 pm – 77 pm = 94 pm. (Note: Radii determined in this manner are approximate and may not exactly match the true *atomic* radii.)

41. Even though positive ions are smaller (often much smaller) than their neutral counterparts, both sodium and potassium are in the same group. Because potassium is lower in the group, it has an additional filled shell, which makes it larger than sodium.

43. If the ionization energies are used as conversion factors, we can complete the calculation in one series of conversions:

$$1.00\,\text{g Na} \times \frac{1\,\text{mol Na}}{22.99\,\text{g Na}} \times \frac{495\,\text{kJ}}{1\,\text{mol Na}} \times \frac{1\,\text{mol K}}{419\,\text{kJ}} \times \frac{39.10\,\text{g K}}{1\,\text{mol K}} = 2.01\,\text{g K}$$

45. a. From lowest to highest: Al, Si, P; the first ionization energy increases with increasing atomic number in a period. For the second ionization energy, you would expect Al to be higher than Si or P, because it would have to remove an electron from a filled 3*s* orbital, while Si and P continue to remove electrons from the unfilled 3*p* orbitals. (Rank: Si, P, Al)

b. From lowest to highest: Kr, Ar, Ne; the first ionization energy decreases with increasing atomic number in a group. For the second ionization energy, you would expect the same order, because the valence configuration remains the same for all three, maintaining the ionization energy trend in the group.

47. Because the initial ionization energy for Element A is low, it is a metal. Additionally, because there is a large jump from the second ionization energy to the third, the element is in Group IIA. (The large third ionization energy is likely from a filled shell and requires much energy.) The ionization energies for Element B are higher and are steadily increasing: It is a nonmetal. There is no discernible jump in the ionization energy that might be expected when encountering a filled *s* orbital; the nonmetal is probably part of Group VIA or VIIA.

49. Element 118 would be part of the noble gases, Group VIIIA. Following the general trend, it should have an ionization energy lower than its neighbor above it in the group. (The first ionization energy decreases with increasing atomic number in a group.)

51. a. Both sodium and potassium are part of Group IA and would lose one electron to reach a filled-shell configuration; therefore, both ions would have a +1 charge

b. K^+ would be the larger ion because it has one more filled shell than Na^+.

c. Potassium would be easier to ionize. (The first ionization energy decreases with increasing atomic number in a group.)

53. Magnesium's first ionization removes an electron from a filled *s* orbital, which requires more energy than removing sodium's electron from an unfilled *s* orbital. The second ionization for sodium requires removing an electron from a completely filled shell, which takes even more energy, while magnesium's second electron is being removed from the now unfilled *s* orbital.

55. a. Magnesium never forms a +3 ion, because removing a third electron would require breaking up a filled shell of electrons.

b. Fluorine, seeking a filled shell, needs to gain an electron to achieve an octet configuration. A +1 fluorine ion is a move in the opposite direction.

c. Hydrogen has only one electron to lose; it can never get to a +2 state.

d. Aluminum, in losing three electrons, achieves an octet (filled-shell) configuration; the octet configuration is more stable than other, nonoctet configurations.

57. In order to form a stable anion, an element needs to have a favorable electron affinity (exothermic / negative). Sulfur's electron affinity is −200 kJ/mol, which is favorable. Xe, a noble gas, already has a filled shell and does not favorably accept another electron. Therefore, S will be more likely to form a stable anion.

59. Chlorine's electron affinity is −349 kJ/mol, which is very favorable and exothermic. Ar, a noble gas, already has a filled shell and does not favorably accept another electron; this makes the process endothermic.

61. You might expect Na, Mg, and Al to follow much the same trend as Li, Be, and B, which lie above Na, Mg, and Al in the periodic table. The electron affinity table in the text shows the trend for Li, Be, and B from least negative to most negative to be Be (>0) to B (−27 kJ/mol) to Li (−60 kJ/mol). We could then predict the trend for Na, Mg, and Al in order from least to most negative as Mg, Al, Na. In order to understand the formation of an anion, we need to consider the electron configuration of each metal. Before adding an electron, magnesium already has a filled *s* orbital, which is more energetically favorable than what would be achieved by placing an added electron in an unfilled set of *p* orbitals. Adding an electron to Na, however, will create a filled *s* orbital, which is better than the case for Al, which achieves no special configuration by adding an electron. None of these three will form stable anions outside of the gas phase, but the ranking of the tendency to form an anion will be the same as given above. The more negative the electron affinity, the more likely is the formation of the anion.

63. Fluorine is the smallest ion-forming element in the second period, which means its electrons are the most tightly packed. Accepting an additional electron into this dense electron arrangement, even though it completes the valence shell, is not as favorable as would otherwise be expected, because of the greater electron–electron repulsions in the small volume.

65. a. Considering only the ionization energy and electron affinity, the total energy for Na + Cl → Na^+ + Cl^- is (+495 kJ/mol) + (−349 kJ/mol) = +146 kJ/mol, indicating an endothermic process.

b. An endothermic process seems at odds with the violent, very exothermic process that actually takes place. The above calculation neglects two additional factors: the energy to dissociate a chlorine atom from the molecule (+242 kJ/mol, half of which could be attributed to a single atom), and the lattice energy of solid NaCl (−769 kJ/mol to form NaCl(*s*) from the ions). Combining all four factors results in −502 kJ/mol NaCl formed, which is a very exothermic process (+146 kJ/mol + (½)×242 kJ/mol + (−769 kJ/mol) = −502 kJ/mol).

67. The general trend is that electronegativity increases toward the upper right of the periodic table, so the order from lowest to highest is Na, Li, As, S, F.

69. The more electronegative element in the pair will attract the electrons to itself: (a) O, (b) S, (c) Br, (d) O

71. The electronegativity difference between Se (2.4) and Ga (1.6) is 0.8. From Zn (1.6) to Sc (1.3), there is a difference of 0.3. This is a much lower difference than might have been predicted on the basis of the electronegativity and proton difference between Se and Ga. If the electronegativity difference were based on the number of protons, you might have expected a difference three times as large (9 protons vs. 3 protons). Electronegativity cannot be solely dependent on the atomic number.

73. The reactivity for metals follows the ease with which they are ionized. Because ionization energy generally decreases toward the left in a period and decreases toward the bottom in a group, magnesium should be less reactive than sodium, which in turn is less reactive than rubidium. Least to most reactive: Mg, Na, Rb.

75. For the metals, the lower the ionization energy, the more reactive the element. For nonmetals the opposite is true, because the nonmetals prefer to add electrons (not lose them) in their reactions. Electron affinity is a better predictor of reactivity than ionization energy for the nonmetals.

77. a. All three coinage metals are found in the *d*-block, Group IB.

b. All three of the coinage metals are found at the bottom of the reactivity series. (They are three of the last four. The fourth, Hg, is a liquid and therefore cannot be used for coins.)

c. Realizing that the most reactive elements are found to the left and right sides of the periodic table (excluding the noble gases) and working from the outside in toward Group IB, the metals in activity series are found in the same order as the periodic table would predict. First listed are the alkali metals (Group IA) and alkaline earth metals (Group IIA), followed by aluminum and zinc from the right and the other transition metals from the left of Group IB.

79. When forming a compound solely of chlorine and oxygen, the oxygen, with its higher electronegativity, will pull shared electrons more toward itself and therefore is the more negative of the two elements.

81. A possible explanation for the relative lack of iron in the mantle compared to the crust is that the during the early formation of the earth, the very dense elements, such as iron, were drawn to the core and are therefore missing from the mantle.

83. a. Argon

b. $$\frac{6.022 \times 10^{23} \text{ particles Air}}{1 \text{ mol Air}} \times \frac{0.00934 \text{ parts Ar}}{1 \text{ part Air}}$$

$$= 5.62 \times 10^{21} \text{ atoms Ar / mol Air}$$

85. To remain uncombined, the elements need to be very nonreactive: copper and gold. All the remaining elements on the list are listed as very reactive and would thus be combined with another element in the crust.

87. a. Johan Dobereiner first identified groups of three elements that shared similar properties. These first "triads" form the basis for the current alkali metals group (Group IA), alkaline earth metals group (Group IIA), and halogens group (Group VIIA).

b. John Newlands first placed elements in order of increasing mass and found that elements eight places apart shared similar properties. He arranged the elements in octaves (periods of eight) that very much mirror the *s*- and *p*-blocks of current periodic tables.

c. Dmitri Mendeleev added additional elements to his table, generally in mass order, that created vertical groups of elements with similar properties. Mendeleev left holes in his table and predicted that elements would be discovered to fill those holes. Mendeleev's use of the table as a predictive tool laid the basis for the modern periodic table.

89. a. The most common form of steel is carbon-steel.

b. The main difference between carbon-steel and other steels is that carbon-steel contains carbon and iron and little else, while the other steels contain a significantly larger amount of elements other than carbon.

91. The length of a C–H bond should be the sum of covalent radii of C and H. The covalent radius of an atom can be calculated by dividing in half the bond length formed between two similar atoms. From the C–C bond we find the covalent radius of C to be 154 pm / 2 = 77 pm. The covalent radius of H is then 75 pm / 2 = 38 pm. We would predict the C–H bond to have a length of 77 pm + 38 pm = 115 pm.

93. a. With the exception of He (Group VIIIA, 2 valence) there is a very good trend between valence electrons and group number. The number of valence electrons is a periodic property.

b. The oxidation number for the first 18 elements also appears to be periodic. Even though there is a large break in the trend, the values repeat with successive periods, making the property periodic and predictable.

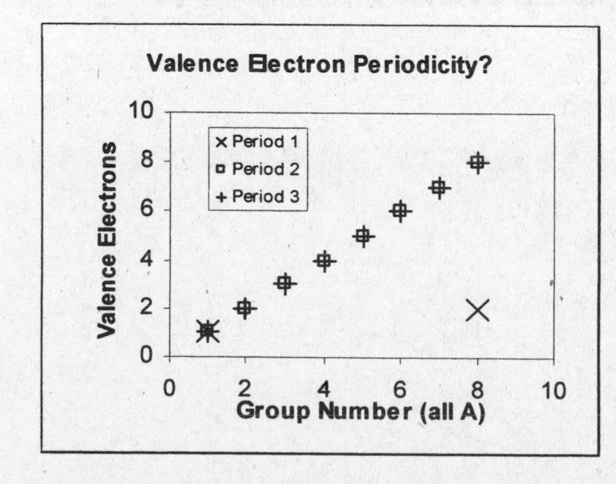

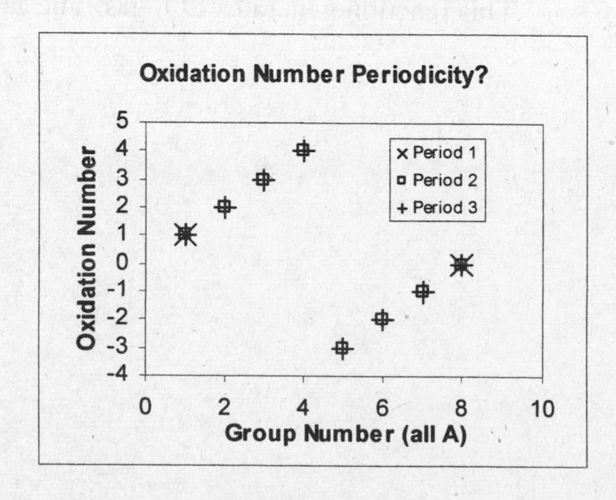

95. There are two competing trends for how the atomic number affects the size of the atom: (1) In a group, size increases with increasing atomic number. (2) In a period, size decreases with increasing atomic number. It is important to account for the relative positions (period and group) of the two atoms to be compared.

97. Even when successive electrons are removed from the same orbital, the remaining ion becomes more positive with each electron removed. The attraction of the increasingly positive ion for the electron raised the energy required to remove another electron. As more electrons are removed, eventually lower-orbital electrons will be ionized that are more strongly bound to the nucleus, which also increases the energy required to ionize the electron.

99. Additional stability results not only from a configuration that completely fills a set of orbitals but also from a configuration that half fills the set of orbitals. Arsenic has the valence configuration $4s^2 4p^3$, half-filling the *p*-orbital set, while selenium has $4s^2 4p^4$ and doesn't have the additional stability. For this reason, it takes slightly more energy to remove an electron from As than from Se.

101. a. A negative electron affinity indicates an exothermic process.

b. The reaction is $Cl + e \rightarrow Cl^- + 350$ kJ/mol.

103. a. $NaHCO_3 + HCl \rightarrow H_2CO_3 + NaCl$

b. H: Period 1, Group 1A; Na, Period 3, Group 1A; C, Period 2, Group IVA; O: Period 2, Group VIA; Cl: Period 3, Group VIIA.

c. Metals: Na only

d. $H_2CO_3 \rightarrow H_2O + CO_2$

e. $$\frac{10.0\,\text{mg NaHCO}_3}{275\,\text{mL}} \times \frac{1\,\text{g}}{1000\,\text{mg}} \times \frac{1\,\text{mol NaHCO}_3}{84.01\,\text{g NaHCO}_3} \times \frac{1\,\text{mol NaCl}}{1\,\text{mol NaHCO}_3} \times \frac{58.44\,\text{g NaCl}}{1\,\text{mol NaCl}} \times \frac{1000\,\text{mg}}{1\,\text{g}} \times \frac{1000\,\text{mL}}{1\,\text{L}}$$

$$= 25.3\,\text{mg/L NaCl} = 25.3\,\text{ppm NaCl}$$

f. $$\frac{25.3\,\text{mg NaCl}}{1\,\text{L}} \times \frac{1\,\text{g}}{1000\,\text{mg}} \times \frac{1\,\text{mol NaCl}}{58.44\,\text{g NaCl}} \times \frac{1\,\text{mol Na}}{1\,\text{mol NaCl}} \times \frac{22.99\,\text{g Na}}{1\,\text{mol Na}} \times \frac{1000\,\text{mg}}{1\,\text{g}}$$

$$= 9.95\,\text{mg/L Na} = 9.95\,\text{ppm Na}$$

g. This reaction generates CO_2 gas. The gas leavens the bread or pastry, making it lighter.

Chapter 8: Bonding Basics

The Bottom Line

- Models are an important tool that chemists use to help them determine the properties of molecules. (Section 8.1)
- Estimation of the properties of a molecule on the basis of the structure of that molecule is only as good as the model of that molecule. (Section 8.1)
- Bonding can range across the full spectrum between equal sharing and complete transfer of electrons. (Section 8.1)
- The three main types of bonds are ionic, polar covalent, and metallic. (Section 8.1)
- An anion is always bigger than the atom from which it is derived. A cation is always smaller than the atom from which it is derived. (Section 8.2)
- The lattice enthalpy is an important expression of the energetic stability of a salt. (Section 8.2)
- We can use bond enthalpy calculations to determine the approximate energy change involved in a reaction. (Section 8.3)
- Lewis dot structures are useful in constructing a simple model showing the location of atoms within a molecule. (Section 8.3)
- Resonance hybrids offer an overall picture of a molecule. Individual resonance structures do not adequately describe a molecule. (Section 8.3)
- VSEPR theory describes the shapes of molecules better than Lewis dot structures. The model drawn using VSEPR provides a three-dimensional picture of the molecule. (Section 8.4)
- The polarity of a molecule is related to the overall forces of the individual bond dipoles in the molecule *and* to the molecule's three-dimensional shape. (Section 8.6)

Note concerning Lewis structures in this key: **In order to maintain clarity, all structures, not just ions, have been placed within brackets when the structure appears within the text. The brackets are not required in normal use for neutral structures.**

Solutions to Practice Problems

8.1. S^{2-} has 18 electrons with the configuration $1s^22s^22p^63s^23p^6$. The 18 electrons make it isoelectronic with Ar. The Lewis dot structure shows only the valence electrons (8): $\left[:\ddot{\underset{\cdot\cdot}{S}}:\right]^{2-}$.

F^- has 10 electrons with the configuration $1s^22s^22p^6$. The 10 electrons make it isoelectronic with Ne. The Lewis dot structure shows only the valence electrons (8): $\left[:\ddot{\underset{\cdot\cdot}{F}}:\right]^{-}$.

Mg^{2+} has 10 electrons with the configuration $1s^22s^22p^6$. The 10 electrons make it isoelectronic with Ne. The Lewis dot structure shows only the valence electrons (0): $[Mg]^{2+}$.

Br^- has 36 electrons with the configuration $1s^22s^22p^63s^23p^64s^23d^{10}4p^6$. The 36 electrons make it isoelectronic with Kr. The Lewis dot structure shows only the valence electrons (8): $\left[:\ddot{\underset{\cdot\cdot}{Br}}:\right]^{-}$.

8.2. We start by drawing the Lewis dot symbols for the calcium and chlorine atoms. Then we show the transfer of an electron from one of the less electronegative sodium atoms to the more electronegative oxygen atom. We show a similar transfer of the other sodium atom's electron to oxygen. We end with three ions (two Na^+ and O^-) that have noble gas electron configurations. Note that the ions end up with either a completely full valence electron configuration (and become isoelectronic with a noble gas) or a completely empty valence shell (and become isoelectronic with a noble gas). This is the octet rule in action.

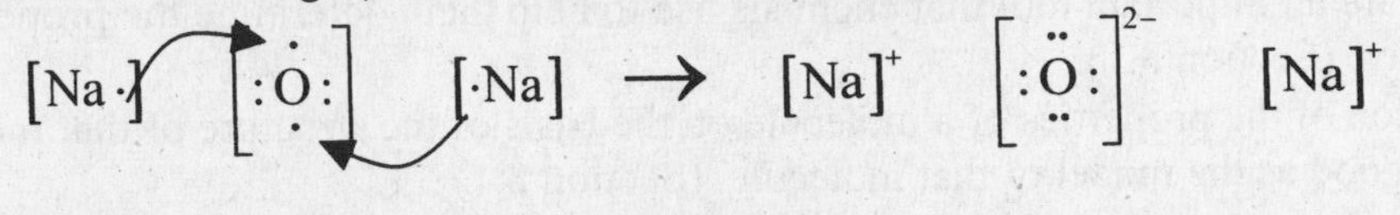

$$[\text{Na}\cdot]\ [:\dot{\text{O}}:]\ [\cdot\text{Na}] \rightarrow [\text{Na}]^+\ [:\ddot{\text{O}}:]^{2-}\ [\text{Na}]^+$$

8.3. We will limit our search to those atoms that have an atomic number close to that of neon—two that have a higher atomic number (Na and Mg) and two that have a smaller atomic number (O and F). Second, the question asks us to organize the ions we've created by size. When two species are isoelectronic, we know that cations are smaller than atoms and that anions are larger (in other words, with the same number of electrons, the species with the most protons is smallest). O and F will have to gain electrons to become isoelectronic with Ne, while Na and Mg will lose electrons. The ions that are created, in order of increasing size, are $Mg^{2+} < Na^+ < Ne < F^- < O^{2-}$.

8.4. For $FeCl_3$ and $FeCl_2$, the ion sizes are generally the same (although Fe^{3+}, with one fewer electron, might be slightly smaller. The most important factor will be the increased charge present on Fe^{3+}, along with the one additional interaction in $FeCl_3$ (three Cl and one Fe versus two Cl and one Fe in $FeCl_2$). $FeCl_3$ should have a larger lattice enthalpy. (It does, with 5359 kJ/mol versus 2631 kJ/mol for $FeCl_2$.)

8.5. NO has a difference in electronegativity of 0.5, which is right on the cutoff between polar covalent and nonpolar covalent bonds. We could call it a weakly polar covalent bond. F_2 has no electronegativity difference between its atoms (the same would be true of any molecule containing only a single type of atom) and has a nonpolar covalent bond. The atoms in MgO have an electronegativity difference of 2.3, placing this compound in the ionic bonding region.

8.6. Before computing the formal charges, we must first create the Lewis diagrams for the molecule.

For OCl^-: $[:\ddot{\text{O}}:\ddot{\text{Cl}}:]^-$. Oxygen normally has 6 valence electrons and has 1 bond and 6 nonbonded electrons in the ion. Its formal charge is −1 (FC = valence – bonds – nonbonded electrons = 6 – 1 – 6 = −1). Chlorine normally has 7 valence electrons and has 1 bond and 6 nonbonded electrons. Its formal charge is 0 (FC = 7 – 1 – 6 = 0).

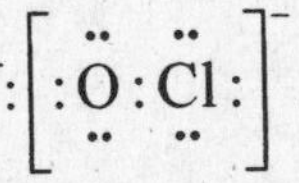

For CH_3NH_2:

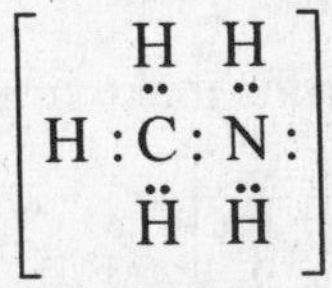

$$\left[\begin{array}{ccc} & \text{H} & \text{H} \\ \text{H}: & \text{C}: & \text{N}: \\ & \text{H} & \text{H} \end{array}\right]$$

. Each hydrogen has 1 bond and no non-bonded electrons, giving each a formal charge of 0 (FC = 1 – 1 – 0 = 0). The carbon has valence of 4 with 4 bonds and no nonbonded electrons, giving it a formal charge of 0 (FC = 4 – 4 – 0 = 0). Nitrogen has 5 valence electrons, 3 bonds, and 2 nonbonded electrons, giving it a formal charge of 0 (FC = 5 – 3 – 2 = 0).

8.7. For C_2H_4, there are not enough electrons for all atoms to obtain an octet (or duet) with only a single bond between the carbons. There are 4 electrons from each carbon and 1 each from the hydrogens, totaling 12.

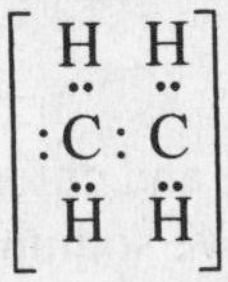

$$\left[\begin{array}{cc} \text{H} & \text{H} \\ :\text{C}: & \text{C} \\ \text{H} & \text{H} \end{array}\right]$$

. Note that the second carbon has only 6 electrons. If 2

electrons from the first carbon are used to create a second bond with the other carbon, $\left[\begin{array}{cc} H & H \\ \ddot{C} & :: \ddot{C} \\ \ddot{H} & \ddot{H} \end{array}\right]$,

we get the correct structure.

For CH_3N, you must create a double bond between carbon and nitrogen to get a structure with filled octets and duets. $\left[\begin{array}{l} H \\ C = \ddot{N} - H \\ \ddot{H} \end{array}\right]$

8.8. Rewriting the reaction in terms of Lewis structure, we see that

$$\begin{array}{c} H \\ H:\ddot{C}:O:H \\ \ddot{H} \end{array} + H:\ddot{\underset{..}{Br}}: \rightarrow \begin{array}{c} H \\ H:\ddot{C}:\ddot{\underset{..}{Br}}: \\ \ddot{H} \end{array} + H:\ddot{\underset{..}{O}}:H$$

To complete the reaction we must break one C–O bond and one H–Br bond, while creating one C–Br bond and one O–H bond. Recall that energy is absorbed (+) when breaking a bond and released (−) when forming a bond.

$$\begin{aligned} 1 \text{ mol C} - \text{O bonds} \times 358 \text{ kJ/mol} &= +358 \text{ kJ} \\ 1 \text{ mol H} - \text{Br bonds} \times 363 \text{ kJ/mol} &= +363 \text{ kJ} \\ \text{Total Bonds Broken} &= +721 \text{ kJ} \end{aligned}$$

$$\begin{aligned} 1 \text{ mol C} - \text{Br bonds} \times (-276 \text{kJ/mol}) &= -276 \text{ kJ} \\ 1 \text{ mol O} - \text{H bonds} \times (-467 \text{kJ/mol}) &= -467 \text{ kJ} \\ \text{Total Bonds Formed} &= -743 \text{ kJ} \end{aligned}$$

$$\begin{aligned} \text{TOTAL Energy Absorbed} &= +721 \text{ kJ} \\ \text{TOTAL Energy Released} &= -743 \text{ kJ} \\ \text{Net Energy Change } (\Delta H) &= -22 \text{ kJ} \end{aligned}$$

8.9. We start with the structures for each molecule: $\left[\begin{array}{c} :\ddot{O}::N:O:H \\ :\ddot{\underset{..}{O}}: \end{array}\right]$; $\left[\begin{array}{c} :\ddot{Cl}: \\ :\ddot{\underset{..}{Cl}}:C:\ddot{\underset{..}{Cl}}: \\ :\ddot{\underset{..}{Cl}}: \end{array}\right]$; $\left[\begin{array}{c} H \\ H:\ddot{N}: \\ \ddot{H} \end{array}\right]$.

In HNO_3, the nitrogen has three electron groups, giving a trigonal planar electron group geometry with 120° bond angles around nitrogen. In CCl_4, the carbon has four electron groups, giving the molecule a tetrahedral electron group and molecular geometry with 109.5° bond angles. In NH_3, the nitrogen has four electron groups, giving the molecule a tetrahedral electron group geometry. With the one lone pair, the molecular geometry is trigonal pyramidal. The lone pair decreases the tetrahedral bond angle slightly to 107°.

8.10. Before completing the question, we note that the structures are $\left[:\ddot{O}::\ddot{S}:\ddot{\underset{..}{O}}:\right]$ and $\left[H:\ddot{\underset{..}{O}}:H\right]$.

With three electron groups, SO_2 will have a trigonal planar geometry. Because there is one lone pair, the O–S–O angle is slightly less than the normal 120° angle. H_2O has four electron groups and a tetrahedral geometry, but the two lone electron pairs reduce the H–O–H angle to slightly less than 109°. SO_2 will have the larger bond angle.

8.11. N_2 has a nonpolar bond and therefore is also a nonpolar molecule. (All molecules containing a single type of atom will have nonpolar bonds and be nonpolar molecules.) NH_3 does have polar bonds because of the electronegativity difference of 0.9. NH_3 has a trigonal pyramidal molecular structure (see Practice 8.9). Nitrogen is the more electronegative of the two, so each bond will have a dipole with the positive end on hydrogen and the negative end on nitrogen. The resulting net dipole is vertical through the center of the pyramid (The horizontal components cancel out.)

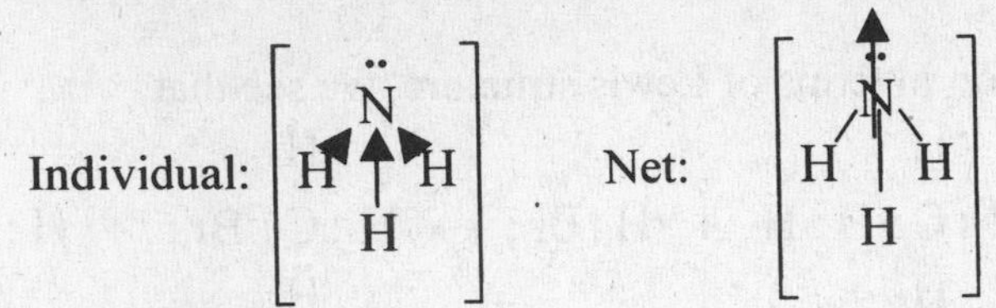

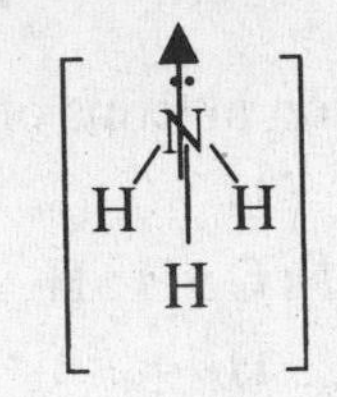

Solutions to Student Problems

1. a. The Lewis structrures are $[:\dot{A}l]$ and $[:\dot{\underset{\cdot}{N}}\cdot]$. Nitrogen (3) has more unpaired electrons than aluminum (1).
   b. Nitrogen (5) has more valence electrons than aluminum (3).
   c. In order to fill an octet by gaining electrons, nitrogen would need 3 electrons and aluminum 5 electrons. Because it needs fewer, nitrogen is more likely to gain electrons to fill its octet.

3. H with 2 electrons is H^- and is isoelectronic with He. B^{3+} is the only one with 2 electrons. C^{2+}, N^{3-}, and H^+ do not have the same configuration.

5. a. Group VIA needs to gain 2 electrons to fill an octet.
   b. Group IIA needs to lose 2 electrons to reduce to an octet.
   c. Group VA needs to gain 2 electrons to fill an octet.

7. a. S is in Group VIA and normally has 6 valence electrons; to get 8 it gained 2 electrons and has a −2 charge.
   b. P is in Group VA and normally has 5 valence electrons; to get 8 it gained 3 electrons and has a −3 charge.
   c. Cl is in Group VIIA and normally has 7 valence electrons; to get 8 it gained 1 electrons and has a −1 charge.

9. a. Se^{2-} has 6 valence to start plus 2 extra = 8 electrons: $[:\ddot{\underset{\cdot\cdot}{Se}}:]^{2-}$
   b. I^- has 7 valence to start plus 1 extra = 8 electrons: $[:\ddot{\underset{\cdot\cdot}{I}}:]^{-}$
   c. Sr^{2+} has 2 valence to start minus 2 electrons = 0 electrons: $[Sr]^{2+}$

d. Sc^{3+} has 3 valence to start minus 3 electrons = 0 electrons: $[Sc]^{3+}$

e. Si^{2+} has 4 valence to start minus 2 electrons = 2 electrons: $[Si:]^{2+}$

11. In order to have the same configuration as Ne, the ion would need to be near the end of Period 2 or the beginning of Period 3. Na^+, F^-, and Al^{3+} are isoelectronic with Ne. ($1s^22s^22p^6$)

13. To see if any pair is isoelectronic, we can look at their configurations: Ca^{2+} ($1s^22s^22p^63s^23p^6$); Sc^+ ($1s^22s^22p^63s^23p^64s^2$); S ($1s^22s^22p^6$); Ar ($1s^22s^22p^63s^23p^6$); Cl^- ($1s^22s^22p^63s^23p^6$). Ca^{2+}, Ar, and Cl^- are isoelectronic.

15. With 4 valence electrons, this atom is in Group IVA (could be C, Si, Ge, Sn, or Pb) or Group IVB (could be Ti, Zr, Hf, or Rf).

17. a. The Lewis structures for Li, Na, and K are [Li•], [Na•], and [K•].

b. In forming a compound, each atom will lose one electron, but oxygen needs two electrons to go from $[:\dot{\underset{\cdot}{O}}:]$ to $[:\ddot{\underset{\cdot\cdot}{O}}:]^{2-}$. Each compound will have two metal ions and a single oxygen anion: Li_2O, Na_2O, and K_2O. (See Practice 8.2.)

19. a. True.

b. False. Electron sharing is typical in covalent bonds.

c. True.

d. True.

21. To reduce its valence to an octet configuration, aluminum goes from the atom to the ion by losing three electrons: $[:\dot{Al}] \rightarrow [Al]^{3+}$ + 3 electrons. Oxygen forms oxide (and creates an octet) by gaining two electrons: $[:\dot{\underset{\cdot}{O}}:]$ + 2 electrons $\rightarrow$ $[:\ddot{\underset{\cdot\cdot}{O}}:]^{2-}$. Chlorine forms chloride (and creates an octet) by gaining one electron: $[:\ddot{\underset{\cdot}{Cl}}:]$ + 1 electrons $\rightarrow$ $[:\ddot{\underset{\cdot\cdot}{Cl}}:]^{-}$ When forming a compound with chlorine, aluminum loses three electrons, which allows for the formation of three chlorides. All electrons have been accounted for, and the charges on the ions balance one another—$AlCl_3$ is the stable compound that forms, creating the 1:3 ratio. When forming a compound with oxygen, two aluminum atoms lose three electrons each (six total), which allows for the formation of three oxides. All electrons have been accounted for, and the charges on the ions balance one another—Al_2O_3 is the stable compound that forms, creating the 2:3 ratio.

23. a. With the same number of protons in each, the ion/atom with the most electrons will be the largest, and the one with the fewest electrons will be the smallest: $I^{5+} < I < I^-$.

b. $S^{6+} < S^{4+} < S^{2-}$

c. $C^+ < C < C^-$

d. $Fe^{3+} < Fe^{2+} < Fe$

25. Coulomb's law states that the attraction between two oppositely charged ions will increase as ionic charge increases and distance between the ions decreases. Because each compound has a +1 metal ion with chloride, the charges will not determine which compound has the larger attraction. The relative sizes of the cations are the determining factor. The periodic trend for the sizes of these alkali metal cations will be that the ions get larger down the group, so $Li^+ < Na^+ < K^+ < Rb^+ < Cs^+$. The smaller the ion, the smaller the distance between the cation and anion, and the larger the attraction. Therefore, the strength of the attraction will be the reverse of the above: LiCl > NaCl > KCl > RbCl > CsCl.

27. In the formation of KCl, potassium loses an electron and chlorine gains an electron, both reaching a stable noble gas configuration:

$$[K\cdot] \quad [:\ddot{\underset{..}{Cl}}\cdot] \rightarrow [K]^+ \quad [:\ddot{\underset{..}{Cl}}:]^-$$

29. Sodium oxide could be represented as $[Na]^+ [:\ddot{\underset{..}{O}}:]^{2-} [Na]^+$. After reacting with water ($Na_2O + H_2O \rightarrow 2\ NaOH$), sodium hydroxide results: $[Na]^+ [:\ddot{\underset{..}{O}}:H]^-$.

31. With the same number of protons in each, the ion/atom with the most electrons will be the largest, and the one with the fewest electrons will be the smallest; therefore, Mn^{5+} will be smaller than Mn^{4+} and will fit into the smaller zeolite hole.

33. Coulomb's law states that the attraction between two oppositely charged ions will increase as ionic charge increases and distance between the ions decreases. Because each compound has the same cation and anion charges, the charges will not determine which compound has the larger attraction. The relative sizes of the anions are the determining factor. Bromide is present in both compounds and will not cause the difference. The periodic trend for the sizes of alkali metal cation will be that the ions get larger down the group, so $K^+ < Cs^+$. The smaller the ion, the smaller the distance between the cation and anion, and the larger the attraction. Therefore, the smaller distance between the K^+ and Br^- gives this compound the higher lattice enthalpy, matching the melting-point data given.

35. Electron affinity gauges how much a particular atom (or molecule) "wants" to wholly gain an electron, with larger electron affinities indicating greater likelihood of gaining an electron. Electronegativity values gauge how much a particular atom will attract electrons from a shared covalent to itself. However, both electron affinity and electronegativity tend to increase toward the upper right of the periodic table (excluding the noble gases), and both can predict the favorability of forming ionic compounds. Electronegativity is more useful as its predictive capability extends to polar covalent bonding and dipoles.

37. Within your table you should have a vertical arrow pointing up, indicating an increase in electronegativity as you go up a group, and a horizontal arrow pointing right, indicating an increase in electronegativity as you go to the right in a period.

39. H_2O: $\left[H-\ddot{\underset{..}{O}}-H\right]$, NO (radical): $\left[:N=\ddot{O}:\right]$, CO: $[:C\equiv O:]$, NO_2 (radical): $\left[:\ddot{\underset{..}{O}}-\dot{N}=\ddot{O}:\right]$,

HCl: $\left[H-\ddot{\underset{..}{Cl}}:\right]$, PCl_2 (radical): $\left[:\ddot{\underset{..}{Cl}}-\dot{\underset{..}{P}}-\ddot{\underset{..}{Cl}}:\right]$, NBr_3: $\left[\begin{array}{c} :\ddot{Br}: \\ | \\ :\ddot{\underset{..}{Br}}-\ddot{N}: \\ | \\ :\underset{..}{Br}: \end{array}\right]$

41. OH^-: $\left[:\ddot{\underset{..}{O}}-H\right]^-$, NO_2^-: $\left[:\ddot{\underset{..}{O}}-\ddot{N}=\ddot{O}:\right]^-$, Br^-: $\left[:\ddot{\underset{..}{Br}}:\right]^-$, PO_4^{3-}: $\left[\begin{array}{c} :\ddot{O}: \\ | \\ \ddot{\underset{..}{O}}=P-\ddot{\underset{..}{O}}: \\ | \\ :\underset{..}{O}: \end{array}\right]^{3-}$,

SO_3^{2-}: $\left[\begin{array}{c} :\ddot{O}: \\ | \\ \ddot{\underset{..}{O}}=S: \\ | \\ :\underset{..}{O}: \end{array}\right]^{2-}$, CO_3^{2-}: $\left[\begin{array}{c} :\ddot{O}: \\ | \\ :\ddot{O}=C \\ | \\ :\underset{..}{O}: \end{array}\right]^{2-}$, BrO_4^-: $\left[\begin{array}{c} :\ddot{O}: \\ || \\ \ddot{\underset{..}{O}}=Br=\ddot{\underset{..}{O}} \\ | \\ :\underset{..}{O}: \end{array}\right]^-$

43. C_2H_6: $\left[\begin{array}{c} \;\;H\;\;H \\ H-C-C-H \\ \;\;H\;\;H \end{array}\right]$, C_3H_6: $\left[\begin{array}{c} \;\;H\;\;H \\ H-C-C=C-H \\ \;\;H\;\;\;\;\;\;H \end{array}\right]$, C_2H_4: $\left[\begin{array}{c} H\;\;H \\ C=C \\ H\;\;H \end{array}\right]$,

C_3H_8: $\left[\begin{array}{c} \;\;H\;\;H\;\;H \\ H-C-C-C-H \\ \;\;H\;\;H\;\;H \end{array}\right]$, C_4H_{10}: $\left[\begin{array}{c} \;\;H\;\;H\;\;H\;\;H \\ H-C-C-C-C-H \\ \;\;H\;\;H\;\;H\;\;H \end{array}\right]$

45. The three resonance structures are

$$\left[\begin{array}{c} :\ddot{O}: \\ || \\ \overset{(-1)}{:\ddot{\underset{..}{O}}}-\overset{(+1)}{N} \\ | \\ \underset{(-1)}{:\underset{..}{O}:} \end{array}\right]^- \leftrightarrow \left[\begin{array}{c} \overset{(-1)}{:\ddot{O}:} \\ | \\ :\ddot{O}=\overset{(+1)}{N} \\ | \\ \underset{(-1)}{:\underset{..}{O}:} \end{array}\right]^- \leftrightarrow \left[\begin{array}{c} \overset{(-1)}{:\ddot{O}:} \\ | \\ \overset{(-1)}{:\ddot{\underset{..}{O}}}-\overset{(+1)}{N} \\ || \\ :O: \end{array}\right]^-$$

Note that each oxygen has a FC = −1 in two of the three structures. The overall hybrid structure is

the average of the three: $\left[\begin{array}{c} :\ddot{O}: \\ \vdots \\ \ddot{\underset{..}{O}}\;\overset{-}{\underset{-}{}}\;N \\ \vdots \\ :O: \end{array}\right]^-$

47. C_4H_4:
$$\left[\begin{array}{cc} H & H \\ | & | \\ C = & C \\ | & | \\ C = & C \\ | & | \\ H & H \end{array}\right] \leftrightarrow \left[\begin{array}{cc} H & H \\ | & | \\ C - & C \\ \| & \| \\ C - & C \\ | & | \\ H & H \end{array}\right]$$
C_4H_6:
$$\left[\begin{array}{ccc} H & H & \\ | & | & \\ C & -C- & H \\ \| & | & \\ C & -C- & H \\ | & | & \\ H & H & \end{array}\right]$$
C_4H_8:
$$\left[\begin{array}{ccccc} & H & & H & \\ & | & & | & \\ H- & C & - & C & -H \\ & | & & | & \\ H- & C & - & C & -H \\ & | & & | & \\ & H & & H & \end{array}\right]$$

49. N_2O: $\left[:N \equiv N - \ddot{\underset{..}{O}}:\right] \leftrightarrow \left[:\ddot{N} = N = \ddot{O}:\right]$, N_2: $\left[:N \equiv N:\right]$, N_2H_4: $\left[\begin{array}{cccc} H- & \ddot{N}- & \ddot{N} & -H \\ & | & | & \\ & H & H & \end{array}\right]$. Because of the resonance structure, N_2O has a N–N bond that is roughly 2.5 bonds, while N_2 has a triple bond and N_2H_6 has a single bond between the nitrogens. Therefore, the bond energies are N_2 = 946 kJ/mol; N_2O = 418 kJ/mol; N_2H_6 = 160 kJ/mol.

51. Na_3PO_4: $[Na]^+$ $\left[\begin{array}{c} :\ddot{O}: \\ | \\ \ddot{\underset{..}{O}} = P - \ddot{\underset{..}{O}}: \\ | \\ :\underset{..}{O}: \end{array}\right]^{3-}$ $[Na]^+$ $[Na]^+$

$CaCO_3$: $[Ca]^{2+}$ $\left[\begin{array}{c} :\ddot{O}: \\ | \\ :\ddot{O} = C \\ | \\ :\underset{..}{O}: \end{array}\right]^{2-}$

$Fe(NO_3)_2$: $\left[\begin{array}{c} :\ddot{O}: \\ | \\ :\ddot{O} = N \\ | \\ :\underset{..}{O}: \end{array}\right]^{-}$ $[Fe]^{2+}$ $\left[\begin{array}{c} :\ddot{O}: \\ | \\ :\ddot{O} = N \\ | \\ :\underset{..}{O}: \end{array}\right]^{-}$

Covalent Bonding

Ionic Bonding

Covalent Bonding

The ionic bonding occurs between the separate ions, while the covalent bonding is happening between the atoms of the polyatomic anions.

53. To answer the question, we calculate the electronegativity difference for each pair of atoms; the pair with the lowest difference is the least polar, and the pair with the highest difference is the most polar. C–C (ΔE = 0) < H–C (ΔE = 0.4) < N–O (ΔE = 0.5) < Ca–H (ΔE = 1.1) < Ca–N (ΔE = 2.0)

55. a. The structure for ethane is $\left[\begin{array}{cccc} & H & H & \\ & | & | & \\ H- & C- & C & -H \\ & | & | & \\ & H & H & \end{array}\right]$. Formal charge = valence electrons – bonds – nonbonded electrons, so for carbon, FC = 4 – 4 – 0 = 0.

b. The structure for ethanol is $\left[\begin{array}{ccccc} & H & H & & \\ & | & | & & \\ H- & C- & C & -\ddot{\underset{..}{O}}- & H \\ & | & | & & \\ & H & H & & \end{array}\right]$. The formal charge on carbon is unchanged (FC = 4 – 4 – 0 = 0).

c. For oxygen, FC = 6 − 2 − 4 = 0.

d. Without the hydrogen on oxygen, the structure is $\left[\begin{array}{c} \text{H} \;\; \text{H} \\ | \;\;\; | \;\;\;\; \ddot{} \\ \text{H}-\text{C}-\text{C}-\ddot{\text{O}}: \\ | \;\;\; | \;\;\;\; \\ \text{H} \;\; \text{H} \end{array}\right]^-$. For oxygen, the formal charge is now FC = 6 − 1 − 6 = −1.

57. The structure of NH_4Cl is $\left[\begin{array}{c} \text{H} \\ | \\ \text{H}-\text{N}-\text{H} \\ | \\ \text{H} \end{array}\right]^+ \left[:\ddot{\underset{\cdot\cdot}{\text{Cl}}}:\right]^-$.

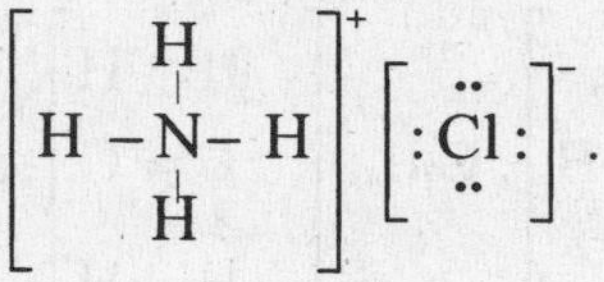

59. The structure of NO_2 is

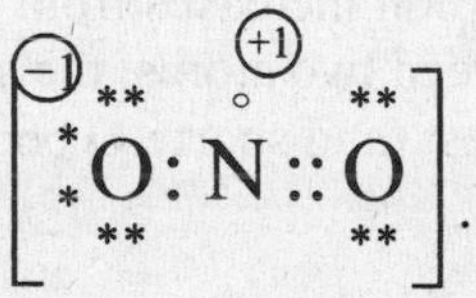

$$\left[\overset{\ominus}{\text{O}}:\overset{(+1)}{\overset{\circ}{\text{N}}}::\text{O}\right]$$

(with ** lone pairs above and below each O, and * lone pair on the left of the first O).

In this diagram, * represents lone-pair electrons, : represents bonding electrons, and ° represents a radical electron. The nonzero formal charges are labeled within the circles.

61. The cyanide (CN^-) structure is $\left[:\overset{\ominus}{\text{C}}\equiv\text{N}:\right]^-$. Carbon has a −1 formal charge and nitrogen has a 0 formal charge. Because the formal charge of −1 is on the carbon, H^+ with its positive charge should attach to carbon and its negative charge.

63. The structure for H_2SO_3 is $\left[\begin{array}{c} :\ddot{\text{O}}-\text{H} \\ | \\ \ddot{\underset{\cdot\cdot}{\text{O}}}=\text{S}: \\ | \\ :\underset{\cdot\cdot}{\text{O}}-\text{H} \end{array}\right]$. The formal charge on sulfur is 0 (FC = 6 − 4 − 2 = 0), while the oxidation number on sulfur is +4. In the calculation of the oxidation number, oxygen is taken at −2 and hydrogen at +1, giving the sulfur a +4 oxidation number, but in the calculation of formal charges both oxygen and hydrogen have a value of zero. This difference is why the oxidation number and the formal charge on sulfur differ.

65. Nitrate (NO_3^-) is $\left[\begin{array}{c} :\ddot{\text{O}}: \\ | \\ :\ddot{\text{O}}=\underset{(+1)}{\text{N}} \\ | \\ :\underset{\cdot\cdot}{\text{O}}: \end{array}\right]^-$, and nitrite ($NO_2^-$) is $\left[:\ddot{\text{O}}-\underset{(0)}{\ddot{\text{N}}}=\ddot{\text{O}}:\right]^-$. The formal charges are labeled on the structures inside the circles.

67. The two structures for glycine (NH_2CH_2COOH) are

$$\left[\begin{array}{c} \quad\quad\quad H \\ H\ \ :\ddot{O}: \\ H-\ddot{N}-C-C=\ddot{O} \\ H\ \ H \end{array}\right] \leftrightarrow \left[\begin{array}{c} \quad\quad\quad H \\ H\ \ \ \ O: \\ H-\ddot{N}-C-C-\ddot{O}: \\ H\ \ H \end{array}\right]$$

69. The structures are C_2H_6: $\left[H-\underset{H}{\overset{H}{C}}-\underset{H}{\overset{H}{C}}-H\right]$, C_2H_4: $\left[\underset{H}{\overset{H}{C}}=\underset{H}{\overset{H}{C}}\right]$, and C_2H_2: $[H-C\equiv C-H]$.

The number of bonds between carbon increases from one to two to three, as shown above. The greater the number of bonds between two atoms, the stronger the bond and the shorter the bond length. C_2H_2 will have the strongest (and shortest) bond, and C_2H_6 will have the longest (and weakest) bond.

71. Rewriting the reaction in terms of Lewis structures, we see that

$$H-\underset{H}{\overset{H}{C}}-\underset{H}{\overset{H}{C}}-\underset{H}{\overset{H}{C}}-\underset{H}{\overset{H}{C}}-H+\frac{13}{2}\ddot{O}=\ddot{O}\rightarrow 4:\ddot{O}=C=\ddot{O}:+5\,H-\ddot{O}-H$$

To complete the reaction we must break 3 C – C bonds, 10 H – C bonds, and 13/2 O=O bonds, while creating 8 C=O bonds and 10 O – H bond. Recall that energy is absorbed (+) when a bond is broken and is released (−) when a bond is formed.

$$\begin{aligned} 3\text{ mol C}-\text{C bonds} \times 347\text{ kJ/mol} &= +1041\text{kJ} \\ 13/2\text{ mol O=O bonds} \times 495\text{kJ/mol} &= +3217.5\text{kJ} \\ 10\text{ mol H}-\text{C bonds} \times 413\text{ kJ/mol} &= +4130\text{kJ} \\ \text{Total Bonds Broken} &= +8388.5\text{kJ} \end{aligned}$$

$$\begin{aligned} 8\text{ mol C}=\text{O bonds} \times (-799\text{kJ/mol}) &= -6392\text{kJ} \\ 10\text{ mol O}-\text{H bonds} \times (-467\text{kJ/mol}) &= -4670\text{kJ} \\ \text{Total Bonds Formed} &= -11062\text{kJ} \end{aligned}$$

$$\begin{aligned} \text{TOTAL Energy Absorbed} &= +8388.5\text{kJ} \\ \text{TOTAL Energy Released} &= -11062\text{kJ} \\ \text{Net Energy Change } (\Delta H) &= -2673.5\text{kJ} = -2674\text{ kJ/mol} \end{aligned}$$

This value is different from the direct determination, because the values for the bond energies are taken from the average of the bond in several molecules and may not be exactly the same as the bond energies in the molecules of this reaction.

73. The reactions are$H_2 + \frac{1}{2} O_2 \rightarrow H_2O$ and $H_2 + O_2 \rightarrow H_2O_2$. For water formation, we break 1 H–H bond and ½ of a O=O bond and form 2 O–H bonds: ΔH = (+432 kJ/mol) + ½(+495 kJ/mol) + 2(−467 kJ/mol) = −254.₅ kJ/mol = −254 kJ/mol. For hydrogen peroxide formation, we break 1 H–H bond and 1 O=O bond and form 2 O–H bonds and 1 O–O bond: ΔH = (+432 kJ/mol) +

(+495 kJ/mol) + 2(−467 kJ/mol) + (−146 kJ/mol) = −153 kJ/mol. Since water releases more energy than hydrogen peroxide in its formation, it will be the more stable of the two molecules.

75. H_2O: $\left[\mathrm{H-\ddot{\underset{\cdot\cdot}{O}}-H}\right]$, has four electron groups (tetrahedral) but two lone pairs, making the molecular geometry bent.

NO_2: $\left[:\ddot{\underset{\cdot\cdot}{O}}-\dot{N}=\ddot{O}:\right]$, has three electron groups (trigonal planar) but one lone radical, making the molecular geometry bent.

PCl_2 (radical): $\left[:\ddot{\underset{\cdot\cdot}{Cl}}-\dot{\underset{\cdot\cdot}{P}}-\ddot{\underset{\cdot\cdot}{Cl}}:\right]$, has four electron groups (tetrahedral) but two lone pairs/radicals, making the molecular geometry bent.

NBr_3: $\left[\begin{array}{c} \mathrm{H} \\ | \\ \mathrm{H-\ N:} \\ | \\ \mathrm{H} \end{array}\right]$, has four electron groups (tetrahedral) but one lone pair, making the molecular geometry trigonal pyramidal.

77. NO_2^-: $\left[:\ddot{\underset{\cdot\cdot}{O}}-\ddot{N}=\ddot{O}:\right]^-$, has three electron groups (trigonal planar) but one lone pair, making the ion geometry bent.

PO_4^{3-}: $\left[\begin{array}{c} :\ddot{O}: \\ | \\ \ddot{\underset{\cdot\cdot}{O}}=\mathrm{P}-\ddot{\underset{\cdot\cdot}{O}}: \\ | \\ :\underset{\cdot\cdot}{O}: \end{array}\right]^{3-}$, has four electron regions and no lone pairs, making the ion geometry tetrahedral.

SO_3^{2-}: $\left[\begin{array}{c} :\ddot{O}: \\ | \\ \ddot{\underset{\cdot\cdot}{O}}=\mathrm{S}: \\ | \\ :\underset{\cdot\cdot}{O}: \end{array}\right]^{2-}$, has four electron regions (tetrahedral) but one lone pair, making the ion geometry trigonal pyramidal.

CO_3^{2-}: $\left[\begin{array}{c} :\ddot{O}: \\ | \\ :\ddot{O}=\mathrm{C} \\ | \\ :\underset{\cdot\cdot}{O}: \end{array}\right]^{2-}$, has three electron regions and no lone pairs, making the ion geometry trigonal planar.

BrO_4^-: $\left[\begin{array}{c} :\ddot{O}: \\ | \\ \ddot{O} = Br = \ddot{O} \\ | \\ :\ddot{O}: \end{array}\right]$, has four electron regions and no lone pairs, making the ion geometry tetrahedral.

79. C_2H_6: $\left[\begin{array}{c} \ \ H\ \ H \\ H-C-C-H \\ \ \ H\ \ H \end{array}\right]$ has a tetrahedral geometry at each carbon atom (four electron groups, no lone pairs).

C_3H_6: $\left[\begin{array}{c} H\ \ H \\ H-C-C=C-H \\ H\qquad H \end{array}\right]$ has a tetrahedral geometry for the leftmost carbon (four electron groups, no lone pairs) and trigonal planar geometry at the right two carbon atoms (three electron groups, no lone pairs).

C_2H_4: $\left[\begin{array}{c} H\ \ H \\ C=C \\ H\ \ H \end{array}\right]$, has a trigonal planar geometry at each carbon atom (three electron groups, no lone pairs).

C_3H_8: $\left[\begin{array}{c} H\ \ H\ \ H \\ H-C-C-C-H \\ H\ \ H\ \ H \end{array}\right]$ has a tetrahedral geometry at each carbon atom (four electron groups, no lone pairs).

C_4H_{10}: $\left[\begin{array}{c} H\ \ H\ \ H\ \ H \\ H-C-C-C-C-H \\ H\ \ H\ \ H\ \ H \end{array}\right]$ has a tetrahedral geometry at each carbon atom (four electron groups, no lone pairs).

81. The Lewis structure of ClO_2 is $\left[:\ddot{O}-\dot{\underset{\cdot\cdot}{Cl}}=\ddot{O}:\right]$ (note the radical electron). At Cl, it has four electron groups (tetrahedral) but two lone pairs/radicals, making the molecular geometry bent.

83. The structures for these molecules are $BeCl_2$: $\left[:\ddot{Cl}-Be-\ddot{Cl}:\right]$, $AlCl_3$: $\left[\begin{array}{c} :\ddot{Cl}: \\ | \\ :\ddot{Cl}-Al-\ddot{Cl}: \end{array}\right]$,

CCl_4: $\left[\begin{array}{c} :\ddot{Cl}: \\ | \\ :\ddot{Cl}-C-\ddot{Cl}: \\ | \\ :\ddot{Cl}: \end{array}\right]$, $XeCl_4$: $\left[\begin{array}{c} :\ddot{Cl}: \\ | \\ :\ddot{Cl}-\ddot{Xe}-\ddot{Cl}: \\ | \\ :\ddot{Cl}: \end{array}\right]$, and NCl_3: $\left[\begin{array}{c} :\ddot{Cl}: \\ | \\ :\ddot{Cl}-\ddot{N}: \\ | \\ :\ddot{Cl}: \end{array}\right]$. The geometries

and bond angles are as follows: $BeCl_2$, linear, 180°; $AlCl_3$, trigonal planar, 120°; CCl_4, tetrahedral, 109°; $XeCl_4$, square planar, 90°; and NCl_3, trigonal pyramid, ~107° (due to repulsion of lone pair). In order from largest to smallest: $BeCl_2 > AlCl_3 > CCl_4 > NCl_3 > XeCl_4$.

85. $SeCl_4$ has the structure $\left[\begin{array}{c} :\ddot{Cl}-\ddot{Se}-\ddot{Cl}: \\ / \quad \backslash \\ :\ddot{Cl}: \quad :\ddot{Cl}: \end{array}\right]$, which has an electron geometry of trigonal bipyramid. To reduce the electron–electron repulsion, the lone-pair electrons take an equatorial position, leaving the rest of the molecule in a see-saw shape.

87. To complete the problem, we need the structures for both species: N_2H_2: $\left[H-\ddot{N}=\ddot{N}-H\right]$ and $N_2H_2^{2+}$: $\left[H-N\equiv N-H\right]^{2+}$. In N_2H_2, each nitrogen has three electron groups (trigonal planar) but one lone pair, so the molecular geometry is bent. The H–N–N bond angle will be slightly less than the normal 120° because of the additional lone-pair repulsion. In $N_2H_2^{2+}$, each nitrogen has only two electron groups and no lone pairs, making the geometry linear. The bond angle is then expanded from 120° in N_2H_2 to 180° in $N_2H_2^{2+}$.

89. For each molecule, the ranking is derived from the electronegativity difference (in parentheses). Ranked from least to most polar: $C\leftarrow H$ (0.4) < $C\rightarrow N$ (0.5) = $C\leftarrow B$ (0.5) = $C\rightarrow Cl$ (0.5) < $C\rightarrow O$ (1.0) < $C\leftarrow Mg$ (1.3).

91. The structures are SF_6 $\left[\begin{array}{c} :\ddot{F}:\ :\ddot{F}: \\ \backslash\ | \\ :\ddot{F}-S-\ddot{F}: \\ |\ \backslash \\ :\ddot{F}:\ :\ddot{F}: \end{array}\right]$ and SF_5 $\left[\begin{array}{c} :\ddot{F}:\ :\ddot{F}: \\ \backslash\ | \\ \cdot S-\ddot{F}: \\ |\ \backslash \\ :\ddot{F}:\ :\ddot{F}: \end{array}\right]$. SF_6 is octahedral (six electron groups around S) and nonpolar because all the S–F dipoles cancel. SF_5 has six electron groups but one lone pair, making the geometry square pyramidal. By removing one S–F dipole that helped in making SF_6 nonpolar, SF_5 is polar because not all the S–F dipoles cancel out.

93. To interpret the table, we need to remember that the higher the melting point, the higher the interactions between ions or molecules in the solid; that water solubility is possible only for polar molecules and hexane solubility only for nonpolar molecules; and that only ionic substances will conduct electricity in a molten state. Therefore, A is an ionic compound, B is a nonpolar molecule, and C is a polar molecule.

95. The structure of H_2O_2 is $\left[H-\ddot{\underset{\cdot\cdot}{O}}-\ddot{\underset{\cdot\cdot}{O}}-H \right]$ and is bent (see Problem 76). The O–H bonds are polar with an electronegativity difference of 1.4. Because the molecule is bent, even though the dipoles point in nearly opposite directions, the dipoles do not cancel and the molecule is polar. If the molecule were linear, the dipoles would cancel out, because they would be aligned and in opposite directions. The linear molecule would be nonpolar.

97. Because each of the possibilities has the same number of protons, the sizes will be determined by the number of electrons. The greater number of electrons will correspond with a larger atom/ion. From left to right the diagrams represent H^+, H^-, and H.

99. Because N_2 has a triple bond, its bond energy is very large (941 kJ/mol). Breaking this bond in order to cause a reaction is very difficult and therefore unlikely. N_2 will emerge just as it entered.

101. There is only a single choice for the anion (−2), so our choice will be based on which cation will have a weaker interaction. The weakest lattice attractions will come from ion pairs that are widely separated and have smaller charges. This set of ions (the small +1 and the larger +3) have these trends in opposition. The ion with the +3 charge has roughly twice the diameter, which is not a large enough increase in size to offset the larger charge. The +3 ion will have a larger ionic attraction. Therefore, the weakest interaction will be between the −2 anion and +1 cation.

103. The highest lattice enthalpy occurs for the ions with the largest attractions; therefore, the +3 cation with −2 anion would have the larger lattice enthalpy.

105. a. The attraction that two charges particles have for one another is larger as the charges on each become larger and as the particles get closer to one another. Because ionic compounds have full (and sometimes multiple) charges and often involve small ions, the Coulomb forces can be quite large. Large attractive forces lead to high melting points. Neutral molecular species at most will have partial charges from the dipoles, limiting the strength of attraction to other molecules. The lower attraction leads to lower melting points.

b. For a high melting point, we should choose species that are small and have high charges. For a low melting point, we should choose low charges and large radii. By far the highest charge and lowest size belong to B^{3+}. For the anion we could choose N^{3-} or O^{2-}. Nitride has the higher charge but is bigger (171 pm versus 140 pm), but it is less than 1.25 times larger while its charge is 1.5 times larger than that of oxide. Highest melting point: BN. The largest ions are I^- and Cs^+; CsI should have the lowest melting point.

107. a. Very few molecules can be found in only a diatomic form allowing for direct measurement of the bond energy. Inside larger molecules, the strength of the bond can vary, depending on the other atoms and how well the electrons are distributed within the molecule. In these cases, several measurements in different molecules are made, and the average value of the bond energy is placed in the table.

b. The ethane molecule (CH_3CH_3) does not have any particular bonding irregularities, but trifluoroethane (CH_3CF_3) could be markedly different. The presence of the three fluorine atoms with their high electronegativity will tend to shift the electrons from the other bonds toward the fluorine atoms. This has the potential to weaken the carbon–carbon bond.

109. a. When the chlorine atoms are placed on opposite carbons, the dipoles will nearly average out. (Recall that the C−C bond is free to rotate, causing the dipoles almost to cancel over long time scales.) The geometry at each carbon is tetrahedral, so the entire molecule could be imagined as two pyramids squashed together at the top points.

$$\begin{array}{c} \mathrm{H}\ \ \mathrm{H} \\ | \quad | \\ \mathrm{Cl}-\mathrm{C}-\mathrm{C}-\mathrm{Cl} \\ | \quad | \\ \mathrm{H}\ \ \mathrm{H} \end{array} \quad \text{or} \quad \mathrm{Cl} \leftarrow \mathrm{CH_2}-\mathrm{CH_2} \rightarrow \mathrm{Cl}$$

b. When both chlorines are placed on the same carbon, the dipoles cannot cancel:

$$\begin{array}{c} \mathrm{H}\ \ \mathrm{Cl} \\ | \quad | \\ \mathrm{H}-\mathrm{C}-\mathrm{C}-\mathrm{Cl} \\ | \quad | \\ \mathrm{H}\ \ \mathrm{H} \end{array} \quad \text{or} \quad \mathrm{H}-\mathrm{CH_2}-\mathrm{CH}(\nearrow\mathrm{Cl}) \rightarrow \mathrm{Cl}$$

c. Wax is non-polar, so we need a nonpolar molecule to dissolve it. Because the dipoles in the compound in part a cancel (at least partially), it should be the better choice.

d. The structure for chlorine is $\left[:\ddot{\mathrm{Cl}}-\ddot{\mathrm{Cl}}:\right]$, and the structure for ethene is $\left[\begin{array}{c} \mathrm{H}\ \ \mathrm{H} \\ | \quad | \\ \mathrm{C}=\mathrm{C} \\ | \quad | \\ \mathrm{H}\ \ \mathrm{H} \end{array}\right]$.

e. The bonds in ethane are nonpolar covalent, and the molecule is symmetric; it should have no net dipole.

f. The net reaction is $Cl_2 + C_2H_4 \rightarrow C_2H_4Cl_2$, so the mass of ethane consumed is

$$2.0\,\mathrm{g\,C_2H_4Cl_2} \times \frac{1\,\mathrm{mol\,C_2H_4Cl_2}}{98.95\,\mathrm{g\,C_2H_4Cl_2}} \times \frac{1\,\mathrm{mol\,C_2H_4}}{1\,\mathrm{mol\,C_2H_4Cl_2}} \times \frac{28.05\,\mathrm{g\,C_2H_4}}{1\,\mathrm{mol\,C_2H_4}} = 0.57\,\mathrm{g\,C_2H_4}$$

Chapter 9: Advanced Models of Bonding

The Bottom Line

- The molecular models that chemists construct increase in complexity from Lewis dot structures, to the VSEPR model, to valence bond theory, to MO theory. This increase in complexity is accompanied by an increase in satisfactory agreement with observed properties for the molecule.
- Valence bond theory, originated by Linus Pauling, defines bonds as the overlap of atomic orbitals. Sigma bonds result from end-on overlap of orbitals; pi bonds result from side-to-side overlap of orbitals. (Section 9.1)
- Hybridization of atomic orbitals gives rise to new orbitals that help explain bonding. Hybridization also gives structures consistent with the VSEPR model. (Section 9.2)
- Mixing n atomic orbitals gives n hybridized orbitals. (Section 9.2)
- Hybridization can be used to explain the existence of sigma (σ) and pi (π) bonds in molecules. (Section 9.2)
- Molecular orbital theory defines bonding with orbitals that are not confined to a single atom. Bonds result from molecular orbitals in a molecule that encompass the atoms. (Section 9.3)
- Mixing n atomic orbitals gives n molecular orbitals. One-half n of these are bonding; the other half are antibonding. Bonding MOs are lower in energy than the atomic orbitals from which they are constructed. Antibonding MOs are higher in energy than the atomic orbitals from which they are constructed.
- The Pauli exclusion principle and Hund's rule must be obeyed when placing electrons in the new MOs. (Section 9.3)
- MO diagrams can be used to identify the number of bonds between atoms. (Section 9.3)
- Paramagnetism results from unpaired electrons in a molecule. Diamagnetism results from complete pairing of all electrons in a molecule. (Section 9.3)
- Electrons in conjugated π orbitals are said to be delocalized, because the electron density can be distributed among more than two atoms. (Section 9.4)

***Note concerning Lewis structures in this key:* In order to maintain clarity, all structures, not just ions, have been placed within brackets when the structure appears within the text. The brackets are not required in normal use for neutral structures.**

Solutions to Practice Problems

9.1. The configuration of the valence electrons in fluorine is $2s^22p^5$. Overlap of one of the $2p$ orbitals from each fluorine allows the electrons to be shared between the two atoms. The valence bond model of F−F shows a $2p$-$2p$ orbital overlap that constitutes the covalent bond.

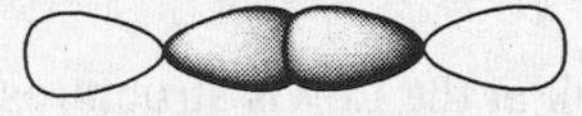

9.2. The bond formed in F_2 is from the overlap of two $2p$ orbitals, in HCl from $1s$ and $3p$ orbitals, and in Cl_2 from two $3p$ orbitals. Because both $2p$ orbitals are smaller than the $3p$ orbitals, the bond in F_2 is shorter than that in Cl_2, and because the $1s$ orbital is much smaller than the $3p$ orbital, the HCl bond is shorter than the Cl_2 bond. Cl_2 will have the longest bond.

9.3. a. The Lewis dot structure model for OF_2, $\left[:\ddot{\underset{\cdot\cdot}{F}}-\ddot{\underset{\cdot\cdot}{O}}-\ddot{\underset{\cdot\cdot}{F}}:\right]$, indicates that the central oxygen has two bonds and two lone pairs. Mixing the four orbitals (one 2*s* and three 2*p*) on the oxygen atom allows us to have two lone pairs of equal energy and two bonds to the adjacent fluorine atoms. The oxygen atom possesses sp^3 hybridized orbitals.

b. The Lewis dot structure model for H_2S, $\left[H-\ddot{\underset{\cdot\cdot}{S}}-H\right]$, indicates that the central sulfur has two bonds and two lone pairs. Mixing the four orbitals (one 3*s* and three 3*p*) on the sulfur atom allows us to have two lone pairs of equal energy and two bonds to the adjacent hydrogen atoms. The sulfur atom possesses sp^3 hybridized orbitals.

c. The Lewis dot structure model for NH_4^+, $\left[\begin{matrix} & H & \\ & | & \\ H- & N & -H \\ & | & \\ & H & \end{matrix}\right]^+$, indicates that the central nitrogen has four bonds. Mixing the four orbitals (one 2*s* and three 2*p*) on the nitrogen atom allows us to have four bonds to the adjacent hydrogen atoms. The nitrogen atom possesses sp^3 hybridized orbitals.

9.4. a. OF_2 geometry: sp^3 hybridized atoms adopt a tetrahedral geometry. Because two of the sp^3 orbitals contain lone pairs, the VSEPR model indicates that the molecule has an overall bent geometry. The bond angles should be less than 109.5° because the lone pairs repel each other more than the bonding pairs.

b. H_2S geometry: sp^3 hybridized atoms adopt a tetrahedral geometry. Because two of the sp^3 orbitals contain lone pairs, the VSEPR model indicates that the molecule has an overall bent geometry. The bond angles should be less than 109.5° because the lone pairs repel each other more than the bonding pairs.

c. NH_4^+ geometry: sp^3 hybridized atoms adopt a tetrahedral geometry. The bond angles should be 109.5°.

9.5. A quick look at the Lewis dot structure for CH_2O, $\left[\begin{matrix} & :O: & \\ & \| & \\ H- & C & -H \end{matrix}\right]$, shows that C will need three sigma bonds (one to O and two to H) and will use sp^2 hybridization to create the bonds. Oxygen also will be sp^2 hybridized to create the sigma bond to carbon and for the two lone pairs. Both carbon and oxygen have half-filled p orbitals that can be used to create a pi bond to complete the double bond (one sigma, one pi) between C and O. Formaldehyde has a total of one pi bond and three sigma bonds.

9.6. To answer the question, a quick look at the Lewis structures will help:

CH_2NOH: $\begin{matrix} H & & & & \\ | & & & & \\ H-C & = & \ddot{N} & - & \ddot{\underset{\cdot\cdot}{O}}-H \end{matrix}$ CH_3NHOH: $\begin{matrix} & H & & & \\ & | & & & \\ H- & C & -\ddot{N}- & \ddot{\underset{\cdot\cdot}{O}} & -H \\ & | & | & & \\ & H & H & & \end{matrix}$

The shorter bond will be found in CH_2NOH as the NO bond is formed from the combination of a sp^2 (on N) and sp^3 (on O) hybridizations. In CH_3NHOH, both N and O have sp^3 hydridization. We

can further explain by seeing that we can write a resonance structure for CH_3NOH in which a double bond exists between N and O, making the bond shorter:

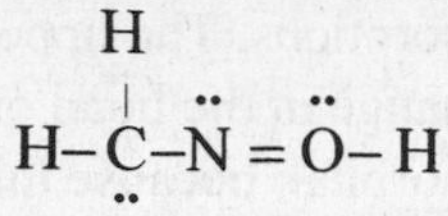

```
    H
    |  ..  ..
  H–C–N = O– H
    ..
```

9.7. **We can get an idea of the shapes by first looking at the Lewis structures:**

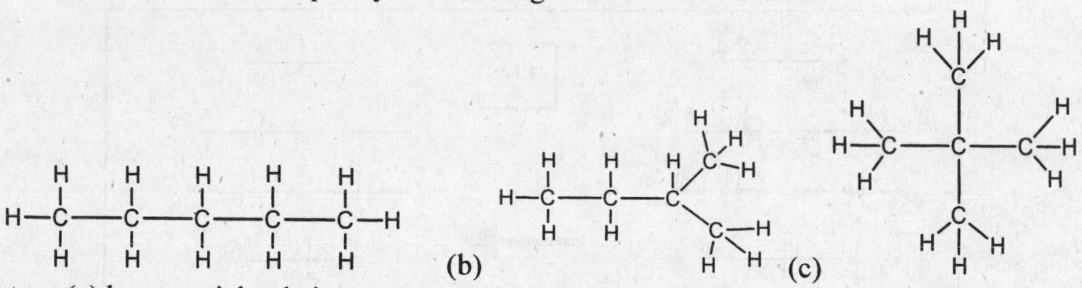

```
   H   H   H   H   H
   |   |   |   |   |
 H–C———C———C———C———C–H
   |   |   |   |   |
   H   H   H   H   H
(a)
```

(b)

(c)

Pentane (a) has a straight-chain structure and will be able to pack more closely together. It should have the most intermolecular attraction and the highest boiling temperature. The other two molecules are progressively more branched and will pack less tightly. They will boil at lower temperatures.

9.8. **The molecular orbital diagram for He_2 is at right, using only the valence electrons ($1s^2$) from each He atom. Note that all the molecular orbitals are filled; equal numbers of bonding and antibonding orbitals are filled. The bond order will be 0, and He_2 is not stable.**

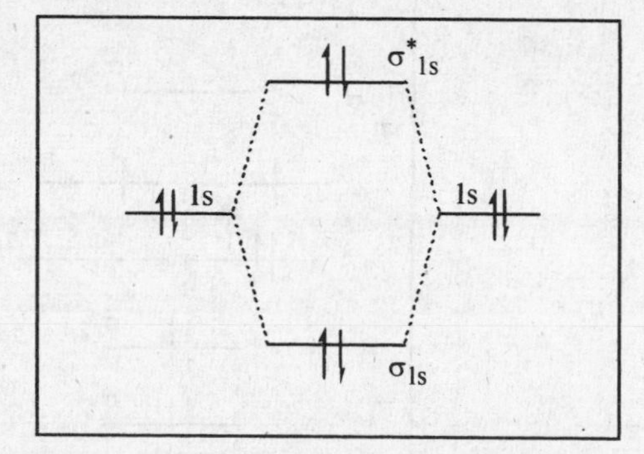

9.9. **The molecular configuration for H_2 shows two valence electrons in the bonding σ_{1s} and none in the antibonding σ^*_{1s}. (See the figure below.) The bond order for H_2 is only 1. Promoting a single bonding electron to an antibonding orbital would make the bond order 0 and cleave the molecular bond.**

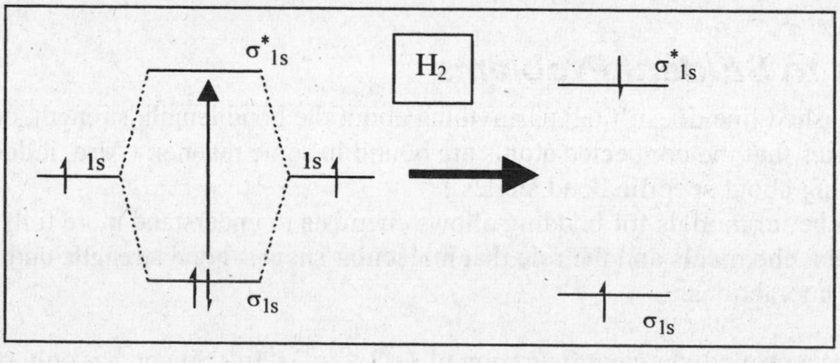

σ*1s
H2
σ*1s
1s
1s
σ1s
σ1s

The molecular configuration for O_2 (using only the valence 2*s* and 2*p* electrons; see the figure below) shows eight bonding electrons and four antibonding electrons, giving a bond order of 2. There are several possibilities for single photon absorptions. The upper diagram shows a transition from the π^* to the σ^* molecular orbital, causing no change in the bond order. The lower diagram shows a transition from the π to the π^* molecular orbital; because this transition moves a bonding electron into an antibonding orbital, the bond order decreases to 1. In either case, the remaining bond order in oxygen remains higher.

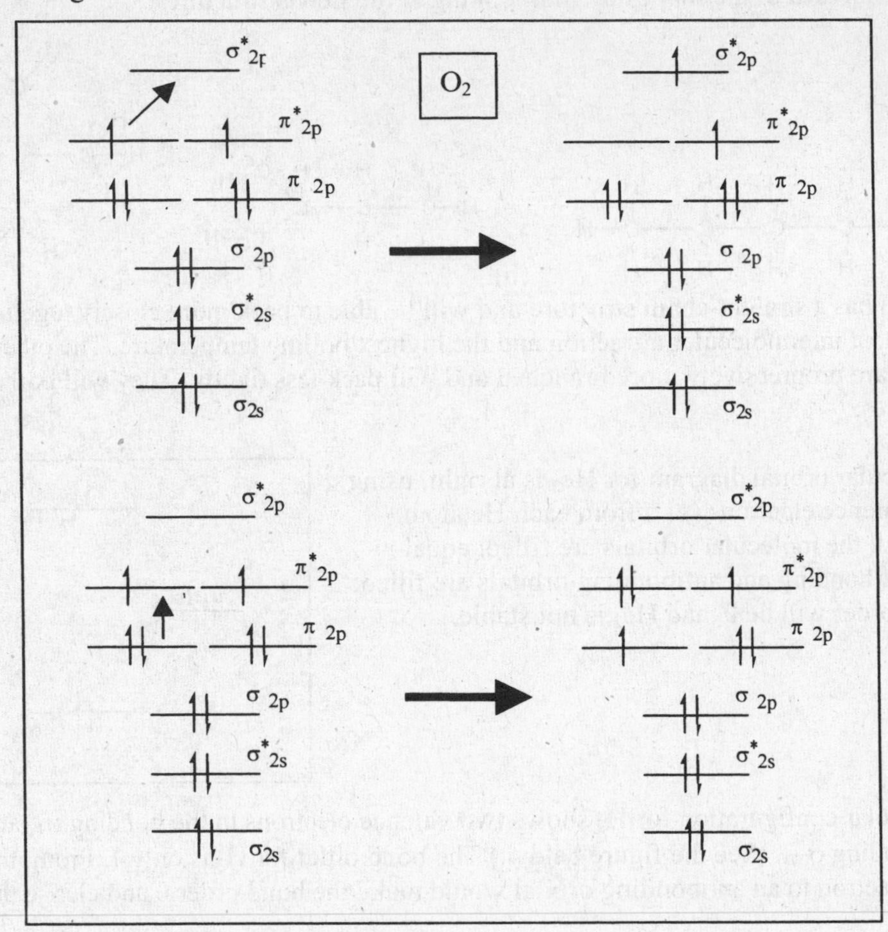

Solutions to Student Problems

1. a. The dashed line doesn't tell us anything about the bond length, strength, or energy; it merely indicates that the connected atoms are bound in some manner. (Also, it doesn't reveal anything about specific bond angles.)

b. Using better models for bonding allows chemists to understand more fully the interaction between chemicals and the role that molecular shapes, bond strength, and polarity play in those interactions.

3. Silicon has a ground-state configuration of $1s^2 2s^2 2p^6 3s^2 3p^2$. Silicon has only two unpaired electrons ($3p^2$) to use in the formation of two covalent bonds. This answer is not consistent with $SiCl_4$, which has four bonds on silicon.

5. The number of bonds that form is predicted from the number of *unpaired* electrons in the valence shell. Cl: $1s^22s^22p^63s^23p^5$, one bond; Se: $1s^22s^22p^63s^23p^64s^23d^{10}4p^4$, two bonds, B: $1s^22s^22p^1$, one bond.

7. Carbon has a ground-state configuration of $1s^22s^22p^2$. Carbon has only two unpaired electrons ($2p_x^{\,1}2p_y^{\,1}$) to use in the formation of two covalent bonds. With only two electrons to form bonds, carbon cannot form CCl_4 and the required four bonds using the ground electron configuration.

9. a. H has a configuration of $1s^1$ and bromine has a configuration of $1s^22s^22p^63s^23p^64s^23d^{10}4p^5$.
b. The $1s$ on hydrogen and the $4p$ on bromine form the overlap that creates the H–Br bond.
c. This bond should be weaker than the H–F bond, because the overlap of the much larger Br orbital with the small H orbital creates a weaker interaction and weaker bond.

11. Because each chlorine atom has a configuration of $1s^22s^22p^63s^23p^5$, one $3p$ orbital from each Cl is involved in creating the covalent bond. These orbitals are exactly the same size and energy, so the amount of overlap in the bond is substantial.

13. Cl_2 has the longer bond since the $3p$ orbitals from Cl are larger than the $1s$ orbitals on H. Br_2 has the longer bond since the $4p$ orbitals from Br are larger than the $3p$ orbitals from Cl. HBr has the longer bond since the $4p$ orbitals from Br are larger than the $3p$ orbitals from Cl, while the $1s$ orbital from H is involved in each molecule.

15. The $1s$ orbital on H ($1s^1$), the $3p$ orbital on Cl ($1s^22s^22p^63s^23p^5$), and the $2p$ orbitals on O ($1s^22s^22p^4$) are likely candidates for bonding in HOCl. (Using the hybridization model, we realize that the sp^3 orbitals on oxygen are used in the molecule.) Because the same bonds are used by O in the bonds with H and Cl, the size of the orbitals on H (smaller) and Cl (larger) predict a larger O–Cl bond.

17. Potassium hydride is KH. If the compound is covalently bound, KH should have the shorter and stronger bond since the 4s orbital on K is smaller than the $6s$ orbital on Cs. (Since K^+ is also smaller than Cs^+, one would expect the ionic attraction of K^+ and H^- to be stronger as well.)

19. The milk represents the s orbital and is involved in all hybridization schemes. The eggs represent the p orbitals and can be mixed with the milk in 1-, 2-, or 3-part proportions. The character of the omelet representing the hybrid orbitals is a little different in each case, but each is still an omelet. Also, the size of the omelet increases (as does the size of the hybrid orbital) as the number of eggs (p orbitals) increases.

21. a. In sp^3d^2 hybridization there are six orbitals involved: one s orbital, three p orbitals, and two d orbitals.
b. We will still finish with generally the same shape for the hybrid orbital; however, the presence of more orbitals causes the angles between each orbital to decrease to 90°, with an overall octahedral shape. Additionally, because the d orbitals are larger than the s or p orbitals, the sp^3d^2 hybrid orbitals will tend to be longer than any of the sp hybrid types.

23. a. The Lewis diagram for Si is $\left[:\dot{\text{Si}}\cdot\right]$. In order to produce four bonds, Si must use sp^3 hybrid orbitals.
b. With four equally spaced sp^3 orbitals, $SiCl_4$ would have a tetrahedral shape.

25. The central atoms differ in the number of lone pairs that are accommodated within the sp^3 hydrid orbitals. Each lone pair removes a vertex from the tetrahedron: CH_4 is tetrahedral, NH_3 loses a vertex and is trigonal pyramidal, and water loses two vertices and is bent. Additionally, the lone-pair electrons have a larger repulsive force than the electrons in a bonding orbital; therefore, the bond angles are reduced for every extra lone pair in the molecule. In CH_4 there are no lone pairs and the bond angle is the normal 109.5° in the tetrahedral geometry. In NH_3 there is a single lone pair, which decreases the H–N–H bond angles from 109.5° to ~107°. In H_2O, there are two lone pairs, further decreasing the bond angle to ~104°.

27. The base geometry of both molecules is tetrahedral; however, although all four of carbon's sp^3 orbitals are the same size and shape, the 3*p* orbitals used in the bonding by Cl are longer and larger than the 1*s* orbitals in H. The lengthening of the three C–Cl bonds relative to the C–H bonds changes the tetrahedral shape. Additionally, the Cl–C–Cl bond angles may increase slightly to accommodate the larger Cl atoms relative to H.

29. Two *sp* orbitals are formed from mixing an *s* orbital with a single *p* orbital.

31. *sp*: 2 bonds; sp^2: three bonds; sp^3: four bonds

33. a. In OF_2, [F–Ö: with F bonded above O, two lone pairs on O], O has two bonds and two lone pairs: sp^3 hybridization

b. In CCl_4, [Cl–C–Cl with Cl above and Cl below C], C has four bonds and no lone pairs: sp^3 hybridization

c. In BCl_3, [Cl–B–Cl with Cl above B], B has three bonds and no lone pairs: sp^2 hybridization

d. In $BeCl_2$, [Cl – Be – Cl], Be has two bonds and no lone pairs: *sp* hybridization

35. a. linear b. trigonal planar c. tetrahedron d. trigonal bypyramid e. octahedron

37. For C_2H_2, the carbon forms two *sp* orbitals, one that forms a sigma bond with the other carbon and one that forms a sigma bond with hydrogen. The remaining two *p* orbitals on each carbon form the two pi bonds. For C_2H_4, the carbon forms three sp^2 orbitals, one that forms a sigma bond with the other carbon and two that form sigma bonds with hydrogen.

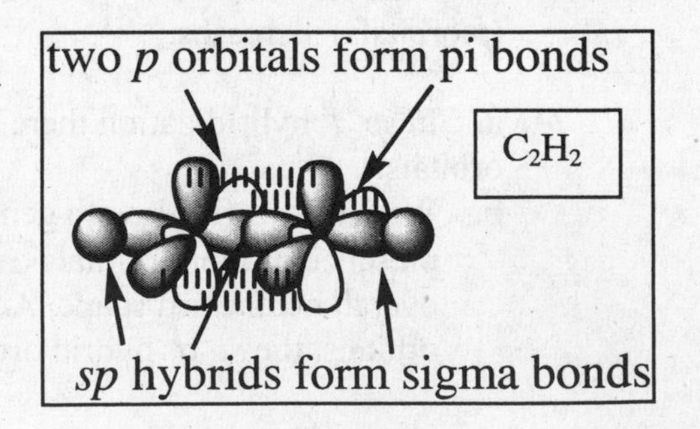

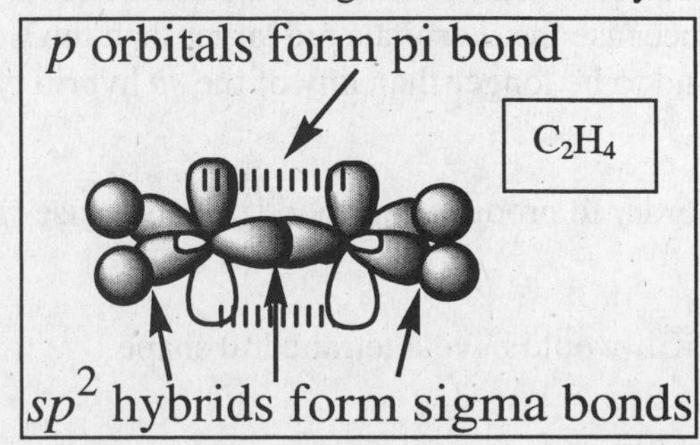

The remaining *p* orbital on each carbon forms the pi bond. For C_2H_6, the carbon forms four sp^3 orbitals, one that forms a sigma bond with the other carbon and three that form sigma bonds with hydrogen. Because C_2H_6 has no pi bonds between the carbons, less energy will be required to break the C–C bond.

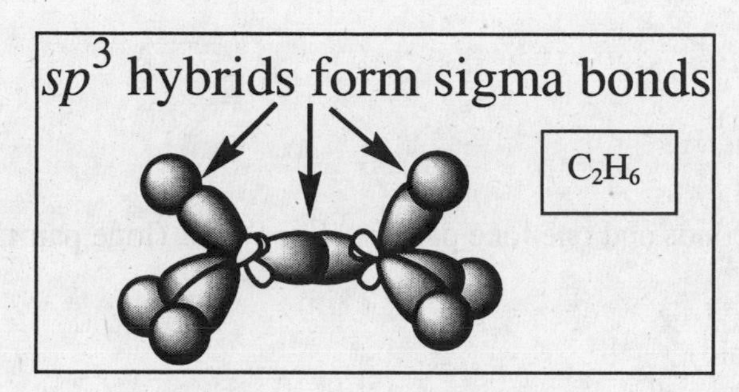

39. a. SO_3: [O=S(=O)=O], forms three sigma bonds: sp^2, trigonal planar

b. SO_3^{2-}: [:Ö–S(:)(=O)–Ö:] $^{2-}$, forms three sigma bonds and has one lone pair: sp^3, trigonal pyramidal

c. SF_6: [S bonded to six :F:], forms six bonds: sp^3d^2, octahedral

d. S_8: is ring structure: (eight-membered S ring), each S has two bonds and two lone pairs, sp^3, bent geometry at each sulfur leading to a crown structure,

41. a. NH_4^+: [H–N(H)(H)–H] $^+$, N has four bonds, sp^3, tetrahedral

b. XeF_4: [Lewis structure], Xe has four bonds and two lone pairs, sp^3d^2, square planar (lone pairs fill top and bottom of octahedron)

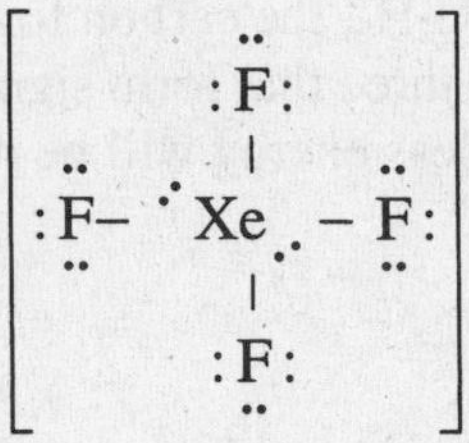

c. SF_4: [Lewis structure], S has four bonds and one lone pair, sp^3d, see-saw, (lone pair takes equatorial position in trigonal bipyramid)

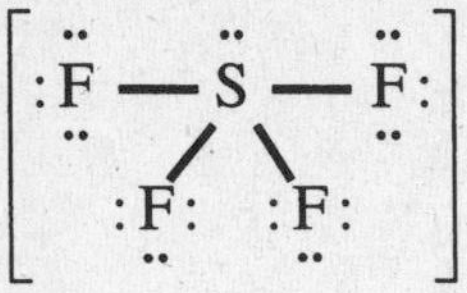

d. NO_2^-; $[:\ddot{O}-\ddot{N}=\ddot{O}:]^-$, N has two sigma bonds and one lone pair, sp^2, bent, (lone pair takes vertex in trigonal planar geometry)

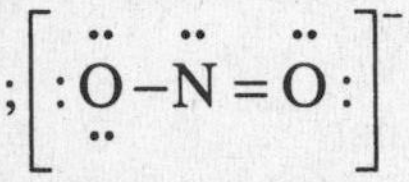

43. For atoms bound to more than one other atom: O, sp^3 (two bonds, two lone pairs); left C, sp^2 (three sigma bonds, no lone pairs); right C, sp^3 (four bonds, no lone pairs); N, sp^3 (three bonds, one lone pair).

45. Both Li_2 and K_2 will have the same bonding (because both have the same valence configuration of one *s* electron); however, the orbitals on K are larger. The overlap of the atomic orbitals from K in forming the covalent bond will be less than that on Li. K_2 would have the weaker bond.

47. The Lewis structure for urea is [H–N(H)–C(=O)–N(H)–H]. With three bonds and a lone pair, nitrogen will be sp^3 hybridized corresponding with tetrahedral geometry. Because of the lone pair, the H-N-C bond angle should be slightly less than the normal 109.5°.

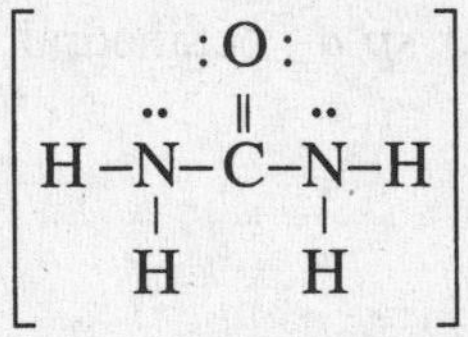

49. There are six pi bonds in the diagram. The carbon in the C=O at the top of the molecule has three sigma bonds and no lone pairs, resulting in sp^2 hybridization, a trigonal planar geometry around the C, and bond angles of 120°.

51. The Lewis structure of the ring portion of the molecule is [ring structure]. Both nitrogen atoms have one lone pair. The left nitrogen has two sigma bonds and a single lone pair. It will be sp^2 hybridized with a 120° bond angle. The right nitrogen has three sigma bonds and a lone pair. It will be sp^3 hybridized with 109.5° bond angles.

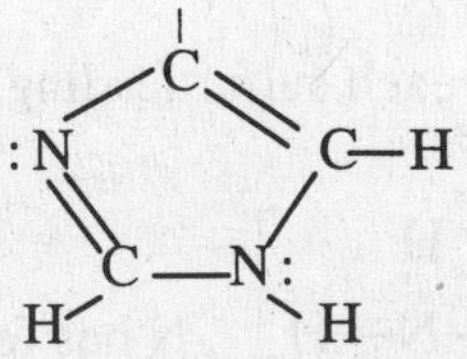

53. Although phosphorus has five total bonds, only four are sigma bonds. P will form sp^3 hybrid orbitals to form those four bonds with 109.5° bond angles.

55. a. The sp^3 hybridized carbons are shown on the structure (all carbons with four sigma bonds).

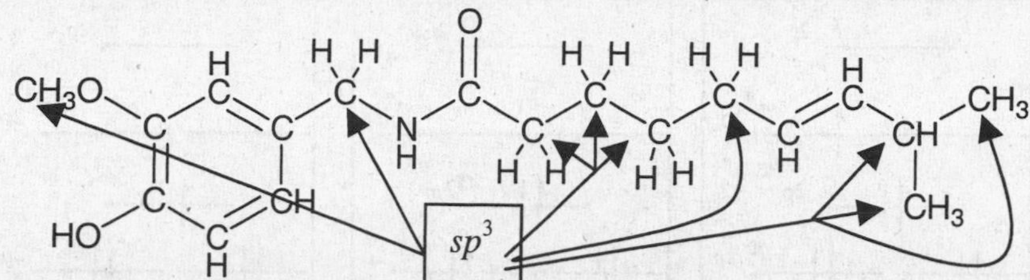

b. In order to create a pi bond, one H from each C atom would have to be removed.
c. There are five pi bonds (one each from the five double bonds).
d. To fill the octet, the oxygen atom needs four electrons in two lone pairs.

57. The table correctly filled out:

MODEL	SCIENTIST	GENERAL SUMMARY
VSEPR	G. Lewis	Distribute bonding and nonbonding electron pairs
VB	L. Pauling	Overlap and hybridization create new orbitals
MO	E. Schrodinger	Subtract and add overlap to create π and σ bonds

59. Because the *p* orbitals on sulfur are larger, the bond distance in S_2 is larger. Because the atoms are farther apart than in oxygen, the *p* orbitals are not able to create a good overlap to form the pi bond.

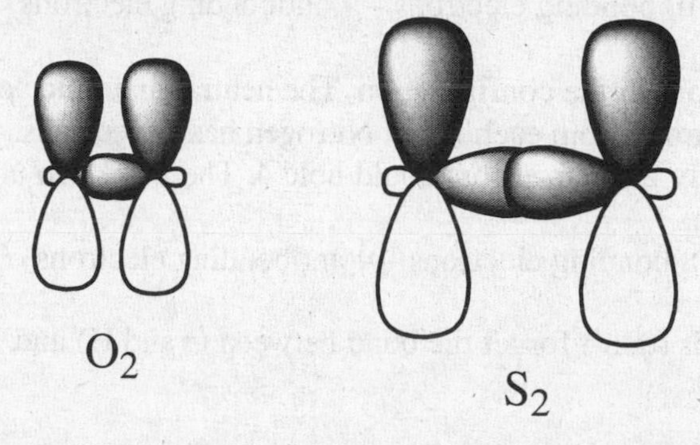

61. The largest difference between bonding and antibonding orbitals is that bonding orbitals do not have a node perpendicular to the bond axis, while antibonding orbitals do. The bonding orbitals also have lower energies than antibonding orbitals made from the same atomic orbitals.

63. a. LCAO – Linear combination of atomic orbitals is the method by which the molecular orbital theory creates its orbitals. The electron densities of the atomic orbitals can be added or subtracted to create the new molecular orbitals.
b. HOMO – The highest occupied molecular orbital is the orbital that has the highest energy among those orbitals that contain electrons.
c. LUMO – The lowest unoccupied molecular orbital is the orbital that has the lowest energy among those orbitals that do not contain electrons.

65. a. The bond orders for the molecules shown below are N_2^- (2.5); N_2 (3); and N_2^+ (2.5).

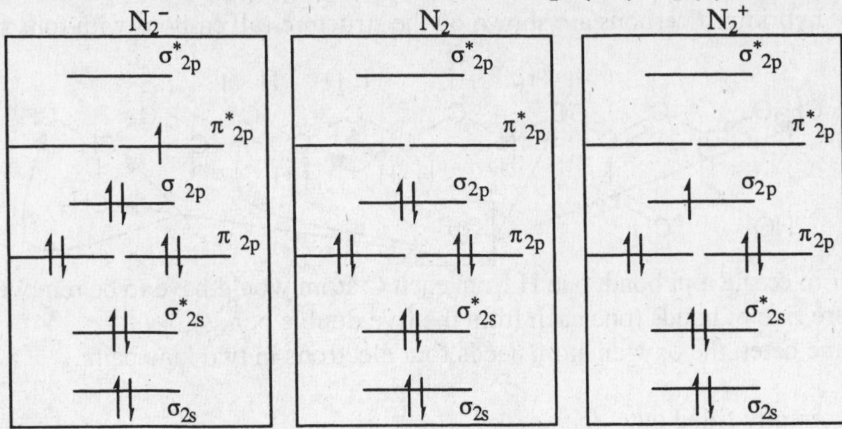

b. N_2^- and N_2^+ have unpaired electrons and are paramagnetic.
c. High bond orders correspond with shorter bond distances, so $N_2 < N_2^- \approx N_2^+$.

67. a. There are 17 electrons in the configuration. The neutral diatomic species would have 20 electrons, or 10 electrons from each atom. Neon has 10 electrons.
b. The HOMO has only 3 but could hold 4. There is an unpaired electron, and the ion is paramagnetic.
c. The bond order is (10 bonding electrons – 7 antibonding electrons) / 2 = 1.5.

68. a. There are 12 electrons in the configuration. The neutral diatomic species would have 14 electrons, or 7 electrons from each atom. Nitrogen has 7 electrons.
b. The HOMO has only 2 electrons but could hold 4. There are two unpaired electrons, and the ion is paramagnetic.
c. The bond order is (8 bonding electrons – 4 antibonding electrons) / 2 = 2.

69. There are 7 sigma bonds (don't forget the bond between O and H) and 2 pi bonds (in the triple bond).

70. There are 12 sigma bonds (don't forget the OH bond, the 2 CH bonds in CH_2 and the CH bonds in the rightmost carbon) and 1 pi bond (in the double bond).

71. a. There is only a single bond present, so the bond order is 1.
b. This is a bond within a ring that has resonant structures (or MOs that are different from the VB orbitals) and is the average of the single and double bonds. It has a bond order of 1.5.
c. There is only a single bond present, so the bond order is 1.

73. a. The diagram is shown at right. The bond order is 3 [(8 bonding electrons – 2 antibonding electrons) / 2].
b. There are no electrons in the highest π^* orbital.

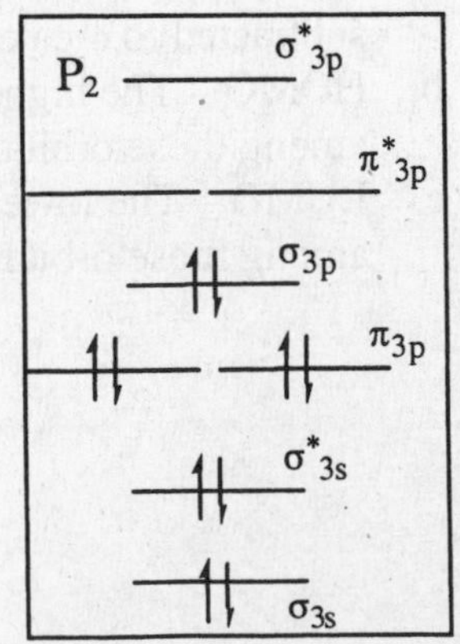

75. a. Orbital overlap occurs when the orbitals on two adjacent atoms share some space in common. When this occurs, the orbitals are free to combine and make bonds (or antibonds).

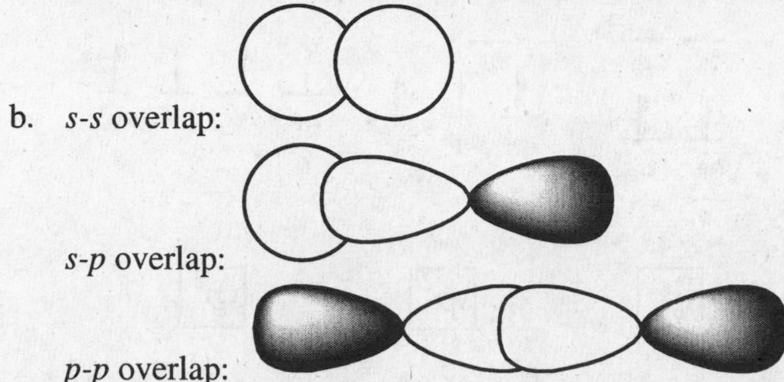

b. *s-s* overlap:

s-p overlap:

p-p overlap:

77. a. The Lewis structures (or skeleton structures) cannot show the single average structure that truly exists for molecules that have resonance structures. Some molecules will have bonds of order 1.5 or 1.33 that are difficult to draw without the resonance structures.

b. This resonance structure has alternate positions for the double bonds inside the ring.

79. The carbon that is second from the right has two sigma bonds and no lone pairs (and two pi bonds.) To form two sigma bonds, the carbon is *sp* hybridized.

80. The bond angles from carbons 1–2–3 and 2–3–4 are both 180° centered around the *sp* hybrid carbons (carbons2 and 3).

81. There are a total of three pi bonds (one in the double bond and two in the triple bond.) There is always one sigma bond between any two connected atoms; there are 13 sigma bonds in the molecule.

82. Both end carbons (carbons 1 and 6) are free to rotate because there are no pi bonds to hold the orientation in a single spot.

87. a. The carbons in the ring and the carbon attached to the ring have delocalized electrons. Although the other carbon with the double bond to O has a pi bond, it is disconnected from the rest.

b. The carbons that have the delocalized electrons all have sp^2 hydridization because they have the three sigma bonds and no lone pairs.

c. The remaining *p* orbital not used in the hybridization is in the pi bond.

89. The structure of naphthalene is shown at the right. All of the carbon atoms have three sigma bonds and are sp^2 hybridized.

91. a. The energy-level diagrams for the four possibilities are shown in the figure below. In the minimum-energy configuration, the ground state has two unpaired electrons.

b. There are no unpaired electrons in the minimum-energy configuration of *sp* hybridization. In the formation of bonds, two electrons are promoted into $2p$ orbitals for pi bonding.

c. There are two unpaired electrons in the minimum-energy configuration of sp^2 hybridization. In the formation of bonds, one electron is promoted into a $2p$ orbital for pi bonding.

d. There are four unpaired electrons in the minimum-energy configuration of sp^3 hybridization.

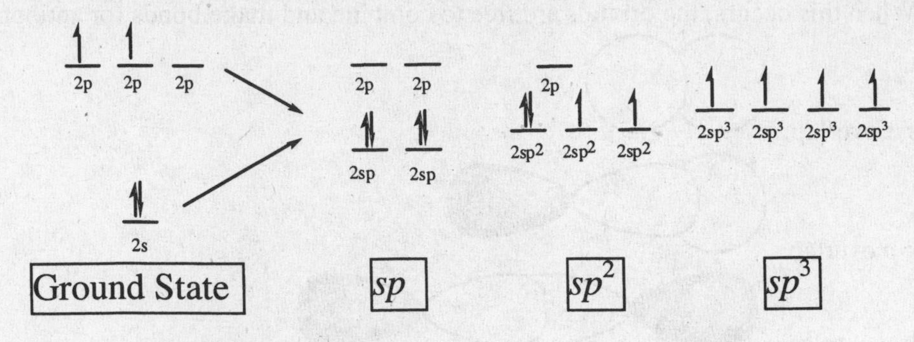

93. a. sp^3 hybridization, tetrahedral shape for orbitals, molecule has trigonal pyramidal geometry.
b. sp^3d^2 hybridization, octahedral shape
c. sp^3 hybridization, tetrahedral shape for orbitals, molecule has bent geometry.
d. sp^3d hybridization, trigonal bipyramid shape

95. NO_3^-, $\left[:\ddot{O}=N(-\ddot{O}:)(-\ddot{O}:) \right]^-$, and NO_2^-, $\left[:\ddot{O}-\ddot{N}=\ddot{O}: \right]^-$, both have sp^2 hybridization that is required to accommodate either the three sigma bonds in nitrate or the two sigma bonds and one lone pair in nitrite. Because nitrite has a lone pair in one of the sp^2 orbitals, and lone pairs of electrons are more repulsive than electron pairs in a bond, the O–N–O bond angle in nitrite is reduced from the normal 120° that is found in the nitrate ion.

97. Each carbon in octane will be sp^3 hybridized with tetrahedral geometry and will form a straight chain of carbons. Each C–C bond will be free to rotate, and the separate molecules can arrange themselves to closely associate with each other, making the formation of a solid more likely. 2,2,3,3-Tetramethylbutane has a branched structure. Each carbon is still sp^3 hybridized, but the center two carbons have three additional carbons attached to each. The molecule will be like a large ball, with little possible change in its geometry. This geometry makes close association between different molecules more difficult and the process of boiling easier. Octane will not boil as easily as 2,2,3,3-tetramethylbutane. (See the structures below.)

$H_3C-CH_2-CH_2-CH_2-CH_2-CH_2-CH_2-CH_3$ $(H_3C)_3C-C(CH_3)_3$

99. The structure of NH_2CHO is $\left[\begin{array}{c} \quad\quad\ \ H \\ \quad\quad\ \ | \\ H-\ddot{N}-C=O \\ | \quad\quad\quad \\ H \quad\quad\quad \end{array}\right]$. The carbon is sp^2 hybridized (three sigma bonds, no lone pairs) and will be trigonal planar. Nitrogen is sp^3 hybridized (three sigma bonds, one lone pair) and will be tetrahedral. Because the carbon is sp^2 hybridized, the atoms connected to carbon lie in a single plane, but nitrogen has tetrahedral geometry, so its two hydrogen atoms are not in the same plane as the rest of the molecule. The overall molecule is not planar and will be puckered.

101. a. The Lewis structure for hydrazine is $\left[\begin{array}{c} H \quad H \\ | \quad\ \ | \\ :N-N: \\ | \quad\ \ | \\ H \quad H \end{array}\right]$.

b. Because each nitrogen must accommodate four pairs of electrons (in bonds or lone pairs), each nitrogen will have a trigonal pyramidal geometry.

c. Each nitrogen has sp^3 hybridization to accommodate the three bonds and the lone pair.

d. This compound in unlikely to have a color, because it does not have a low-lying orbital to which an electron may make a transition. The sp^3 hybridization scheme has all four orbitals used in bond making or holding an electron pair. No empty pi or pi* levels are present.

e. The combustion of hydrazine is written $N_2H_4 + O_2 \rightarrow N_2 + 2\ H_2O$. This reaction requires the breaking of a N–N bond, four N – H bonds, and an O – O bond. Then an N – N triple bond and four H – O bonds are formed:

$$\begin{array}{lr} 1 \text{ mol N}-\text{N bonds} \times 160 \text{ kJ/mol} = & +160 \text{ kJ} \\ 1 \text{ mol O=O bonds} \times 495 \text{ kJ/mol} = & +495 \text{ kJ} \\ 4 \text{ mol H}-\text{N bonds} \times 391 \text{ kJ/mol} = & \underline{+1564 \text{ kJ}} \\ \text{Total Bonds Broken} = & +2219 \text{ kJ} \end{array}$$

$$\begin{array}{lr} 1 \text{ mol N} \equiv \text{N bonds} \times (-941 \text{ kJ/mol}) = & -941 \text{ kJ} \\ 4 \text{ mol O}-\text{H bonds} \times (-467 \text{ kJ/mol}) = & \underline{-1868 \text{ kJ}} \\ \text{Total Bonds Formed} = & -2809 \text{ kJ} \end{array}$$

$$\begin{array}{lr} \text{TOTAL Energy Absorbed} = & +2219 \text{ kJ} \\ \text{TOTAL Energy Released} = & \underline{-2809 \text{ kJ}} \\ \text{Net Energy Change } (\Delta H) = & -590 \text{ kJ/mol} \end{array}$$

Chapter 10: The Behavior and Applications of Gases

The Bottom Line

- Industrial gases are used in manufacturing, medicine, and other industries. (Section 10.1)
- Gases present at low density all behave in about the same way. This permits us to generalize about their behavior in a manner that is not possible with solids and liquids. (Section 10.1)
- Gases that behave as though each particle has no interactions with any other are called ideal gases. (Section 10.1)
- High pressure, low temperature, and intermolecular forces of attraction between gas molecules cause gases to deviate from ideal behavior. (Section 10.1)
- Real gases deviate from ideal behavior, especially at low temperatures and high pressures. (Section 10.1)
- Pressure is a measure of force per unit area. Pressure is the result of collisions of gas molecules with the walls of a container. (Section 10.2)
- Several units of pressure exist, including the pascal (the SI unit), atmosphere, mm Hg, in Hg, torr, bar, psi, and psig. (Section 10.2)
- In current usage, the standard atmosphere (1 atm) is different from standard pressure (1 bar, or about 0.987 atm). (Section 10.2)
- Dalton's law of partial pressures is concerned with the contribution of each gas to the pressure of the entire gas mixture. (Section 10.3)
- Avogadro's law deals with the relationship between the volume and number of moles of ideal gases, at constant temperature and pressure. (Section 10.4)
- Boyle's law expresses the relationship between volume and pressure of an ideal gas, at constant number of moles and temperature. (Section 10.4)
- Charles's law expresses the relationship between the volume and temperature of an ideal gas, at constant number of moles and pressure. (Section 10.4)
- The ideal gas equation combines the gas laws to interrelate the pressure, volume, amount, and temperature of an ideal gas. (Section 10.5)
- The ideal gas equation can be applied to find the molar mass and the density of an ideal gas. (Section 10.5)
- Several scientists have made mathematical models to account for ideal gas behavior. J. D. van der Waals's model is the most commonly used because it is relatively simple and takes into account corrections for pressure and volume. (Section 10.5)
- The molar mass and the density of an ideal gas are directly proportional. (Section 10.6)
- The kinetic-molecular theory shows how it is possible to start from some elementary constructs about gas behavior and derive the gas laws, therefore showing the consistency between theory and experiment. (Section 10.7)
- Graham's law of effusion shows the inverse relationship between the speed of a gas and its molar mass. (Section 10.8)
- The effects of industrialization on the atmosphere are currently being debated. Ozone levels and the greenhouse effect are two issues of greatest social concern. (Section 10.9)

Solutions to Practice Problems

10.1. Although we often think of pounds and kilograms as both referring to mass, the first is actually a force and only the second is a true mass. We use the equation $F = m \times a$ to begin the problem. Rearranging the formula to solve for the acceleration gives

$$a = \frac{F}{m} = \frac{15\,\text{lb} \times \frac{4.47\,\text{N}}{1\,\text{lb}} \times \frac{1\,\frac{\text{kg m}}{\text{s}^2}}{1\,\text{N}}}{10.0\,\text{kg}} = 6.7\,\text{m/s}^2 .$$

10.2. We can treat each as a simple unit conversion problem:

$$29.3\,\text{in Hg} \times \frac{1\,\text{atm}}{29.921\,\text{in Hg}} = 0.979\,\text{atm}$$

$$0.979\,\text{atm} \times \frac{760\,\text{torr}}{1\,\text{atm}} = 744\,\text{torr}$$

$$0.979\,\text{atm} \times \frac{1.01325\,\text{bar}}{1\,\text{atm}} = 0.992\,\text{bar}$$

$$0.979\,\text{atm} \times \frac{101325\,\text{Pa}}{1\,\text{atm}} = 9.92 \times 10^4\,\text{Pa}$$

10.3. Assuming that the composition of the air in the plane is the same as the normal composition of dry air, nitrogen is 78.08% of all the gas. Before calculating the pressure of the nitrogen, we need to determine the total pressure of air in the wheel well. Because O_2 is 21% of the total, and the partial pressure of O_2 is given (32 torr), we can find the total pressure:

$$P_{O_2} = (0.21) \times P_{TOT}$$

$$P_{TOT} = \frac{P_{O_2}}{0.21} = \frac{32\,\text{torr}}{0.21} = 152\,\text{torr} = 1.5_2 \times 10^2\,\text{torr}$$

The pressure of the nitrogen in the wheel well is

$$P_{N_2} = (0.7808) \times P_{TOT} = (0.7808)\left(1.5_2 \times 10^2\,\text{torr}\right) = 119\,\text{torr} = 1.2 \times 10^2\,\text{torr}$$

10.4. We apply Avogadro's law to solve the problem.

$$\frac{V_{\text{final}}}{n_{\text{final}}} = \frac{V_{\text{initial}}}{n_{\text{initial}}}$$

$$\text{so } n_{\text{final}} = V_{\text{final}} \times \frac{n_{\text{initial}}}{V_{\text{initial}}} = 1.5\,\text{L} \times \frac{6.4\,\text{mol}}{12.8\,\text{L}} = 0.75\,\text{mol}$$

10.5. We use Boyle's law to solve the problem (using the final values from the prior problem: 1.3 atm and 0.94 L).

$$P_{\text{final}} V_{\text{final}} = P_{\text{initial}} V_{\text{initial}}$$

$$\text{so } V_{\text{final}} = \frac{P_{\text{initial}} V_{\text{initial}}}{P_{\text{final}}} = \frac{(1.3\,\text{atm})(0.94\,\text{L})}{(0.975\,\text{atm})} = 1.3\,\text{L}$$

10.6. We can use Charles's law to solve for the temperature (using the final values from the problem: 1.62 L at 423 K).

$$\frac{V_{final}}{T_{final}} = \frac{V_{initial}}{T_{initial}}$$

$$\text{so } T_{final} = V_{final} \times \frac{T_{initial}}{V_{initial}} = 2.00\,\text{L} \times \frac{423\,\text{K}}{1.62\,\text{L}} = 522\,\text{K}$$

10.7. HBr is a polar molecule (electronegativity difference of 0.7). Ideal gases have little intermolecular attractions. Because HBr is polar, it will have at least some attraction to other HBr molecules. (This attraction actually would decrease the volume of the gas, at sufficiently low pressures and temperatures, and eventually leads to its condensation into a liquid.)

10.8. Using the combined gas equation, we can rearrange the formula to solve for the final volume.

$$\frac{P_{initial}V_{initial}}{n_{initial}T_{initial}} = \frac{P_{final}V_{final}}{n_{final}T_{final}}$$

$$V_{final} = \frac{P_{initial}V_{initial}n_{final}T_{final}}{n_{initial}T_{initial}P_{final}}$$

$$V_{final} = \frac{P_{initial}V_{initial}T_{final}}{T_{initial}P_{final}} \quad \text{since } n_{final} = n_{initial}$$

$$V_{final} = \frac{(1.00\,\text{atm})(0.300\,\text{L})\{(273.15-11.0)\text{K}\}}{(273\,\text{K})(0.85\,\text{atm})} = 0.34\,\text{L}$$

10.9. Using (and rearranging) the ideal gas equation, we can solve for the moles.

$$PV = nRT$$

$$n = \frac{PV}{RT} = \frac{(0.35\,\text{atm})(2.50\,\text{L})}{(0.08206\,\text{L}\cdot\text{atm/mol}\cdot\text{K})\{(273+25)\text{K}\}}$$

$$n = 0.036\,\text{mol}$$

10.10. Using (and rearranging) the ideal gas equation, we can solve for the moles.

$$PV = nRT$$

$$P = \frac{nRT}{V} = \frac{(0.250\,\text{mol})(0.08206\,\text{L}\cdot\text{atm/mol}\cdot\text{K})\{(273.15+28.7)\text{K}\}}{(8.44\,\text{L})}$$

$$P = 0.734\,\text{atm}$$

10.11. For O_2, the van der Waals (vdW) constants are $a = 1.36$ atm L^2 mol^{-2} and $b = 0.0318$ L mol^{-1}. For NH_3, the van der Waals constants are $a = 4.17$ atm L^2 mol^{-2} and $b = 0.0371$ L mol^{-1}. The ideal gas equation gives the same solution for both:

$$P = \frac{nRT}{V} = \frac{(1.00\,\text{mol})(0.08206\,\text{L}\cdot\text{atm/mol}\cdot\text{K})(298\text{K})}{(1.00\text{L})} = 24.4\,\text{atm}$$

For O_2, vdW gives

$$P = \frac{nRT}{V - nb} - \frac{an^2}{V^2}$$

$$P = \frac{(1.00\,\text{mol})(0.08206\,\text{L}\cdot\text{atm/mol}\cdot\text{K})(298\,K)}{1.00\,\text{L} - (1.00\text{mol})(0.0318\,\text{L/mol})} - \frac{(1.36\,\text{atm}\cdot\text{L}^2/\text{mol}^2)(1.00\,\text{mol})^2}{(1.00\,\text{L})^2}$$

$$P = 23.9\,\text{atm}$$

For NH_3, vdW gives

$$P = \frac{nRT}{V - nb} - \frac{an^2}{V^2}$$

$$P = \frac{(1.00\,\text{mol})(0.08206\,\text{L}\cdot\text{atm/mol}\cdot\text{K})(298\,\text{K})}{1.00\,\text{L} - (1.00\text{mol})(0.0371\,\text{L/mol})} - \frac{(4.17\,\text{atm}\cdot\text{L}^2/\text{mol}^2)(1.00\,\text{mol})^2}{(1.00\,L)^2}$$

$$P = 21.2\,\text{atm}$$

O_2 varies only a little from the ideal gas law, but ammonia, with its large intermolecular attractions, deviates significantly from the ideal gas law.

10.12. We first use the ideal gas law to solve for the number of moles in 1.0 L of gas at 289 K (273 + 16) and 1.75 atm.

$$n = \frac{PV}{RT} = \frac{(1.75\,\text{atm})(1.00\text{L})}{(0.08206\,\text{L}\cdot\text{atm/mol}\cdot\text{K})(289\,\text{K})} = 0.0738\,\text{mol}$$

The mass of 1.00 L of this gas is 3.40 g, so the molecular weight is

$$\underline{M} = \frac{3.40\,\text{g}}{0.738\,\text{mol}} = 46.1\,\text{g/mol}$$

10.13. We will first solve for the number of moles of N_2 from the ideal gas law, then use the stoichiometric ratio from the balance equation to convert to moles of NaN_3, and finally use the molecular weight of NaN_3 to get the mass from the moles of NaN_3.

$$n = \frac{PV}{RT} = \frac{(1.00\,\text{atm})(30.9\text{L})}{(0.08206\,\text{L}\cdot\text{atm/mol}\cdot\text{K})(304\,\text{K})} = 1.24\,\text{mol}\,\text{N}_2$$

$$1.24\,\text{mol}\,\text{N}_2 \times \frac{2\,\text{mol}\,\text{NaN}_3}{3\,\text{mol}\,\text{N}_2} \times \frac{65.0\,\text{g}\,\text{NaN}_3}{1\,\text{mol}\,\text{NaN}_3} = 53.7\,\text{g}\,\text{NaN}_3$$

10.14. To solve the problem, we first need to solve for the moles of C_2H_2 produced and then use the stoichiometry and percent yield to get the mass of methane used. (We assume the same storage conditions for C_2H_2 of 1.00 atm and 38.0 °C.)

$$n = \frac{PV}{RT} = \frac{(1.00\,\text{atm})(5000\text{L})}{(0.08206\,\text{L}\cdot\text{atm/mol}\cdot\text{K})(311\text{K})} = 196\,\text{mol}\,\text{C}_2\text{H}_2$$

$$196\,\text{mol}\,\text{C}_2\text{H}_2\ \text{actual} \times \frac{100\,\text{mol}\,\text{C}_2\text{H}_2\ \text{theor.}}{34\,\text{mol}\,\text{C}_2\text{H}_2\ \text{actual}} \times \frac{2\,\text{mol}\,\text{CH}_4}{1\,\text{mol}\,\text{C}_2\text{H}_2} \times \frac{16.05\,\text{CH}_4}{1\,\text{mol}\,\text{CH}_4}$$

$$= 18505\,\text{g}\,\text{CH}_4 = 18.5\,\text{kg}\,\text{CH}_4$$

10.15. The kinetic energy is given by $KE = \frac{1}{2} mv^2$. Note that the kinetic energy increases both for increases in mass and increases in velocity. If the masses of two particles are different, but the kinetic energy is the same, the velocity must be different to offset the change in mass. As the mass increases, the velocity must decrease to retain the same kinetic energy.

10.16. The average velocity of each is in a constant proportion with the root-mean-square velocity. Because the root-mean-square velocity is inversely proportional to the square root of the molar mass, an increase in molar mass decreases the root-mean-square velocity (assuming the temperature is constant.) F_2 has less mass and will therefore have the higher velocity.

10.17. Because H_2 has less mass than He, it will have the higher velocity and therefore a higher rate of effusion. The ratio of effusion rates is $\frac{re_{H_2}}{re_{He}} = \sqrt{\frac{M_{He}}{M_{H_2}}} = \sqrt{\frac{4.00\,g/mol}{2.02\,g/mol}} = 1.41$. Hydrogen will effuse at a rate 1.4 times greater than helium.

Solutions to Student Problems

1. The forces that most gases experience are very small (because of the large distance between the molecules and the rapid velocities). These small forces are collectively known as van der Waals forces.

3. Any conditions that limit interaction of the gas particles: high temperatures, which make the molecules move quickly past one another, and low pressures, which spread the molecules apart.

5. We can use the relation $F = m \times a$ to find the force:

$$F = m \times a = (0.89\,kg)(9.81\,m/s^2) = 8.7\,kg \cdot m/s^2 = 8.7\,N$$

7. $F = m \times a = (86\,kg)(9.81\,m/s^2) = 844\,kg \cdot m/s^2 = 8.4x10^2\,N$

9. The pressure conversions are

$$797\,mm\,Hg \times \frac{1\,atm}{760\,mm\,Hg} \times \frac{101325\,Pa}{1\,atm} = 1.06 \times 10^5\,Pa$$

$$1.06 \times 10^5\,Pa \times \frac{1\,kPa}{1000\,Pa} = 106\,kPa$$

$$797\,mm\,Hg \times \frac{1\,atm}{760\,mm\,Hg} = 1.05\,atm$$

$$797\,mm\,Hg \times \frac{1\,torr}{1\,mm\,Hg} = 797\,torr$$

$$797\,mm\,Hg \times \frac{1\,atm}{760\,mm\,Hg} \times \frac{1.01325\,bar}{1\,atm} = 1.06\,bar$$

$$797\,mm\,Hg \times \frac{1\,in}{25.4\,mm} = 31.4\,in\,Hg$$

11. The mass on each planet is the same, so we can rearrange the force equation: (Note that it is not necessary to convert the mass from pounds, because the acceleration units cancel directly.)

$$F = m \times a$$

$$m = \frac{F}{a}$$

$$m_{\text{Earth}} = m_{\text{planet}}$$

$$\frac{F_{\text{Earth}}}{a_{\text{Earth}}} = \frac{F_{\text{planet}}}{a_{\text{planet}}}$$

$$F_{\text{planet}} = \frac{F_{\text{Earth}}}{a_{\text{Earth}}} \times a_{\text{planet}} = \frac{145\text{ lb}}{9.8\text{ m/s}^2} \times 0.33\text{ m/s}^2 = 4.9\text{ lb}$$

13. The area in square inches is $567\,\text{cm}^2 \times \left(\frac{1\,\text{in}}{2.54\,\text{cm}}\right)^2 = 87.9\,\text{in}^2$. The pressure given is a gauge pressure (psig), so the true pressure is greater by 1 atm (14.7 psi):

$$P = 22.3 + 14.7 = 37.0\text{ psi}$$

Pressure is force per unit area, so the force is pressure times area:

$$F = (37.0\text{ lb/in}^2)(87.9\text{ in}^2) = 3.25 \times 10^3\text{ lb}$$

15. The pressure given is a gauge pressure (psig) so the true pressure is greater by 1 atm (14.7 psi):

$$P = 17.5 + 14.7 = 32.2\text{ psi}$$

$$\text{In atmospheres: } 32.2\,\text{psi} \times \frac{1\,\text{atm}}{14.7\,\text{psi}} = 2.19\,\text{atm}$$

17. We can apply Dalton's law to solve this problem: $P_{\text{A}} = X_{\text{A}} P_{\text{TOT}}$, where X_A is the fraction of the gas out of the total. Solving for P_{TOT} yields

$$P_{\text{TOT}} = \frac{P_{\text{A}}}{X_{\text{A}}} = \frac{115\,\text{mm Hg}}{0.21} = 5.5 \times 10^2\text{ mm Hg}$$

19. We can get the partial pressures of gases by looking at the ratio of moles of each to the total moles (a mole fraction). There are 12.4 g / (28.02 g/mol) = 0.443 mol N_2 and 12.4 g / (32.0 g/mol) = 0.388 mol O_2. The mole fractions are

$$X_{\text{N}_2} = \frac{\text{moles N}_2}{\text{moles total}} = \frac{0.443\,\text{mol}}{0.443\,\text{mol} + 0.388\,\text{mol}} = 0.533$$

$$X_{\text{O}_2} = \frac{\text{moles O}_2}{\text{moles total}} = \frac{0.388\,\text{mol}}{0.443\,\text{mol} + 0.388\,\text{mol}} = 0.467$$

So the partial pressures are

For N_2: $P_{\text{N}_2} = X_{\text{N}_2} P_{\text{TOT}} = 0.533 \times 1.23\,\text{atm} = 0.656\,\text{atm}$

For O_2: $P_{\text{O}_2} = X_{\text{O}_2} P_{\text{TOT}} = 0.467 \times 1.23\,\text{atm} = 0.574\,\text{atm}$

21. The gas that has been collected is a mixture of water and oxygen. The oxygen pressure will be the difference between the total pressure and the pressure of the water:

$$P_{\text{O}_2} = P_{\text{TOT}} - P_{\text{H}_2\text{O}} = 785\text{ mm Hg} - 27\,\text{mm Hg} = 758\,\text{mm Hg}$$

23. Using Dalton's law:

For Ne: $P_{Ne} = X_{Ne}P_{TOT} = 0.0895 \times 3.42\,\text{mm Hg} = 0.306\,\text{mm Hg}$

For He: $P_{He} = X_{He}P_{TOT} = 0.9115 \times 3.42\,\text{mm Hg} = 3.12\,\text{mm Hg}$

25. We apply Avogadro's law to solve the problem:

$$\frac{V_{\text{final}}}{n_{\text{final}}} = \frac{V_{\text{initial}}}{n_{\text{initial}}}$$

For 5.0 L: $n_{\text{final}} = V_{\text{final}} \times \frac{n_{\text{initial}}}{V_{\text{initial}}} = 5.0\,\text{L} \times \frac{4.5\,\text{mol}}{2.5\,\text{L}} = 9.0\,\text{mol}$

For 1.0 L: $n_{\text{final}} = V_{\text{final}} \times \frac{n_{\text{initial}}}{V_{\text{initial}}} = 1.0\,\text{L} \times \frac{4.5\,\text{mol}}{2.5\,\text{L}} = 1.8\,\text{mol}$

27. We use Boyle's law, $P_{\text{final}}V_{\text{final}} = P_{\text{initial}}V_{\text{initial}}$.

For 760 torr: $V_{\text{final}} = \frac{P_{\text{initial}}V_{\text{initial}}}{P_{\text{final}}} = \frac{(861\,\text{torr})(4.70\,\text{L})}{(760\,\text{torr})} = 5.32\,\text{L}$

For 400 torr: $V_{\text{final}} = \frac{P_{\text{initial}}V_{\text{initial}}}{P_{\text{final}}} = \frac{(861\,\text{torr})(4.70\,\text{L})}{(400\,\text{torr})} = 10.1\,\text{L}$

29. We can use Charles's law to solve for the volume:

$$\frac{V_{\text{final}}}{T_{\text{final}}} = \frac{V_{\text{initial}}}{T_{\text{initial}}}$$

First we convert the temperatures to the Celsius scale:

$$-10°\text{F} \Rightarrow \left(°\text{C} = \frac{(-10°\text{F} - 32)}{1.8} = -23°\text{C}\right);\ 72°\text{F} \Rightarrow \left(°\text{C} = \frac{(72°\text{F} - 32)}{1.8} = 22°\text{C}\right):$$

For −10°F: $V_{\text{final}} = T_{\text{final}} \times \frac{V_{\text{initial}}}{T_{\text{initial}}} = (-23 + 273)\text{K} \times \frac{1.00\,\text{L}}{(22 + 273)\text{K}} = 0.847\,\text{L}$

For 250°F => $\left(°\text{C} = \frac{(250°\text{F} - 32)}{1.8} = 121°\text{C}\right):$

$$V_{\text{final}} = T_{\text{final}} \times \frac{V_{\text{initial}}}{T_{\text{initial}}} = (121 + 273)\text{K} \times \frac{1.00\,\text{L}}{(22 + 273)\text{K}} = 1.34\,\text{L}$$

31. We can use the combined gas laws to solve for the volume:

$$\frac{P_{initial}V_{initial}}{n_{initial}T_{initial}} = \frac{P_{final}V_{final}}{n_{final}T_{final}}$$

$$V_{final} = \frac{P_{initial}V_{initial}n_{final}T_{final}}{n_{initial}T_{initial}P_{final}}$$

$$V_{final} = \frac{P_{initial}V_{initial}T_{final}}{T_{initial}P_{final}} \text{ since } n_{final} = n_{initial}$$

$$V_{final} = \frac{(2839\,\text{mm Hg})(3.20\,\text{L})(273\,\text{K})}{\{(273+37)\text{K}\}(760\,\text{mm Hg})} = 10.5\,\text{L}$$

33. Using our alternative form of Charless' law yields

$$T_{final} = \frac{T_{initial}P_{final}}{P_{initial}} = \frac{(273\,\text{K})(32.0\,\text{psi})}{14.7\,\text{psi}} = 594\,\text{K}$$

35. Because the compound remains the same, we can use the masses in place of the moles in Avogadro's law:

$$\frac{V_{final}}{n_{final}} = \frac{V_{initial}}{n_{initial}}$$

$$\text{so } V_{final} = n_{final} \times \frac{V_{initial}}{n_{initial}} = 0.33\,\text{g} \times \frac{0.217\,\text{L}}{0.58\,\text{g}} = 0.12\,\text{L}$$

37. We can use Charles's law to solve for the volume:

$$\frac{V_{final}}{T_{final}} = \frac{V_{initial}}{T_{initial}}$$

$$\text{For 25°C: } V_{final} = T_{final} \times \frac{V_{initial}}{T_{initial}} = (25+273)\text{K} \times \frac{0.150\,\text{mL}}{(37+273)\text{K}} = 0.144\,\text{mL}$$

39. a. We can use the ideal gas equation to find the value for R in each case:

In atm: $$R = \frac{PV}{nT} = \frac{(1.00\,\text{atm})(22.4\,\text{L})}{(1.00\,\text{mol})(273\,\text{K})} = 0.08205\,\text{L}\cdot\text{atm/mol}\cdot\text{K}$$

In Pa: $$R = \frac{PV}{nT} = \frac{(101325\,\text{Pa})(22.4\,\text{L})}{(1.00\,\text{mol})(273\,\text{K})} = 8314\,\text{L}\cdot\text{Pa/mol}\cdot\text{K}$$

b. In mm Hg: $$R = \frac{PV}{nT} = \frac{(760\,\text{mm Hg})(22.4\,\text{L})}{(1.00\,\text{mol})(273\,\text{K})} = 62.36\,\text{L}\cdot\text{mm Hg/mol}\cdot\text{K}$$

c. In psi: $$R = \frac{PV}{nT} = \frac{(14.7\,\text{psi})(22.4\,\text{L})}{(1.00\,\text{mol})(273\,\text{K})} = 1.206\,\text{L}\cdot\text{psi/mol}\cdot\text{K}$$

41. Solving the ideal gas equation for the volume, we get

$$V = \frac{nRT}{P} = \frac{\left(6.0\times10^{6}\,\text{g}\times\frac{1\,\text{mol He}}{4.00\,\text{g}}\right)(0.08206\,\text{L}\cdot\text{atm/mol}\cdot\text{K})(22.6+273.15\text{K})}{0.955\,\text{atm}} = 3.8\times10^{7}\,\text{L}$$

43. Solving the ideal gas equation for the volume, we get

For 44.0 g at STP: $$V = \frac{nRT}{P} = \frac{\left(44.0\,\text{g} \times \dfrac{1\,\text{mol}\,CO_2}{44.01\,\text{g}}\right)(0.08206\,\text{L}\cdot\text{atm}/\text{mol}\cdot\text{K})(273\,\text{K})}{1.00\,\text{atm}} = 22.4\,\text{L}$$

At 39°C and 0.500 atm: $$V = \frac{\left(44.0\,\text{g} \times \dfrac{1\,\text{mol}\,CO_2}{44.01\,\text{g}}\right)(0.08206\,\text{L}\cdot\text{atm}/\text{mol}\cdot\text{K})(39+273\,\text{K})}{0.500\,\text{atm}} = 51.2\,\text{L}$$

45. Solving the ideal gas equation for the number of moles, we get:

$$n = \frac{PV}{RT} = \frac{(1.00\,\text{atm})(4.33\,\text{L})}{(0.08206\,\text{L}\cdot\text{atm}/\text{mol}\cdot\text{K})(273\,\text{K})} = 0.193\,\text{mol}$$

47. We can rearrange the ideal gas law to solve directly for the molar mass from the density:

$$d = \frac{\text{mass}}{\text{volume}} = \frac{m}{V} = \frac{m}{\dfrac{nRT}{P}} = \left(\frac{m}{n}\right)\frac{P}{RT} = (\underline{M})\frac{P}{RT}$$

$$\text{so } \underline{M} = \frac{dRT}{P}$$

For this gas: $$\underline{M} = \frac{dRT}{P} = \frac{(0.600\,\text{g/L})(0.08206\,\text{L}\cdot\text{atm}/\text{mol}\cdot\text{K})(66+273\,\text{K})}{\left(743\,\text{mm Hg} \times \dfrac{1\,\text{atm}}{760\,\text{mm Hg}}\right)} = 17.1\,\text{g/mol}$$

49. For CH_4 at 25°C and 1.15 atm:

$$d = (\underline{M})\frac{P}{RT} = (16.05\,\text{g/mol})\left(\frac{1.15\,\text{atm}}{(0.08206\,\text{L}\cdot\text{atm}/\text{mol}\cdot\text{K})(298\,\text{K})}\right) = 0.755\,\text{g/L}$$

51. We first need to find the number of moles of oxygen in that breath and then convert to molecules. Note that we multiply the pressure by 0.21 because O_2 makes up only about 21% of air at sea level.

$$n = \frac{PV}{RT} = \frac{(0.922\,\text{atm} \times 0.21)(0.450\,\text{L})}{(0.08206\,\text{L}\cdot\text{atm}/\text{mol}\cdot\text{K})(37+273\,\text{K})} = 0.00342_{51}\,\text{mol}$$

$$0.00342_{51}\,\text{mol}\,O_2 \times \frac{6.022\times10^{23}\,\text{molecules}}{1\,\text{mol}} = 2.06\times10^{21}\,\text{molecules}\,O_2$$

53. For this gas, the molar mass (see Problem 47 for derivation of the equation) is

$$\underline{M} = \frac{dRT}{P} = \frac{(4.48\,\text{g/L})(0.08206\,\text{L}\cdot\text{atm}/\text{mol}\cdot\text{K})(301\,\text{K})}{(0.97\,\text{atm})} = 114\,\text{g/mol}$$

55. We will solve for the molar mass directly (see Problem 47 for derivation of the equation).

$$\underline{M} = \frac{dRT}{P} = \frac{\left(\dfrac{5.00\,\text{g}}{0.0850\,\text{L}}\right)(0.08206\,\text{L}\cdot\text{atm}/\text{mol}\cdot\text{K})(20.0+273.15\,\text{K})}{(1.00\,\text{atm})} = 1.42\times10^{3}\,\text{g/mol}$$

The molar mass should not change as raising the temperature will also lower the density.

57. We will solve for the temperature using the ideal gas equation.

$$T = \frac{PV}{nR} = \frac{\left(745\,\text{mm Hg} \times \dfrac{1\,\text{atm}}{760\,\text{mm Hg}}\right) \times (0.478\,\text{L})}{\left(0.675\,\text{g}\,O_2 \times \dfrac{1\,\text{mol}\,O_2}{32.00\,\text{g}\,O_2}\right) \times (0.08206\,\text{L}\cdot\text{atm/mol}\cdot\text{K})} = 271\,\text{K}$$

59. We will first use the stochiometry to solve for the number of moles of CO_2 produced and then find its volume.

$$10.0\,\text{g}\,C_6H_{12}O_6 \times \frac{1\,\text{mol}\ C_6H_{12}O_6}{180.15\,\text{g}\,C_6H_{12}O_6} \times \frac{6\,\text{mol}\,CO_2}{1\,\text{mol}\ C_6H_{12}O_6} = 0.333\,\text{mol}\,CO_2$$

$$V = \frac{nRT}{P} = \frac{(0.333\,\text{mol})(0.08206\,\text{L}\cdot\text{atm/mol}\cdot\text{K})(37 + 273\,\text{K})}{1.00\,\text{atm}} = 8.47\,\text{L}$$

61. We can solve for the density of ethanol (46.07 g/mol) directly (see Problem 47 for derivation of the equation).

$$d = (M)\frac{P}{RT} = (46.07\,\text{g/mol})\left(\frac{6.8\times10^6\ \text{Pa} \times \dfrac{1.00\,\text{atm}}{101325\,\text{Pa}}}{(0.08206\,\text{L}\cdot\text{atm/mol}\cdot\text{K})(3.00\times10^3\ \text{K})}\right) = 12.6\,\text{g/L}$$

63. $$P = \frac{nRT}{V} = \frac{\left(0.100\,\text{g} \times \dfrac{1\,\text{mol Ar}}{39.95\,\text{g}}\right)(0.08206\,\text{L}\cdot\text{atm/mol}\cdot\text{K})(298\,\text{K})}{0.2000\,\text{L}} = 0.306\,\text{atm}$$

65. a. Because the gases are under different conditions, we need to calculate the number of moles of each to determine which is the limiting reagent.

$$n_{H_2} = \frac{PV}{RT} = \frac{\left(1120\,\text{mm Hg} \times \dfrac{1\,\text{atm}}{760\,\text{mm Hg}}\right)(2.00\,\text{L})}{(0.08206\,\text{L}\cdot\text{atm/mol}\cdot\text{K})(21.0 + 273.15\,\text{K})} = 0.122\,\text{mol}\,H_2$$

$$n_{N_2} = \frac{PV}{RT} = \frac{(1.00\,\text{atm})(2.00\,\text{L})}{(0.08206\,\text{L}\cdot\text{atm/mol}\cdot\text{K})(273\,\text{K})} = 0.0893\,\text{mol}\,H_2$$

Because the reaction consumes 3 moles of H_2 for every mole of N_2, the amount of N_2 required to just react with 0.122 mole of H_2 is

$$0.122\,\text{mol}\,H_2 \times \frac{1\,\text{mol}\,N_2}{3\,\text{mol}\,H_2} = 0.0407\,\text{mol}\ N_2$$

There is an excess of N_2 available to react; therefore, H_2 is the limiting reagent.

b. When the pressure and temperature are the same, the number of moles of a gas is proportional to the volume. Additionally, we can use the stoichiometric coefficients instead of moles as relative volumes when comparing two gases within a reaction. 2.00 L of N_2 would require 6.00 L of H_2 for full reaction, so the H_2 gas limits the reaction. At STP, 2.0 L of H_2 would react fully to produce

$$2.00\,\text{L}\,H_2 \times \frac{2\,\text{L}\,NH_3}{3\,\text{L}\,H_2} = 1.33\,\text{L}\,NH_3$$

67. Using the equation derived above yields

$$\underline{M} = \frac{dRT}{P} = \frac{\left(\frac{1.60\,\text{g}}{1.00\,\text{L}}\right)(0.08206\,\text{L}\cdot\text{atm}/\text{mol}\cdot\text{K})((89.0+273.15)\text{K})}{\left(1065\,\text{torr}\times\frac{1\,\text{atm}}{760\,\text{torr}}\right)} = 33.9\,\text{g}/\text{mol}$$

69. Solving for the molar mass yields

$$\underline{M} = \frac{dRT}{P} = \frac{(1.70\,\text{g}/\text{L})(0.08206\,\text{L}\cdot\text{atm}/\text{mol}\cdot\text{K})(273\text{K})}{(1.00\,\text{atm})} = 38.1\,\text{g}/\text{mol}$$

Because all the halogens form diatomic gases, the atomic mass is ½(38.1 g.mol) = 19.05 g/mol. The gas is F_2.

71. $$V = \frac{nRT}{P} = \frac{\left(1.00\,\text{lb Ne}\times\frac{1\,\text{kg}}{2.205\,\text{lb}}\times\frac{1000\,\text{g}}{1\,\text{kg}}\times\frac{1\,\text{mol Ne}}{20.18\,\text{g Ne}}\right)(0.08206\,\text{L}\cdot\text{atm}/\text{mol}\cdot\text{K})(25+273)\text{K})}{\left(789\,\text{torr}\times\frac{1\,\text{atm}}{760\,\text{torr}}\right)}$$

$$V = 529\,\text{L}$$

73. According to the kinetic molecular theory, the average kinetic energy of a particle is directly proportional to the temperature. The pressure of a gas inside the cylinder depends on both the number of collisions with the walls of the container and the speed of the molecules. As the temperature increases, the speed increases and both the number and velocity of the collisions increase; therefore, the pressure is directly proportional to the temperature.

75. a. According to the kinetic molecular theory, the average kinetic energy of the particles is proportional to the temperature; because both gases are at the same temperature, the average kinetic energy per particle is the same for each.

b. ClO_2 has less mass than Cl_2, giving it a higher velocity: Because $KE = ½\,mv^2$ and the KE of both gases is the same, the smaller mass of ClO_2 requires the velocity to be more to maintain the constant KE.

c. Although both have the same kinetic energy, there are more particles of ClO_2 in the same 5.0-g mass than there are of Cl_2. More particles will give a higher pressure. ClO_2 will have the higher pressure.

d. Even in the same container, the greater number of particles of ClO_2 will give it the greater partial pressure.

77. a. For each, the mass must be converted to kilograms per particle. The kinetic energies given by $KE = ½\,mv^2$ are as follows (for a single particle).

For CO_2: $$KE = \frac{1}{2}mv^2 = \frac{1}{2}\left(\frac{44.02\,\text{g}/\text{mol}}{6.022\times10^{23}/\text{mol}}\times\frac{1\,\text{kg}}{1000\,\text{g}}\right)(1100\,\text{m}/\text{s})^2 = 4.4\times10^{-20}\,\text{J}$$

b. For NH_3: $$KE = \frac{1}{2}mv^2 = \frac{1}{2}\left(\frac{17.03\,\text{g}/\text{mol}}{6.022\times10^{23}/\text{mol}}\times\frac{1\,\text{kg}}{1000\,\text{g}}\right)(1100\,\text{m}/\text{s})^2 = 1.7\times10^{-20}\,\text{J}$$

c. For Ne: $KE = \frac{1}{2}mv^2 = \frac{1}{2}\left(\frac{20.18\,\text{g/mol}}{6.022\times10^{23}\,/\text{mol}} \times \frac{1\,\text{kg}}{1000\,\text{g}}\right)(1100\,\text{m/s})^2 = 2.0\times10^{-20}\,\text{J}$

79. Because at the same temperature, all particles have the same average kinetic energy, the velocity will be inversely proportional to the mass. In other words, the lower the mass, the higher the velocity. From lowest to highest: Ar < C_2H_6 < H_2

81. Calculation of the root-mean-square velocity is as follows. (Note that the mass needs to be in units of kilograms per mole.)

$$u_{\text{rms}} = \sqrt{\frac{3RT}{M}} = \sqrt{\frac{3(8.3145\,\text{J/mol}\cdot\text{K})(298\,\text{K})}{\left(44.09\,\text{g/mol} \times \frac{1\,\text{kg}}{1000\,\text{g}}\right)}} = 411\,\text{m/s}$$

83. For CO_2: $u_{\text{rms}} = \sqrt{\frac{3RT}{M}} = \sqrt{\frac{3(8.3145\,\text{J/mol}\cdot\text{K})(250\,\text{K})}{\left(44.01\,\text{g/mol} \times \frac{1\,\text{kg}}{1000\,\text{g}}\right)}} = 376\,\text{m/s}$

For H_2O: $u_{\text{rms}} = \sqrt{\frac{3RT}{M}} = \sqrt{\frac{3(8.3145\,\text{J/mol}\cdot\text{K})(250\,\text{K})}{\left(18.02\,\text{g/mol} \times \frac{1\,\text{kg}}{1000\,\text{g}}\right)}} = 588\,\text{m/s}$

85. Point 1: *Gases are composed of particles that are negligibly small compared to their container and to the distance between other molecules.*

The marbles do not have negligible size relative either to the container or to the distance between marbles.

Point 2: *Therefore, intermolecular attractions, which are exhibited at small distances, are assumed to be nonexistent.*

Unless the marbles are dirty and sticky, this should be true, because the marbles will not be strongly attracted to each other.

Point 3: *Gases are in constant, random motion*, colliding with the walls of the container and with each other.

If the marbles are continuously shaken during the demonstration, this should be true.

Point 4: *Pressure is the force per unit area caused by the molecules colliding per unit time with the walls of the container.*

This should also be true during the demonstration if the marbles are continuously shaken. As the piston is moved, the number of particle collisions per surface area will change. A smaller volume will correspond to larger pressures.

Point 5: Because pressure is constant in a container over time, *molecular collisions are assumed to be perfectly elastic*. That is, no energy of any kind is lost upon collision.

This will be approximately correct. There will be some small energy loss (due to friction and collisional heating) as the marbles move and collide with each other and the container.

Point 6: *The average kinetic energy of the molecules in a system is linearly proportional to the absolute (Kelvin) temperature.*

This will not be true in the analogy, but more violent shaking could be applied to show higher temperatures.

87. The rate of effusion is proportional to the square root of the mass, so the lighter the molecule, the faster it will diffuse. From lowest to highest rates: $Xe < SO_3 < CO_2 < N_2$

89. Using Graham's law:

$$\frac{re_1}{re_2} = \sqrt{\frac{M_2}{M_1}}$$

$$M_2 = M_1 \times \left(\frac{re_1}{re_2}\right)^2$$

$$M_2 = 2.02\,\text{g/mol} \times \left(\frac{1.800\,\text{m/s}}{0.300\,\text{m/s}}\right)^2 = 72.7\,\text{g/mol}$$

91. Using Graham's law:

$$M_2 = M_{\text{He}} \times \left(\frac{re_{\text{He}}}{re_2}\right)^2$$

$$M_2 = 4.00\,\text{g/mol} \times \left(\frac{1}{0.317}\right)^2 = 39.8\,\text{g/mol}$$

The gas is argon (39.95 g/mol).

93. Because the masses are not exactly the same (CO_2 = 44.01 g/mol and C_3H_8= 44.09 g/mol), there will be a slight difference in the effusion rates. It would be possible, though impractical, to try to separate these molecules via effusion. (A gas chromatograph (GC) operates on similar principles and can separate gases with very close masses.)

95. If we assume that the diffusion rate behaves as the effusion rate (a close approximation, but not exact), we can estimate the position. The relative rates of NH_3 and HCl are

$$\frac{re_{NH_3}}{re_{HCl}} = \sqrt{\frac{M_{HCl}}{M_{NH_3}}} = \sqrt{\frac{36.46\,\text{g/mol}}{17.03\,\text{g/mol}}} = 1.46:1$$

We divide the distance by the total relative rate (each side is moving, so the total is 1.46 + 1 = 2.46):

$$39\,\text{cm} \div 2.46 = 15.9\,\text{cm}$$

Each unit is 15.9 cm, which is the distance the slower HCl has moved at the time the two chemicals collide and precipitate. The precipitate forms about 16 cm from the HCl end of the tube.

97. Using Graham's law, we can predict the ratio of the effusion rates:

$$\frac{re_{^{235}UF_6}}{re_{^{238}UF_6}} = \sqrt{\frac{M_{^{238}UF_6}}{M_{^{235}UF_6}}} = \sqrt{\frac{352\,\text{g/mol}}{349\,\text{g/mol}}} = 1.0057:1$$

$^{235}UF_6$ effuses only 0.57% faster.

99. a. There were 10.8 million tons of NH_3, which has an ideal volume, at SATP, of

$$V=\frac{nRT}{P}=\frac{\left(10.8\times10^{6}\,\text{m.ton}\times\frac{1000\,\text{kg}}{1\,\text{m.ton}}\times\frac{1000\,\text{g}}{1\,\text{kg}}\times\frac{1\,\text{mol}}{17.03\,\text{g}}\right)(0.08206\,\text{L}\cdot\text{atm/mol}\cdot\text{K})(25+273\text{K})}{1.00\,\text{bar}\times\frac{1.00\,\text{atm}}{1.01325\,\text{bar}}}$$

$V=1.57\times10^{13}\,\text{L}$

b. There were 25.5 million tons of O_2, which has an ideal volume, at SATP, of

$$V=\frac{nRT}{P}=\frac{\left(25.5\times10^{6}\,\text{m.ton}\times\frac{1000\,\text{kg}}{1\,\text{m.ton}}\times\frac{1000\,\text{g}}{1\,\text{kg}}\times\frac{1\,\text{mol}}{32.00\,\text{g}}\right)(0.08206\,\text{L}\cdot\text{atm/mol}\cdot\text{K})(25+273\text{K})}{1.00\,\text{bar}\times\frac{1.00\,\text{atm}}{1.01325\,\text{bar}}}$$

$V=1.97\times10^{13}\,\text{L}$

c. There were 17.7 x 10^9 m^3 of H_2 produced; the number of metric tons is

$$n=\frac{PV}{RT}=\frac{\left(1.00\,\text{bar}\times\frac{1.00\,\text{atm}}{1.01325\,\text{bar}}\right)\left(17.7\text{x}10^{9}\,\text{m}^{3}\times\frac{1000\text{L}}{1\,\text{m}^{3}}\right)}{(0.08206\,\text{L}\cdot\text{atm/mol}\cdot\text{K})((25+273)\text{K})}=7.14\times10^{11}\,\text{mol}\,\text{H}_2$$

$$7.14\times10^{11}\,\text{mol}\,\text{H}_2\times\frac{2.002\,\text{g}\,\text{H}_2}{1\,\text{mol}\,\text{H}_2}\times\frac{1\,\text{kg}}{1000\,\text{g}}\times\frac{1\,\text{m.ton}}{1000\,\text{kg}}=1.43\times10^{6}\,\text{metric tons}$$

d. There were 30.3 × 10^6 metric tons of N_2 produced; the number of moles is

$$30.6\times10^{6}\,\text{m.tons}\times\frac{1000\,\text{kg}}{1\,\text{m.ton}}\times\frac{1000\,\text{g}}{1\,\text{kg}}\times\frac{1\,\text{mol}\,\text{N}_2}{28.02\,\text{g}\,\text{N}_2}=1.09\times10^{12}\,\text{mol}\,\text{N}_2$$

101. To solve the problem, we can use the data given as conversion constants to get the volume, and then we can use the volume in the ideal gas law to get moles and then molecules.

$$\frac{75\,\text{mL}}{1\text{kg}\cdot\text{min}}\times86.8\,\text{kg}\times30\,\text{min}\times\frac{1\,\text{L}}{1000\,\text{mL}}=195.3\,\text{L}$$

$$\text{n}=\frac{\text{PV}}{\text{RT}}=\frac{\left(1.00\,\text{bar}\times\frac{1.00\,\text{atm}}{1.01325\,\text{bar}}\right)(195.3\,\text{L})}{(0.08206\,\text{L}\cdot\text{atm/mol}\cdot\text{K})(298\text{K})}=7.88\,\text{mol}$$

$$7.88\,\text{mol}\times\left(6.022\times10^{23}\,\text{molecules/mol}\right)=4.75\times10^{24}\,\text{molecules}$$

103. For H_2, the van der Waals constants are $a = 0.244$ atm L^2 mol^{-2} and $b = 0.0266$L mol^{-1}. The ideal gas equation gives:

$$P=\frac{nRT}{V}=\frac{\left(45.0\,\text{g}\times\frac{1\,\text{mol}}{2.02\,\text{g}}\right)(0.08206\,\text{L}\cdot\text{atm/mol}\cdot\text{K})(298\text{K})}{(25.00\,\text{L})}=21.8\,\text{atm}$$

The van der Waals equation gives

$$P = \frac{nRT}{V - nb} - \frac{an^2}{V^2}$$

$$P = \frac{\left(45.0\,\text{g} \times \frac{1\,\text{mol}}{2.02\,\text{g}}\right)(0.08206\,\text{L}\cdot\text{atm/mol}\cdot\text{K})(298\,\text{K})}{25.00\,\text{L} - \left(45.0\,\text{g} \times \frac{1\,\text{mol}}{2.02\,\text{g}}\right)(0.0266\,\text{L/mol})} - \frac{(0.244\,\text{atm}\cdot\text{L}^2/\text{mol}^2)\left(45.0\,\text{g} \times \frac{1\,\text{mol}}{2.02\,\text{g}}\right)^2}{(25.00\,\text{L})^2}$$

$$P = 22.1\,\text{atm}$$

105. Because the volume (at constant pressure) is proportional to the temperature, you would need to quadruple the Kelvin temperature to 1192K in order to quadruple the volume. The temperature change would be 894 K.

Lowering the temperature to 273 K would reduce the original volume to

$$V_{\text{final}} = T_{\text{final}} \times \frac{V_{\text{initial}}}{T_{\text{initial}}} = (273\text{K}) \times \frac{V_{\text{initial}}}{(298\text{K})} = 0.916 V_{\text{initial}}$$

The volume change is 0.084 V_{initial}, or a decrease of 8.4% of the original volume.

107. To solve the problem, we first need to find the number of moles of He in the tank:

$$n = \frac{PV}{RT} = \frac{(21.0\,\text{atm})(27.0\,\text{L})}{(0.08206\,\text{L}\cdot\text{atm/mol}\cdot\text{K})((29.5 + 273.15)\text{K})} = 22.8\,\text{mol}$$

$$22.8\,\text{mol} \times 4.00\,\text{g/mol} = 91.4\,\text{g He}$$

Each balloon would hold

$$n = \frac{PV}{RT} = \frac{(1.11\,\text{atm})(1.50\,\text{L})}{(0.08206\,\text{L}\cdot\text{atm/mol}\cdot\text{K})((27.5 + 273.15)\text{K})} = 0.0675\,\text{mol}$$

The total number of balloons that could be filled is

$$\frac{22.8\,\text{mol He}}{1\,\text{tank}} \times \frac{1\,\text{balloon}}{0.0675\,\text{mol He}} = 338\,\text{balloons/tank}$$

109. One of the considerations is which property is easier to manage, a temperature of only 20 K or a high pressure of 400 atm. Each requires a different infrastructure to handle: High pressures require thick walls with excellent welding, while low temperatures require vacuum dewers (like a thermos) and any variation in temperature causes the hydrogen to boil and be lost.

110. a. Hydrocarbon gases are nonpolar molecules and are subject only to dispersion or van der Waals forces.

b. Because they have only very weak interactions, these gases tend to act ideally, except at very low temperatures and high pressures.

c. $n = \frac{PV}{RT} = \frac{(1.00\,\text{atm})(0.300\,\text{L})}{(0.08206\,\text{L}\cdot\text{atm/mol}\cdot\text{K})(273\text{K})} = 0.0134\,\text{mol}$

d. During the combustion reaction (C_xH_y + excess $O_2 \rightarrow x\ CO_2 + y/2\ H_2O$), we get predictable relations from the stoichiometry. If the gases are all measured under the same conditions of pressure and temperature, those volumes will behave exactly as moles would behave in the balanced reaction above. For each volume of hydrocarbon burned, there are three volumes of CO_2 produced. (The ratio of volumes is 300 / 100 = 3:1). We can conclude, then, that there are three carbons in the original hydrocarbon. Also, because the ratio of water to hydrocarbon is 4:1, the coefficient in the equation preceding H_2O is 4. Because there is only one molecule of water produced for every two hydrogen atoms in the hydrocarbon, there must be eight atoms of hydrogen in the molecule. The equation of the hydrocarbon is C_3H_8.

e. Only the moles are constant in this change; we can use the combined gas law to solve for the new volume.

$$\frac{P_{\text{initial}}V_{\text{initial}}}{n_{\text{initial}}T_{\text{initial}}} = \frac{P_{\text{final}}V_{\text{final}}}{n_{\text{final}}T_{\text{final}}}$$

$$V_{\text{final}} = \frac{P_{\text{initial}}V_{\text{initial}}n_{\text{final}}T_{\text{final}}}{n_{\text{initial}}T_{\text{initial}}P_{\text{final}}}$$

$$V_{\text{final}} = \frac{P_{\text{initial}}V_{\text{initial}}T_{\text{final}}}{T_{\text{initial}}P_{\text{final}}} \text{ since } n_{\text{final}} = n_{\text{initial}}$$

$$V_{\text{final}} = \frac{(1.00\,\text{atm})(0.400\,\text{L})\{(273+500)\text{K}\}}{(273\,\text{K})(2.5\,\text{atm})} = 0.453\,\text{L}$$

Chapter 11: Chemistry of Water and the Nature of Liquids

The Bottom Line

- Water and other liquids have in common intermolecular forces called van der Waals forces that hold their molecules together. These forces are weaker than covalent or ionic bonds. (Section 11.2)
- Among these forces is dipole–dipole interactions. (Section 11.2)
- A hydrogen bond is an especially strong interaction that occurs when a hydrogen attached to an oxygen, nitrogen, or fluorine atom is close to a different atom of oxygen, nitrogen, or fluorine. (Section 11.2)
- Weaker, but collectively important intermolecular interactions in large molecules are known as London forces. (Section 11.2)
- Water is special among liquids because of intermolecular hydrogen bonds. (Section 11.3)
- Intermolecular interactions can be used to explain many physical properties of water, including its viscosity, surface tension, and capillary action. (Section 11.3)
- The pressure exerted by the evaporation of a liquid in equilibrium with its surroundings is called its vapor pressure. The boiling point is reached when the vapor pressure is equal to the surrounding pressure. (Section 11.3)
- A heating curve describes the changes in phase and temperature that occur as a substance is heated at constant pressure. (Section 11.3)
- A phase diagram describes the changes in phase as pressure and temperature are changed. (Section 11.4)
- Water is known as the universal solvent because of its ability to form solutions with many substances. (Section 11.6)
- Solution formation can be understood in terms of specific types of energy changes. (Section 11.6)
- Solution concentration can be expressed in a variety of units, including molarity, molality, and parts per million. (Section 11.7)
- Raoult's law describes the change in vapor pressure of a solvent as solute is added to a solution. (Section 11.7)
- Solubility is affected by pressure and temperature in, as a first approximation, predictable ways. (Section 11.8)
- Colligative properties are based on the number of particles in a solution and can be understood in terms of Raoult's law. (Section 11.9)
- Colligative properties include vapor pressure lowering, boiling-point elevation, freezing-point depression, and osmosis. (Section 11.9)
- It is possible that reverse osmosis will become a vital process for supplying clean water to large cities. (Section 11.9)

Solutions to Practice Problems

11.1. The size of the London dispersion force between the molecules increases as the size of the molecule increases. Therefore, the forces between the molecules of dodecane are much larger than the forces between hexane molecules, which in turn are much larger than the forces between propane molecules. The larger the force, the more closely and strongly the molecules will be held together. Dodecane is so strongly held that it forms a solid, and hexane is held strongly enough to cause it to condense to a liquid at room temperature. The forces between propane molecules are not large enough at room temperature to cause condensation.

11.2. Propylene glycol has two –OH group just as ethylene glycol does, and it should have similar hydrogen-bonding properties. Because of its larger size (due to the extra CH_2 group in the center), the dispersion forces should be slightly larger. Therefore, the overall forces in propylene glycol are larger, and the expected boiling point is higher. (The larger the forces, the higher the boiling or melting temperature will be.)

11.3. Any molecule with larger total forces should have a vapor pressure that is lower than that of water or methanol. Either propylene glycol or ethylene glycol (as discussed in Practice 11.2) would be suitable examples. Additionally, any molecule with enough forces to solidify should also have a lower vapor pressure.

11.4. Because acetone is a polar molecule, there will be dipole–dipole interactions as well as dispersion forces holding the liquid together. Hexane is a slightly larger molecule than acetone (6-carbon chain compared with acetone's 3-carbon chain) and will have slightly larger dispersion forces, but it has no dipole–dipole forces. Hexane should have lower overall forces than acetone and is likely to have the higher vapor pressure.

11.5. The total heat needed is the sum of five processes: heating ice from –40°C to 0°C, melting ice, heating water from 0°C to 100°C, boiling water, and heating steam from 100°C to 160°C.

Heating ice (c_{ice} = 2.05 J/g·°C): $q = mc_{ice}\Delta T = (130.0\,\text{g})(2.05\,\text{J/g}\cdot°\text{C})[0° - (-40°\text{C})] = 10660\,\text{J}$

Melting ice ($\Delta_{fus}H$ = 334 J/g): $q = m\Delta_{fus}H = (130.0\,\text{g})(334\,\text{J/g}) = 43420\,\text{J}$

Heating water (c_{liq} = 4.184 J/g·°C): $q = mc_{liq}\Delta T = (130.0\,\text{g})(4.184\,\text{J/g}\cdot°\text{C})(100° - 0°\text{C}) = 54392\,\text{J}$

Boiling water ($\Delta_{fus}H$ = 2440 J/g): $q = m\Delta_{fus}H = (130.0\,\text{g})(2440\,\text{J/g}) = 317200\,\text{J}$

Heating steam (c_{vap} = 1.84 J/g·°C): $q = mc_{vap}\Delta T = (130.0\,\text{g})(1.84\,\text{J/g}\cdot°\text{C})(160° - 100°\text{C}) = 14352\,\text{J}$

Total heat required: 10660 + 43420 + 54392 + 317200 + 14352 = 440024 J = **440 kJ**

11.6. The boiling point of water at 0.70 atm is 80°C, lower than at 1.0 atm. This does agree, because a lower temperature is needed to boil water at high altitudes.

11.7. In order to be soluble in hexane, which is a nonpolar molecule, a molecule should also be nonpolar (and therefore should not hydrogen-bond). Both iodine (I_2, nonpolar bond) and CCl_4 (polar bonds but nonpolar molecule) will be soluble in hexane. Water with its polar bonds and hydrogen bonding would "rather" stay with itself than mix into hexane.

11.8. We use the molality and volume to determine the moles of KOH required, which can then be converted to a mass:

$$600.0\,\text{mL soln} \times \frac{1\,\text{L}}{1000\,\text{mL}} \times \frac{1.40\,\text{mol KOH}}{1\,\text{L soln}} = 0.840\,\text{mol}$$

$$0.840\,\text{mol} \times \frac{56.11\,\text{g KOH}}{1\,\text{mol KOH}} = 47.1\,\text{g KOH}$$

If the density of the solution is 1.07 g/mL, there are 642 g of solution (600.0 mL × 1.07 g/mL). The mass of water is 642 g – 47.1 g = 594.9 g water. The number of moles of water is

$$594.9\,\text{g H}_2\text{O} \times \frac{1\,\text{mol H}_2\text{O}}{18.02\,\text{g H}_2\text{O}} = 33.01\,\text{mol H}_2\text{O}$$

The mole fraction of water is

$$\chi_{H_2O} = \frac{\text{mol water}}{\text{mol total}} = \frac{33.01\,\text{mol}}{33.01\,\text{mol} + 0.84\,\text{mol}} = 0.975$$

11.9. For dilute solutions, we can use the conversion factor 1 ppb = 1 μg/L and then use dimensional analysis to convert the mass to moles:

$$\frac{10\,\mu\text{g As}}{1\,\text{L soln}} \times \frac{1\,\text{g}}{10^6\,\mu\text{g}} \times \frac{1\,\text{mol As}}{74.92\,\text{g As}} = \frac{1.33\times10^{-7}\,\text{mol As}}{1\,\text{L soln}} = 1.33\times10^{-7}\,M\,\text{As}$$

11.10. The concentration, in units of g/100 g, is $\frac{22.7\,\text{g NaCl}}{55.0\,\text{g}\,H_2O} \times \frac{100\,\text{g}\,H_2O}{(100\,\text{g}\,H_2O)} = 41.3\,\text{g NaCl}/100\,\text{g}\,H_2O$.

According to the figure, the solubility of NaCl at 25°C is ~37g NaCl/100g H_2O, so the solution is saturated, with some solid NaCl left over. At 37 g NaCl/100 g water, the total amount of water needed to dissolve 22.7 g NaCl is $22.7\,\text{g NaCl} \times \frac{100\,\text{g}\,H_2O}{37\,\text{g NaCl}} = 61.4\,\text{g NaCl}/100\,\text{g}\,H_2O$. The additional amount of water needed is 61.4 g − 55.0 g = 6.4 g H_2O ≈ 6.4 mL water.

11.11. To find the vapor pressure, we first need to find the mole fraction of ethanol in the solution:

$$\text{mol}\,C_2H_5OH = 127.9\,\text{g}\,C_2H_5OH \times \frac{1\,\text{mol}\,C_2H_5OH}{46.07\,\text{g}\,C_2H_5OH} = 2.776\,\text{mol}\,C_2H_5OH$$

$$\text{mol}\,C_3H_8O_3 = 26.8\,\text{g}\,C_3H_8O_3 \times \frac{1\,\text{mol}\,C_3H_8O_3}{92.09\,\text{g}\,C_3H_8O_3} = 0.291\,\text{mol}\,C_3H_9O_2$$

$$\chi_{C_2H_5OH} = \frac{\text{mol}\,C_2H_5OH}{\text{mol}\,C_2H_5OH + \text{mol}\,C_3H_9O_2} = \frac{2.776\,\text{mol}}{2.776\,\text{mol} + 0.291\,\text{mol}} = 0.9051$$

The vapor pressure, according to Raoult's law, is

$$P_{C_2H_5OH} = \chi_{C_2H_5OH} P^\circ_{C_2H_5OH} = (0.9051)(135.3\,\text{torr}) = 122.5\,\text{torr}$$

11.12. a. We first use the equation for the boiling-point elevation to find the temperature change, which is added to the normal boiling point to get the solution boiling point.

For NaCl (i = 2):

$$\Delta T = iK_b m = (2)(0.51°\text{C}/m)(1.00) = 1.02°\text{C}$$
$$T_b = 100.0° + 1.02°\text{C} = 101.0°\text{C}$$

b. For $FeCl_3$ (i = 4):

$$\Delta T = iK_b m = (4)(0.51°\text{C}/m)(0.35) = 0.71°\text{C}$$
$$T_b = 100.0° + 0.71°\text{C} = 100.7°\text{C}$$

c. For KCl (i = 2):

$$\Delta T = iK_b m = (2)(0.80°\text{C}/m)(1.5) = 2.4°\text{C}$$
$$T_b = 64.7° + 2.4°\text{C} = 67.1°\text{C}$$

11.13. a. For sucrose (i = 1):

$$\Pi = iM\,RT = (1)(0.100\,\text{mol/L})(0.08206\,\text{L}\cdot\text{atm/mol}\cdot\text{K})(298\,\text{K})$$
$$\Pi = 2.45\,\text{atm}$$

b. For $CaCl_2$ (i = 3):

$$\Pi = iM\,RT = (3)(0.375\,\text{mol/L})(0.08206\,\text{L}\cdot\text{atm/mol}\cdot\text{K})(298\,\text{K})$$
$$\Pi = 27.5\,\text{atm}$$

c. For $(NH_4)SO_4$ (i = 3)

$$\Pi = iM\,RT = (3)(1.250\,\text{mol/L})(0.08206\,\text{L}\cdot\text{atm/mol}\cdot\text{K})(298\,\text{K})$$
$$\Pi = 91.7\,\text{atm}$$

Solutions to Student Problems

1. The main difference between the two is that *inter*molecular forces occur between molecules, and *intra*molecular forces occur within a molecule and consist of some type of bonding. The intramolecular forces are much stronger than intermolecular forces.

3. Intermolecular forces arise from the interaction of charges, positive on one and negative on the other. These charges are due to the full ionic charges or to smaller ones resulting from permanent dipoles within the molecule or from induced dipoles that are much smaller in magnitude and temporary in nature.

5. From least to most polar: S–H = C–H < N–H < O–H

7. a. Water freezes when intermolecular attractions are forming.
b. The production of carbon and sulfur from CS_2 indicates intramolecular bonds have been broken.
c. When dry ice sublimes, intermolecular bonds are broken.
d. In general, intramolecular bonds are stronger than intermolecular bonds.

9. The shorter the distance between two atoms, the stronger must be the force holding them together. The 0.101-nm distance is due to the intramolecular bond, and the 0.175-nm distance is due to the intermolecular attraction. The 0.101-nm distance in the bond corresponds to the stronger attraction.

11. a. In general, those molecules with larger intermolecular forces will have the higher boiling point. Additionally, for those molecules without large dipoles or hydrogen bonding, the larger and less compact molecules will interact over a greater area, leading to larger forces. Because pentane is "stretched out" relative to the same-sized and more compact 2,2-dimethlpropane, pentane will have the larger forces and higher boiling point.
b. Another general point is that cyclic structures will have stronger interactions than noncyclic structures of similar size. Cyclohexane will have the higher boiling point.
c. Hexane and pentane are similar in shape and types of forces, but hexane is the larger of the two. Hexane will have the stronger forces and the higher boiling point.
d. Although water is smaller than pentane, it will have much stronger hydrogen bonding, while pentane has only dispersion forces. Water has the higher boiling point.
e. Both pentane and methane will have only dispersion forces, but pentane is the larger of the two and will have the stronger attractive force, giving pentane the higher boiling point.

13. a. Based on electronegativity differences (the arrow shows the direction of electron density shift pointing toward the negative end of the dipole): B→O

b. P→Cl
c. H→O

15. a. CH_3CH_2OH must overcome dispersion (all molecules have dispersion forces), dipole–dipole attractions, and hydrogen bonding.
b. CO_2 must overcome only dispersion forces.
c. SF_6 is nonpolar and must overcome only dispersion forces.
d. HF must overcome dispersion, dipole–dipole attractions and hydrogen bonding.

17. a. Since F is the smaller atom, with fewer electrons than Cl, CF_4 will be less polarizable.
b. Since O is the smaller atom, with fewer electrons than Se, H_2O will be less polarizable.

19. a. CH_3OH will have hydrogen bonding.
b. NH_3 will have hydrogen bonding.
c. C_6H_6 will not have hydrogen bonding as it lacks an O, F, or N.
d. $C_2H_5OC_2H_5$ will not have hydrogen bonding as it lacks a hydrogen bound to O, F, or N.

21. a. The bonds between C and O will be the most polar.
b. 2-Methylfuran will have dispersion forces and dipole–dipole forces (due to polar C–O bonds)

23. Hexane is a nonpolar molecule with only dispersion forces, and hexanol is roughly the same size but also has hydrogen bonding and dipolar forces. Because the forces on hexane are smaller, it will have the higher vapor pressure and will evaporate first.

25. As the atoms become larger and have more electrons, they are more polarizable. Larger London forces will result from the more polarizable atoms making liquefaction easier. Kr, the largest in the group, should have the strongest London forces.

27. a. As the central atom becomes larger and more polarizable, the intermolecular forces increase. Higher forces lead to increasing boiling point, which is the trend seen down the group (excluding NH_3).
b. NH_3 deviates from the trend because it is a polar molecule capable of dipole–dipole interactions and can also hydrogen-bond.

29. Even though the masses of the two are nearly the same (18 g/mol for water and 16 g/mol for methane), giving them nearly equivalent dispersion forces, water also has dipole–dipole attractions and hydrogen bonding, which must be overcome to separate the molecules during boiling. Overcoming the additional interactions requires additional energy in the form of heat.

31. The heat absorbed by the water is

$$q = mc_{\text{liq}}\Delta T = \left(12.0\,\text{kg} \times \frac{1000\,\text{g}}{1\,\text{kg}}\right)(4.184\,\text{J/g}\cdot{}^\circ\text{C})(75^\circ - 25^\circ\text{C}) = 2.5\times10^6\,\text{J}$$

33. a. Propane (C_3H_8) consists entirely of nonpolar bonds and will then only have dispersion forces.
b. Because propane is a small molecule with only very weak forces, it should be a gas under the given conditions.
c. In order to liquefy propane at normal temperatures, it is pressurized.

35. Te has an electronegativity of 2.1, which is the same as that of hydrogen. It will form a nonpolar covalent bond with hydrogen. Oxygen has an electronegativity of 3.5, causing it to form a very polar bond with hydrogen. This polarization leaves hydrogen with a significant positive partial charge, causing it to interact strongly with negative charges—forming the "hydrogen bond." Because there is no polarization in the H–Te bond, there can be no hydrogen bonding.

37. The total heat needed is the sum of four processes: heating ice from −5°C to 0°C, melting ice, heating water from 0°C to 100°C, and boiling water.

Heating ice (c_{ice} = 2.05 J/g·°C): $q = mc_{ice}\Delta T = (15.0\,g)(2.05\,J/g\cdot°C)[0° - (-5°C)] = 153.75\,J$

Melting ice ($\Delta_{fus}H$ = 334 J/g): $q = m\Delta_{fus}H = (15.0\,g)(334\,J/g) = 5010\,J$

Heating water (c_{liq} = 4.184 J/g·°C): $q = mc_{liq}\Delta T = (15.0\,g)(4.184\,J/g\cdot°C)(100° - 0°C) = 6276\,J$.

Boiling water ($\Delta_{vap}H$ = 2440 J/g): $q = m\Delta_{vap}H = (15.0\,g)(2440\,J/g) = 36600\,J$

Total heat required: 154 + 5010 + 6276 + 36600 = 48040 J = **48.0 kJ**

39. a. The temperature is read where the gas–liquid phase line crosses 1 atm: 231 K
b. At the crossing of all three phase lines: 0.75 atm and 224 K
c. At −25°C (248 K) and 1 atm: gas

41. Along the vertical line at 220 K from 0.5 atm to 3 atm: gas to solid

43. Along the horizontal line at 1.0 atm from −45°C (228 K) to −53°C (220 K): from liquid to solid

45. Yes, at 240 K, gas will become a liquid at 1.4 atm.

47. The diagram is at right. Note that the pressure scale is logarithmic to allow for all points to be seen.

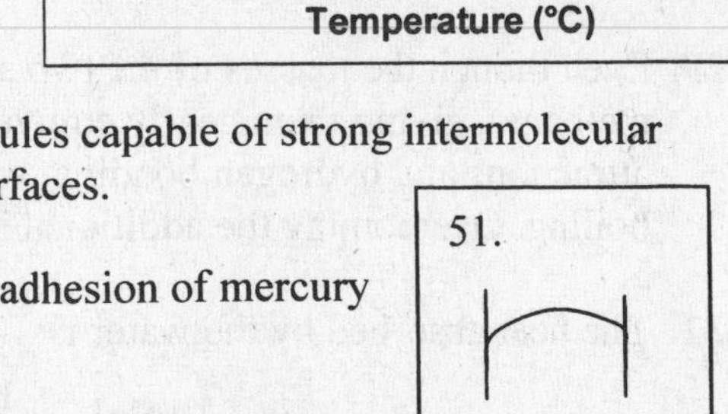

49. Surface tension is a measure of how strongly the molecules of a substance interact and "pull" the surface molecules toward the center. Viscosity is a measure of how strongly the molecules interact and prevent each other from flowing past one another. Capillary action is a balance of two interactions: that of the substance with its container and that of the substance with itself. Molecules capable of strong intermolecular interactions often have strong interactions with other surfaces.

51. The cohesion of the atoms in mercury is larger than the adhesion of mercury to glass. (See figure at right.)

53. From least to most viscous: gasoline, water, honey. Gasoline is nonpolar and will have the lowest intermolecular attractions. Honey, with its dissolved sugars that have many hydrogen-bonding locations, has the largest intermolecular interactions and therefore the highest viscosity.

55. Even without the strong dipole–dipole forces or hydrogen bonding, these molecules have large intermolecular forces. The strength of the dispersion forces is proportional to the size of the molecule. The molecules in the oil are large (and long) and can have significant interaction that keeps the molecules from easily moving past one another.

57. a. Because the formation of the products (the solution) increases with the increase in temperature, the overall reaction must be endothermic (solute + solvent + heat → solution) such that the increase in the "reactant" heat increases the amount of solute that dissolves.

b. Interpolating between the two values, we can set up the proportion

$$\frac{(68-39)\text{g}/100\text{g}}{(80-20)°\text{C}} = \frac{(x-39)\text{g}/100\text{g}}{(40-20)°\text{C}}$$

$$x = 39\text{g}/100\text{g} + \frac{(68-39)\text{g}/100\text{g}}{(80-20)°\text{C}} \times (40-20)°\text{C} = 49\text{g}/100\text{g}$$

Approximately 49 g of NH_4Cl will dissolve in 100 g of water at 40°C.

59. Water is able to dissolve many different sizes and shapes of molecules, from very small salts to very large protein molecules. Any molecule that possesses at least some type of charge (from full charges on ions, to permanent partial charges in dipoles, to temporary charges that it induces in nonpolar molecules) across the molecule will be somewhat soluble in water, because water has both positive and negative regions that can interact strongly and allow for dissolution.

61. a. Yes, if the interactions between two different molecules were the same as those between separate types of molecules.

b. There is additional stability derived from mixing of solutions even if the enthalpy change is not favorable. This is the same reason why some solutions whose formation is slightly endothermic still are possible.

63. The meniscus in A is much more curved, which is indicative of stronger interactions with the glass. Water has stronger intermolecular forces than heptane and matches buret A. Buret B must be heptane.

65. a. A substance that is only soluble has a limit to the amount that can be dissolved in water, while a substance that is miscible forms a solution in any proportion.

b. Miscible substances must share similar size and types of intermolecular interactions. Both water and methanol are small, polar, hydrogen-bonding substances.

67. a.
$$\frac{42.0\,\text{g NaOH} \times \frac{1\,\text{mol NaOH}}{40.00\,\text{g NaOH}}}{0.500\,\text{L}} = 2.10\,M\ \text{NaOH}$$

b.
$$\frac{10.0\,\text{g}\,C_6H_{12}O_6 \times \frac{1\,\text{mol}\,C_6H_{12}O_6}{180.16\,\text{g}\,C_6H_{12}O_6}}{0.250\,\text{L}} = 0.222\,M\,C_6H_{12}O_6$$

c.
$$\frac{25.0\,\text{g}\,NH_2CONH_2 \times \frac{1\,\text{mol}\,NH_2CONH_2}{60.06\,\text{g}\,NH_2CONH_2}}{100.0\,\text{mL} \times \frac{1\,\text{L}}{1000\,\text{mL}}} = 4.16\,M\,NH_2CONH_2$$

69. a. $$\frac{12.5\,\text{g}\,CH_2OHCH_2OH \times \dfrac{1\,\text{mol}\,CH_2OHCH_2OH}{62.07\,\text{g}\,CH_2OHCH_2OH}}{0.100\,\text{kg}} = 2.01\,m\,CH_2OHCH_2OH$$

b. $$\frac{53.0\,\text{g}\,C_{12}H_{22}O_{11} \times \dfrac{1\,\text{mol}\,C_{12}H_{22}O_{11}}{342.30\,\text{g}\,C_{12}H_{22}O_{11}}}{500.0\,\text{g} \times \dfrac{1\,\text{kg}}{1000\,\text{g}}} = 0.310\,m\,C_{12}H_{22}O_{11}$$

c. $$\frac{4.55\,\text{g}\,NaHCO_3 \times \dfrac{1\,\text{mol}\,NaHCO_3}{84.01\,\text{g}\,NaHCO_3}}{250.0\,\text{g} \times \dfrac{1\,\text{kg}}{1000\,\text{g}}} = 0.217\,m\,NaHCO_3$$

71. a. **To solve for the mole fraction, we calculate the number of moles of each and then place those values in the equation for the mole fraction.**

For benzene:

$$\text{mol}\,C_6H_6 = 22.7\,\text{g}\,C_6H_6 \times \frac{1\,\text{mol}\,C_6H_6}{78.11\,\text{g}\,C_6H_6} = 0.2906\,\text{mol}\,C_6H_6$$

$$\text{mol}\,C_6H_{12} = 67.5\,\text{g}\,C_6H_{12} \times \frac{1\,\text{mol}\,C_6H_{12}}{84.16\,\text{g}\,C_6H_{12}} = 0.8020\,\text{mol}\,C_6H_{12}$$

$$\chi_{C_6H_6} = \frac{\text{mol}\,C_6H_6}{\text{mol}\,C_6H_6 + \text{mol}\,C_6H_{12}} = \frac{0.2906\,\text{mol}}{0.2906\,\text{mol} + 0.8020\,\text{mol}} = 0.266$$

b. **For formic acid:**

$$\text{mol}\,CH_2O_2 = 15.0\,\text{g}\,CH_2O_2 \times \frac{1\,\text{mol}\,CH_2O_2}{46.03\,\text{g}\,CH_2O_2} = 0.3259\,\text{mol}\,CH_2O_2$$

$$\text{mol}\,H_2O = 100.0\,\text{g}\,C_6H_{12} \times \frac{1\,\text{mol}\,H_2O}{18.02\,\text{g}\,H_2O} = 5.549\,\text{mol}\,H_2O$$

$$\chi_{CH_2O_2} = \frac{\text{mol}\,CH_2O_2}{\text{mol}\,CH_2O_2 + \text{mol}\,H_2O} = \frac{0.3259\,\text{mol}}{0.3259\,\text{mol} + 5.549\,\text{mol}} = 0.0555$$

c. **For acetaldehyde:**

$$\text{mol}\,C_2H_4O = 0.195\,\text{g}\,C_2H_4O \times \frac{1\,\text{mol}\,C_2H_4O}{44.05\,\text{g}\,C_2H_4O} = 0.004427\,\text{mol}\,C_2H_4O$$

$$\text{mol}\,H_2O = 25.0\,\text{g}\,C_6H_{12} \times \frac{1\,\text{mol}\,H_2O}{18.02\,\text{g}\,H_2O} = 1.387\,\text{mol}\,H_2O$$

$$\chi_{C_2H_4O} = \frac{\text{mol}\,C_2H_4O}{\text{mol}\,C_2H_4O + \text{mol}\,H_2O} = \frac{0.004427\,\text{mol}}{0.00427\,\text{mol} + 1.387\,\text{mol}} = 0.00318$$

73. Fill in the missing information for the following table:

Compound	Grams of Compound	Grams of Water	Mole Fraction of Solute	Molality
NH_4Cl	12.5 grams	95.0 grams	**0.0424**	**2.46**
KNO_3	**1.96 grams**	125 grams	**0.00278**	0.155
$C_6H_{12}O_6$	**325 grams**	250.0 grams	0.115	**7.21**

Calculations for Problem 73:

For NH_4Cl:

$$\text{mol NH}_4\text{Cl} = 12.5\,\text{g NH}_4\text{Cl} \times \frac{1\,\text{mol NH}_4\text{Cl}}{53.49\,\text{g NH}_4\text{Cl}} = 0.2337\,\text{mol NH}_4\text{Cl}$$

$$\text{mol H}_2\text{O} = 95.0\,\text{g C}_6\text{H}_{12} \times \frac{1\,\text{mol H}_2\text{O}}{18.02\,\text{g H}_2\text{O}} = 5.272\,\text{mol H}_2\text{O}$$

$$\chi_{\text{NH}_4\text{Cl}} = \frac{\text{mol NH}_4\text{Cl}}{\text{mol NH}_4\text{Cl} + \text{mol H}_2\text{O}} = \frac{0.2337\,\text{mol}}{0.2337\,\text{mol} + 5.272\,\text{mol}} = 0.0424$$

$$m_{\text{NH}_4\text{Cl}} = \frac{\text{mol NH}_4\text{Cl}}{\text{kg H}_2\text{O}} = \frac{0.2337\,\text{mol}}{0.095\,\text{kg}} = 2.46\,m$$

For KNO_3:

$$\text{mol KNO}_3 = m_{\text{KNO}_3} \times \text{kg}_{\text{H}_2\text{O}} = 0.155\,m \times 0.125\,\text{kg} = 0.0194\,\text{mol KNO}_3$$

$$\text{g KNO}_3 = 0.0194\,\text{mol KNO}_3 \times \frac{101.11\,\text{g KNO}_3}{1\,\text{mol KNO}_3} = 1.96\,\text{g KNO}_3$$

$$\text{mol H}_2\text{O} = 125.0\,\text{g C}_6\text{H}_{12} \times \frac{1\,\text{mol H}_2\text{O}}{18.02\,\text{g H}_2\text{O}} = 6.937\,\text{mol H}_2\text{O}$$

$$\chi_{\text{KNO}_3} = \frac{\text{mol KNO}_3}{\text{mol KNO}_3 + \text{mol H}_2\text{O}} = \frac{0.01938\,\text{mol}}{0.01938\,\text{mol} + 6.937\,\text{mol}} = 0.00279$$

For $C_6H_{12}O_6$:

$$\text{mol H}_2\text{O} = 250.0\,\text{g C}_6\text{H}_{12} \times \frac{1\,\text{mol H}_2\text{O}}{18.02\,\text{g H}_2\text{O}} = 13.873\,\text{mol H}_2\text{O}$$

$$\chi_{\text{C}_6\text{H}_{12}\text{O}_6} = \frac{\text{mol C}_6\text{H}_{12}\text{O}_6}{\text{mol C}_6\text{H}_{12}\text{O}_6 + \text{mol H}_2\text{O}} = \frac{\text{mol C}_6\text{H}_{12}\text{O}_6}{\text{mol C}_6\text{H}_{12}\text{O}_6 + 13.873} = 0.115$$

$$\text{mol C}_6\text{H}_{12}\text{O}_6 = 0.115 \times (\text{mol C}_6\text{H}_{12}\text{O}_6 + 13.873) = 0.115(\text{mol C}_6\text{H}_{12}\text{O}_6) + 1.595$$

$$(1 - 0.115)\text{mol C}_6\text{H}_{12}\text{O}_6 = 0.885\,\text{mol C}_6\text{H}_{12}\text{O}_6 = 1.595$$

$$\text{mol C}_6\text{H}_{12}\text{O}_6 = \frac{1.595}{0.885} = 1.802\,\text{mol C}_6\text{H}_{12}\text{O}_6$$

$$\text{mass C}_6\text{H}_{12}\text{O}_6 = 1.802\,\text{mol C}_6\text{H}_{12}\text{O}_6 \times \frac{180.16\,\text{g C}_6\text{H}_{12}\text{O}_6}{1\,\text{mol C}_6\text{H}_{12}\text{O}_6} = 325\,\text{g C}_6\text{H}_{12}\text{O}_6$$

$$m_{\text{C}_6\text{H}_{12}\text{O}_6} = \frac{\text{mol C}_6\text{H}_{12}\text{O}_6}{\text{kg H}_2\text{O}} = \frac{1.802\,\text{mol}}{0.250\,\text{kg}} = 7.21\,m$$

75. Using a density of water of 1.0 g/mL the mass percent is

$$\%\text{C}_8\text{H}_{10}\text{O}_2\text{N}_4 = \frac{\text{mass C}_8\text{H}_{10}\text{O}_2\text{N}_4}{\text{mass C}_8\text{H}_{10}\text{O}_2\text{N}_4 + \text{mass H}_2\text{O}} \times 100\% = \frac{0.075\text{g}}{0.075\text{g} + 200\text{g}} \times 100\% = 0.037\%$$

The total mass of the solution is 200.075 g, so the volume of the solution is

$$200.075\text{g} \times \frac{1\,\text{mL}}{1.09\,\text{g}} = 183._6\,\text{mL} = 184\,\text{mL}$$

The molarity would be

$$\text{Molarity} = \frac{\text{mol}\,C_8H_{10}O_2N_4}{\text{L soln}} = \frac{0.075\,\text{g} \times \dfrac{1\,\text{mol}\,C_8H_{10}O_2N_4}{194.20\,\text{g}\,C_8H_{10}O_2N_4}}{183._6\,\text{mL} \times \dfrac{1\,\text{L}}{1000\,\text{mL}}} = 2.1 \times 10^{-3}\,M\,C_8H_{10}O_2N_4$$

The molality would be

$$\text{Molality} = \frac{\text{mol}\,C_8H_{10}O_2N_4}{\text{kg}\,H_2O} = \frac{0.075\,\text{g} \times \dfrac{1\,\text{mol}\,C_8H_{10}O_2N_4}{194.20\,\text{g}\,C_8H_{10}O_2N_4}}{200.0\,\text{g} \times \dfrac{1\,\text{kg}}{1000\,\text{g}}} = 1.9 \times 10^{-3}\,m\,C_8H_{10}O_2N_4$$

77. Let us assume a total mass of the solution of 100.0 g. Therefore, the solution is made from 5.00 g NaCl and 95.00 g H_2O. The volume of the solution is $100.0\,\text{g} \times \dfrac{1\,\text{mL}}{1.02\,\text{g}} = 98.04\,\text{mL}$.

$$\text{mol}\,Na^+ = 5.00\,\text{g NaCl} \times \frac{1\,\text{mol NaCl}}{58.44\,\text{g NaCl}} \times \frac{1\,\text{mol}\,Na^+}{1\,\text{mol NaCl}} = 0.08556\,\text{mol}\,Na^+$$

$$M_{Na^+} = \frac{\text{mol}\,Na^+}{\text{L soln}} = \frac{0.08556\,\text{mol}\,Na^+}{0.09804\,\text{L}} = 0.873\,M$$

79. Each measure of relative mass has at its heart the mass fraction, which is equal to the mass of the solute over the total mass of the solution. If we take F to represent

$$F = \frac{\text{mass solute}}{\text{mass solute} + \text{mass solvent}}$$

then we can define mass percent, ppm, and ppb as follows:

$$F = \frac{\text{mass solute}}{\text{mass solute} + \text{mass solvent}}$$

$$\text{mass\%} = \frac{\text{mass solute}}{\text{mass solute} + \text{mass solvent}} \times 10^2\% = F \times 10^2\%$$

$$\text{ppm} = \frac{\text{mass solute}}{\text{mass solute} + \text{mass solvent}} \times 10^6\,\text{ppm} = F \times 10^6\,\text{ppm}$$

$$\text{ppb} = \frac{\text{mass solute}}{\text{mass solute} + \text{mass solvent}} \times 10^9\,\text{ppm} = F \times 10^9\,\text{ppm}$$

$$\text{so } F = \frac{\text{mass\%}}{10^2\%} = \frac{\text{ppm}}{10^6} = \frac{\text{ppb}}{10^9}$$

a. For 130.0 ppm Ca^{2+}, $F = \dfrac{130.0\,\text{ppm}}{10^6\,\text{ppm}} = 1.300 \times 10^{-4}$, and

$$\%Ca^{2+} = F \times 100\% = 1.300 \times 10^{-4} \times 100\% = 0.01300\%$$

b. $\text{ppb}\,Ca^{2+} = F \times 10^9\,\text{ppb} = 1.300 \times 10^{-4} \times 10^9\,\text{ppb} = 1.300 \times 10^5\,\text{ppb}$

81. a. An increase in pressure will force more O_2 into solution, shifting the reaction to the left.
b. Because the solubility of all gases decreases with increasing temperature, the reaction will shift toward the right.

83. The partial pressure of nitrogen is 0.78(1 atm) = 0.78 atm. Using Henry's law to solve for the concentration:

$$P_{gas} = kC_{gas}$$

$$C_{gas} = \frac{P_{gas}}{k} = \frac{0.78\,\text{atm}}{1540\,\text{atm}/M} = 5.1\times10^{-4}\,M$$

85. $CaCl_2$ dissolves into three ions, so the concentration of ions would be 3(0.00100 *M*) = 0.00300 *M*.

87. In general, those molecules with the larger intermolecular forces have lower vapor pressures. Pentane has only dispersion forces, and water has strong dipole and hydrogen bonding, but glycerol has larger dispersion forces than water and also has strong dipole and hydrogen bonding. The expected order from highest to lowest vapor pressure is pentane > water > glycerol.

89. a. We first use the equation for the boiling-point elevation to find the temperature change, which is added to the normal boiling point to get the solution boiling point.

For NaCl (i = 2): $\Delta T = iK_b m = (2)(0.51°\text{C}/m)(0.75) = 0.765°\text{C}$

$$T_b = 100.00° + 0.765°\text{C} = 100.76°\text{C}$$

b. For $C_6H_{12}O_6$ (i = 1): $\Delta T = iK_b m = (1)(0.51°\text{C}/m)(0.040) = 0.020°\text{C}$

$$T_b = 100.00° + 0.020°\text{C} = 100.02°\text{C}$$

c. For 0.250 *M* $CaCl_2$ (i = 3), the solution mass is 1000 g/L and the mass of $CaCl_2$ per liter is

$$0.250\,\text{mol/L} \times 1.00\,\text{L} \times 110.98\,\text{g/mol} = 27.7\,\text{g}\,CaCl_2$$

$$\text{Mass water} = 1000\,\text{g} - 27.7\,\text{g} = 972.3\,\text{g} = 0.9723\,\text{kg}\,H_2O$$

$$\text{Molality}\,CaCl_2 = \frac{0.250\,\text{mol}}{0.9723\,\text{kg}} = 0.257\,m\,CaCl_2$$

$$\Delta T = iK_b m = (3)(0.51°\text{C}/m)(0.257) = 0.39°\text{C}$$

$$T_b = 100.00° + 0.39°\text{C} = 100.39°\text{C}$$

d. For $\chi_{naphthalene}$ = 0.25, there are 25 mol of naphthalene (i = 1) and 75 mol of benzene for each 100 mol total. The mass of benzene is 75 mol (78.11 g/mol) = 5858.25 g = 5.858 kg. The molality is

$$\frac{25\,\text{mol naphthalene}}{5.858\,\text{kg benzene}} = 4.27\,m\,\text{naphthalene}$$

For benzene, K_{bp} = 2.6°C/*m* and T_{bp}=80.1°C, so

$$\Delta T = iK_b m = (1)(2.1°\text{C}/m)(4.27) = 8.97°\text{C}$$

$$T_b = 80.1° + 8.97°\text{C} = 89.1°\text{C}$$

91. a. We first use the equation for the freezing-point depression to find the temperature change, which is added (the change is negative) to the normal freezing point to get the solution freezing/melting point.

For glucose (i = 1): $\Delta T = -iK_f m = (1)(1.86°\text{C}/m)(0.500m) = -0.93°\text{C}$

$$T_f = 0.00° + (-0.93°\text{C}) = -0.93°\text{C}$$

b. For LiOH (i = 2): $\Delta T = -iK_f m = (2)(1.86°\text{C}/m)(0.055m) = -0.20°\text{C}$

$$T_f = 0.00° + (-0.20°\text{C}) = -0.20°\text{C}$$

c. For methanol (i = 1) in phenol: $\Delta T = -iK_f m = (1)(7.4°\text{C}/m)(0.125m) = -0.92°\text{C}$

$$T_f = 43.0° + (-0.925°\text{C}) = 42.1°\text{C}$$

d. For benzoic acid ($i = 1$): $\Delta T = -iK_f m = (1)(1.86°C/m)(1.20m) = -2.23°C$

$$T_f = 0.00° + (-2.23°C) = -2.23°C$$

93. a. Water at 0°C has a vapor pressure of 4.58 torr. To solve the problem, we need to know the mole fraction of the solvent when *all particles* are considered in calculating the mole fraction. For $\chi_{glucose} = 0.200$, $\chi_{water} = 0.800$; then $P_{water} = \chi_{water} P°_{water} = (0.800)(4.58\,torr) = 3.66\,torr$

b. $CoCl_2$ dissolves into three ions in water. In 10 mol of solution, there are 9.45 mol of water and 0.55 mol of $CoCl_2$, which breaks into 0.55 mol of Co^{2+}, and 1.10 mol of Cl^-. The actual mole fraction of water is

$$\chi_{water} = \frac{\text{mol water}}{\text{mol water} + \text{mol}\,Co^{2+} + \text{mol}\,Cl^-} = \frac{9.45\,\text{mol}}{9.45\,\text{mol} + 0.55\,\text{mol} + 1.10\,\text{mol}} = 0.85$$

Then $\quad P_{water} = \chi_{water} P°_{water} = (0.85)(4.58\,torr) = 3.9\,torr$

c. For $\chi_{benzoic\ acid} = 0.125$, $\chi_{water} = 0.875$; then $P_{water} = \chi_{water} P°_{water} = (0.875)(4.58\,torr) = 4.01\,torr$

d. HCl dissolves into two ions in water. In 1 kg of water, there are 1.20 mol of HCl, which is 1.20 mol of H^+ and 1.20 mol of Cl^-. The 1000 g of water is (1000g) / (18.02 g/mol) = 55.5 mol water. The actual mole fraction of water is

$$\chi_{water} = \frac{\text{mol water}}{\text{mol water} + \text{mol}\,H^+ + \text{mol}\,Cl^-} = \frac{55.5\,\text{mol}}{55.5\,\text{mol} + 1.2\,\text{mol} + 1.2\,\text{mol}} = 0.959$$

Then $\quad P_{water} = \chi_{water} P°_{water} = (0.959)(4.58\,torr) = 4.39\,torr$

95. The vapor pressure is the equilibrium position at a specific temperature and will not depend on the surface area of the liquid. The area of the liquid will affect how quickly that equilibrium could be reached, but the final value of the vapor pressure will be the same whether the liquid is in a cup or forms an ocean—as long as the container is closed. In an open container, the partial pressure of the water cannot build up to a high value where the equilibrium pressure may lie, because the vapor is often removed (by wind or other air currents) from close proximity to the liquid.

97. Boiling-point elevation is a colligative property, which means that the amount of elevation in the boiling point is dependent on the number of particles that are dissolved in the solvent. All nonelectrolytes remain intact in the solution, and so all have the same concentration of particles. Electrolytes dissolve and produce more than one particle; the exact number of particles depends on the electrolyte. Therefore, the number of particles per volume will vary with different electrolytes, as will the elevation of the boiling point.

99. In order to have a freezing point of −32°C, the concentration of the solution would have to be

$$\Delta T = -iK_f m$$

$$m = -\frac{\Delta T}{iK_f} = -\frac{-32°C}{(1)(1.86°C/m)} = 17.2m$$

For 5.0 kg of solvent, the mass of antifreeze would be

$$\frac{17.2\,\text{mol HOCH}_2\text{CH}_2\text{OH}}{1\,\text{kg water}} \times 5.0\,\text{kg water} = 86.0\,\text{mol}$$

$$86.0\,\text{mol} \times \frac{62.07\,\text{g HOCH}_2\text{CH}_2\text{OH}}{1\,\text{mol HOCH}_2\text{CH}_2\text{OH}} = 5340\,\text{g HOCH}_2\text{CH}_2\text{OH}$$

Therefore, 5.340 kg of antifreeze is needed. (This is roughly a 50:50 mix by mass, compared to the normal 50:50 mix by volume.)

101. a. A solvent will move across a semipermeable membrane from an area of low concentration to an area of higher concentration. The water inside the cell will migrate out of the cell. Highly simplified:

b. If the osmotic pressure of the fluid added to the patient does not match the osmotic pressure of the tissue, the cells will either shrink (as above) in a hypertonic solution or rupture as a consequence of too much fluid entering the cells from a hypotonic solution.

103. To create the structure, a hydrogen bond is formed in pairs between a hydrogen on O or N and a second O or N without a hydrogen.

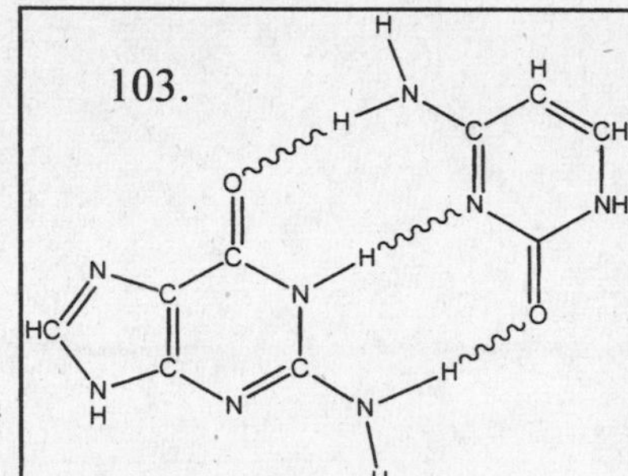

105. The energy needed in Problem 37 is 48.0 kJ. The combustion of methane provides −55.5 kJ/g of heat. Therefore, the mass of methane required is 48.0 kJ / (55.5 kJ/g) = 0.865 g CH_4.

107. The carbon–halogen bonds in CCl_4 and $CHBr_3$ are both polar, but the dipoles for each bond cancel in CCl_4, while a dipole remains in $CHBr_3$. Therefore, $CHBr_3$ can have dipole–dipole intermolecular attractions. Additionally, Br is larger and more polarizable than Cl, so $CHBr_3$ should also have larger dispersion forces between the molecules. Overall, the total forces in $CHBr_3$ are stronger, which gives it the greater surface tension.

109. The process of boiling is an equilibrium process that occurs when the vapor pressure of the liquid equals the external pressure. Because the pressure in the mountains is lower, the temperature of the boiling will be lower as well. The only way to increase the temperature of the boiling process would be to increase the external pressure (as in a pressure cooker.) Turning up the gas will cause the boiling to happen more forcefully, but still at the same temperature.

111. The easiest place to start a conversion from mass percent to molarity is with the density:

$$\frac{1.58\,\text{g soln}}{1\,\text{mL soln}} \times \frac{1000\,\text{mL soln}}{1\,\text{L soln}} \times \frac{35.6\,\text{g H}_2\text{SO}_4}{100\,\text{g soln}} \times \frac{1\,\text{mol H}_2\text{SO}_4}{98.09\,\text{g H}_2\text{SO}_4} = \frac{5.73\,\text{mol H}_2\text{SO}_4}{1\,\text{L soln}} = 5.73\,M\ \text{H}_2\text{SO}_4$$

112. a. The Lewis dot structure is

```
   H   H   H   H
   |   |   |   |
H--C---C---C---C--H
   |   |   |   |
   H   H   H   H
```

b. Because all the bonds in the molecule are nonpolar, the molecule is a nonpolar molecule. Only dispersion forces are possible.

c. Because the molecule is relatively small, and only dispersion forces are possible, the low boiling point is expected.
d. Because this molecule does not have a permanent dipole and cannot hydrogen-bond, the solubility of butane in water should be very low.
e. Butane burns according to the equation 2 C_4H_{10} + 13 $O_2 \rightarrow$ 8 CO_2 + 10 H_2O. The number of moles of water created is $3.50\,\text{g}\,C_4H_{10} \times \frac{1\,\text{mol}\,C_4H_{10}}{58.12\,\text{g}\,C_4H_{10}} \times \frac{10\,\text{mol}\,H_2O}{2\,\text{mol}\,C_4H_{10}} = 0.301\,\text{mol}\,H_2O$.
f. The butane inside the lighter is at higher pressures such that the boiling point is higher (above room temperature.
g. Because pentane is a larger molecule that should have more dispersion forces than butane, it should have a boiling point that is higher than that of butane (+36.1 °C). Because propane is a smaller molecule than butane that would have less dispersion forces, it should have a boiling point that is lower than that of butane (−42.1°C).

Chapter 12: Carbon

The Bottom Line

- Organic chemistry is the study of carbon-containing compounds, particularly those derived from living things or fossil fuels. (Section opener)
- The element carbon occurs in the form of three allotropes: the diamond, graphite, and fullerene forms. (Section 12.1)
- Crude oil is our major source of carbon-containing compounds, which are processed by the chemical industry into a wide range of products. (Section 12.2)
- Hydrocarbons are compounds composed of carbon and hydrogen only. They include the alkanes, alkenes, alkynes, and aromatic hydrocarbons. (Section 12.3)
- Organic compounds can exist as structural and geometric isomers. (Section 12.3)
- Hydrocarbons are separated into fractions by fractional distillation of crude oil. (Section 12.4)
- Cracking and reforming are two important methods of preparing specific organic molecules. (Section 12.5)
- Alkanes can undergo substitution and dehydrogenation reactions. (Section 12.6)
- All organic compounds can be regarded as carrying zero, one, or more distinctive functional groups on a hydrocarbon frame. (Section 12.7)
- Each functional group is associated with a distinctive set of chemical characteristics. (Section 12.7)
- Many modern materials are polymers composed of thousands of organic monomer units linked together. (Section 12.8)
- Molecules may exhibit chirality, which may lead to different chemical properties in two stereoisomers. (Section 12.14)
- Organic chemists can isolate and purify the active constituents of complex mixtures in order to make products such as pharmaceuticals. (Section 12.15)

A Note on Structural Drawings

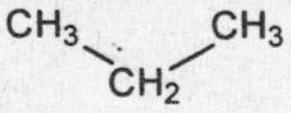

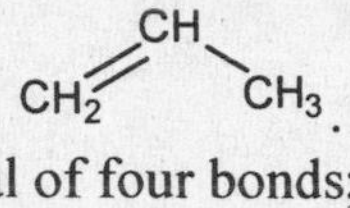

Chemists often use many types of shorthand notations to make drawing chemicals structures quicker and easier. One of these types of notation is the line drawing of organic structures. This notation is characterized by the elimination of the symbols for carbon (C), and hydrogen (H) from the drawings. One example is propane, CH_3 CH_2 CH_3, whose structure has three carbons connected in a line with three hydrogen atoms attached to the first and third carbons. The second carbon has two hydrogen atoms bound to it. The line drawing for the structure is therefore  . Note that the basic backbone is exactly the same. We infer from the diagram that there is a carbon at the end of each line and at the vertex in the middle. We further infer that there are enough hydrogen atoms on each carbon so that the carbon will have a total of four bonds. When there is a multiple bond (a double or triple bond) in the structure, there is more than one line between the carbon atoms. For example, propene (CH_2CHCH_3) can be drawn as 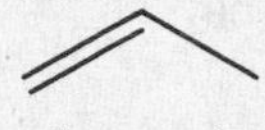 , which is the same as CH_2 CH CH_3. Because the first carbon has a double bond, it needs two hydrogens to have a total of four bonds; the second carbon has three bonds to other carbon atoms and needs only a single hydrogen; and the third carbon needs three additional hydrogens because it has only one bond to another carbon atom. When a structure includes an atom other than carbon or hydrogen, that atom is explicitly shown in the line drawing. For example, the common chemical acetone, CH_3COCH_3, is $CH_3C(=O)CH_3$ and would be drawn [line drawing] as a line drawing. Both structural and line drawings will be used in this answer key. Because both are commonly used by chemists, it is important to be able to understand both conventions.

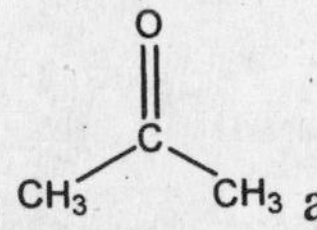

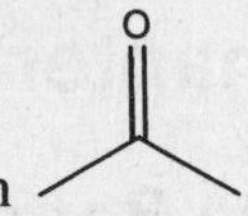

Solutions to Practice Problems

12.1. The isomers are

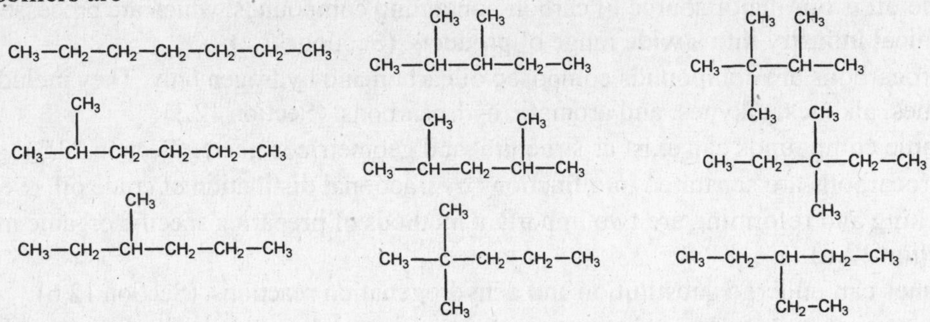

$CH_3-CH_2-CH_2-CH_2-CH_2-CH_2-CH_3$

$CH_3-CH(CH_3)-CH(CH_3)-CH_2-CH_3$

$CH_3-C(CH_3)_2-CH(CH_3)-CH_3$

$CH_3-CH(CH_3)-CH_2-CH_2-CH_2-CH_3$

$CH_3-CH(CH_3)-CH_2-CH(CH_3)-CH_3$

$CH_3-CH_2-C(CH_3)_2-CH_2-CH_3$

$CH_3-CH_2-CH(CH_3)-CH_2-CH_2-CH_3$

$CH_3-C(CH_3)_2-CH_2-CH_2-CH_3$

$CH_3-CH_2-CH(CH_2-CH_3)-CH_2-CH_3$

12.2. Hexene has the formula C_6H_{12}, and the cyclic structure is cyclohexane:

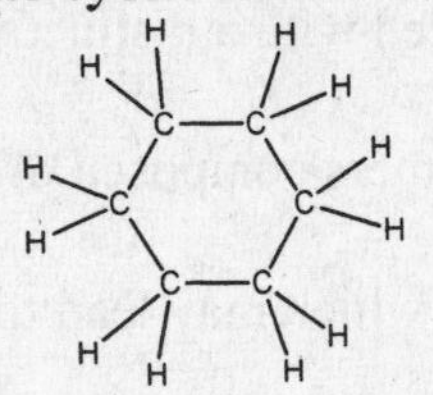

[cyclohexane structure: six C atoms in a ring, each bonded to two H atoms]

12.3. For C_5H_{10}, there are only six structural isomers, only one of which has separate *cis* and *trans* isomers:

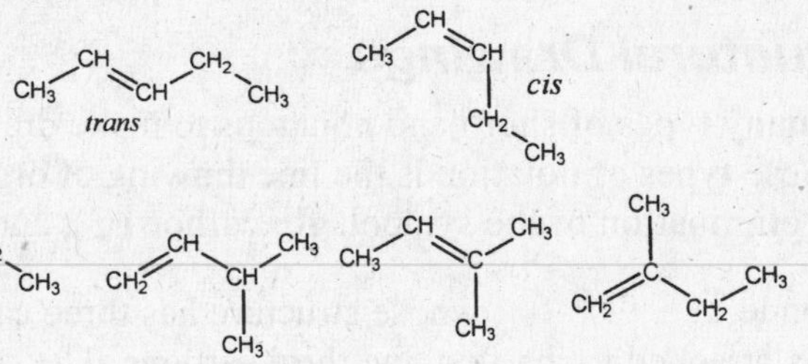

$CH_3-CH=CH-CH_2-CH_3$ (*trans*)

$CH_3-CH=CH-CH_2-CH_3$ (*cis*)

$CH_2=CH-CH_2-CH_2-CH_3$

$CH_2=CH-CH(CH_3)-CH_3$

$CH_3-CH=C(CH_3)-CH_3$

$CH_2=C(CH_3)-CH_2-CH_3$

12.4. Octane has at its base an eight-member carbon chain, with two single carbon (methyl) groups on the second carbon, and a three-carbon group (propyl) on the fourth carbon:

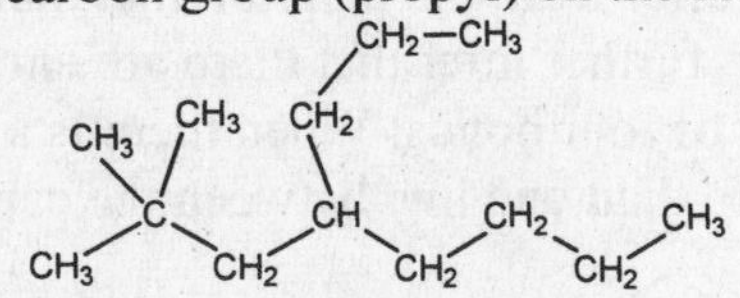

$CH_3-C(CH_3)_2-CH_2-CH(CH_2-CH_2-CH_3)-CH_2-CH_2-CH_2-CH_3$

12.5.

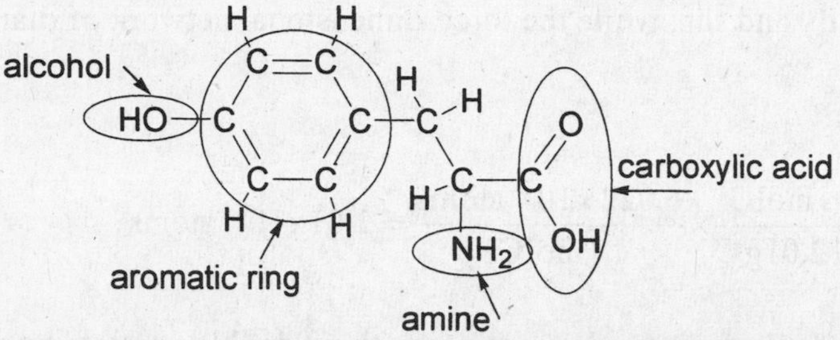

alcohol, aromatic ring, amine, carboxylic acid

12.6. Ethyl benzoate has two carbons on the alcohol side and the benzyl group on the carboxylic acid side. Benzyl ethanoate has the benzyl group on the alcohol side and a two-carbon group on the carboxylic acid side.

ethyl benzoate

benzyl ethanoate

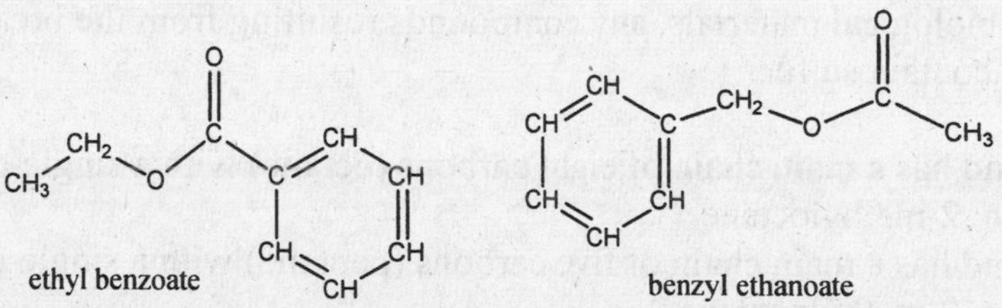

12.7. In the left molecule, the carbon second from the left in the chain is bound to four different groups and is chiral (H, Br, CH_3, and C_2H_5). In the right molecule, the carbon on the lower right is bound to four different groups and is chiral (H, F, CH_2, CF_2).

12.8. m-Amsacrine is very likely to have (serious) side effects. Nearly every cell has DNA in the nucleus of the cell. The DNA of healthy cells, along with that of the cancer cells, would be fragmented, leading to the death of both types of cells.

12.9. Of those listed in Table 12.6, there are one alkene, one ether, two alcohols, one ketone, and one ester:

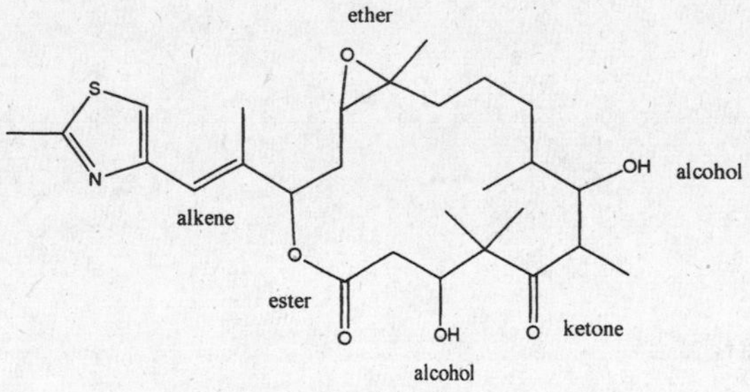

ether, alcohol, alkene, ester, alcohol, ketone

Solutions to Student Problems

1. Three allotropes of carbon are diamond (sp^3), graphite (sp^2 and planar network of six-member rings), fullerenes (sp^2 and curved due to five-member rings with six-member rings similar to the pattern on many soccer balls).

3. In diamond, all of the carbons are sp^3 hybridized (tetrahedral) and form an extensive three-dimensional network of carbon atoms bound together. The carbon atoms in graphite are all sp^2 (recall that sp^2 atoms form trigonal planes) hybridized and form extended planar surfaces that are only weakly bound via van der Waals forces between the planes. Because the planes are only

weakly held, they can fracture easily and slip, while the three-dimensional network of diamond is very hard.

5. Using dimensional analysis:

$$2.00\,\text{carats} \times \frac{200\,\text{mg}}{1\,\text{carat}} \times \frac{1\,\text{g}}{1000\,\text{mg}} \times \frac{1\,\text{mol C}}{12.01\,\text{g C}} \times \frac{6.022\times10^{23}\ \text{atoms}}{1\,\text{mol C}} = 2.01\times10^{22}\ \text{atoms}$$

7. The main difference is that the aliphatic compounds are those with no double or triple bonds, while the aromatics are characterized by structures with several multiple bonds, often in a cyclic arrangement.

9. Because many biological compounds (proteins and others) contain sulfur and the source of the crude oil is those biological materials, any compounds resulting from the breakdown of this material will also contain sulfur.

11. a. This compound has a main chain of eight carbons (octane) with a single-carbon group on the second carbon: 2-methyloctane.

b. This compound has a main chain of five carbons (pentane) with a single-carbon group on the second carbon: 2-methylpentane.

c. This compound has a main chain of five carbons (pentane) with a two-carbon group on the third carbon: 3-ethylpentane.

d. This compound has a single chain of five carbons: pentane.

13. a.

CH_3–CH_2–CH_2–CH_2–CH_2–CH_2–$CH(CH_3)$–CH_3

b.

CH_3–CH_2–CH_2–$CH(CH_3)$–CH_3

c.

CH_3–CH_2–$CH(CH_2CH_3)$–CH_2–CH_3

d.

CH_3–CH_2–CH_2–CH_2–CH_3

CH_3–$CH(CH_3)$–CH_3

15.

	Alkane	Alkene	Alkyne
Type of carbon-carbon bond	C–C	C=C	C≡C
Type of hybridization	sp^3	sp^2	sp

17. a.

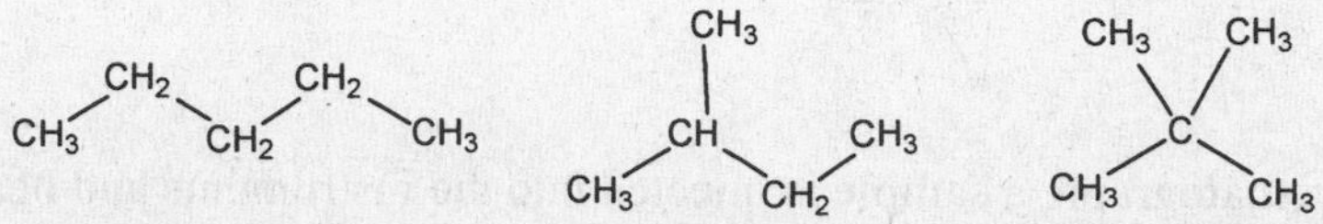

$CH_3CH_2CH_2CH_2CH_3$ — Pentane

$(CH_3)_2CHCH_2CH_3$ — 2-methybutane

$(CH_3)_4C$ — 2,2-dimethylpropane

Each molecule is C_5H_{12}.

b. The more branched a structure is, the lower the intermolecular forces and the lower the boiling point. Therefore: pentane (36°C), 2-methylbutane (28°C), 2,2-dimethylpropane (9.5°C).

19. All completely saturated, noncyclic structures have the formula C_nH_{2n+2}, so the formula is $C_{28}H_{58}$.

21. There are seven possible isomers of octane containing a six-carbon continuous chain. Each vertex and each line endpoint is a carbon with the appropriate number of hydrogens:

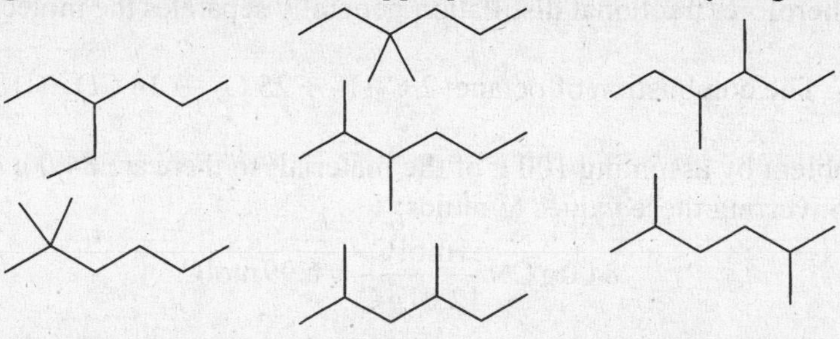

23. A cycloalkane will have the general formula C_nH_{2n} if there is only a single ring. The only possibilities that match are (b) C_6H_{12} and (d) C_7H_{14}. (The other two match the formula C_nH_{2n+2} of a noncyclic alkane.)

25. The left structure has the **nonhydrogen** groups on opposite sides of the double bond and is *trans*, while the right structure has the nonhydrogen groups on the same side of the double bond and is *cis*.

27. $CaC_2 + 2\ H_2O \rightarrow Ca(OH)_2 + C_2H_2$

29. a. The structures for cyclohexane and benzene have 12 and 6 hydrogen atoms, respectively:

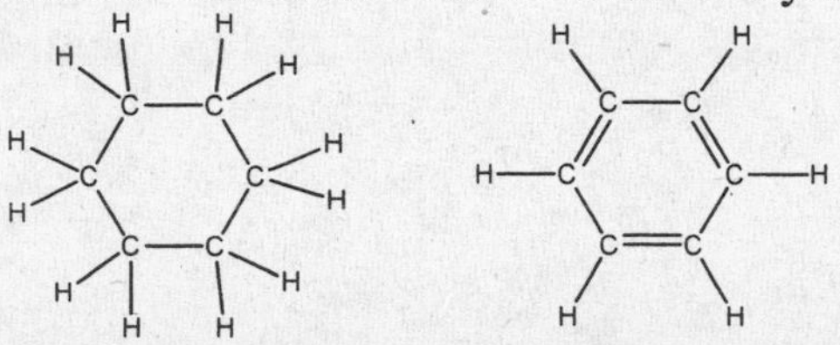

b. In cyclohexane, the carbons have four bonds, making the carbons sp^3 hybridized, while the benzene carbons have three sigma bonds (one pi bond) and are sp^2 hybridized.

c. The bond angles in cyclohexane are 109.5°, matching sp^3 tetrahedral geometry, while benzene has 120° angles, matching sp^2 trigonal planar geometry.

d. Benzene has three double bonds; cyclohexane has none.
e. Only benzene has an alternative resonance structure:

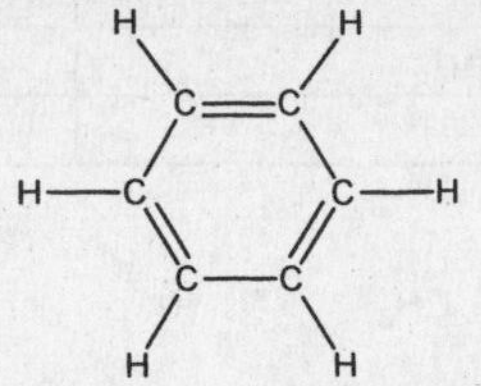

31. During a "run" of the gas chromatograph, a sample is injected into the instruments and heated to vaporize the sample. A carrier gas takes the sample into and through a column. Inside the column, the different molecules interact to varying extents with the materials inside the column. The greater the interaction, the more slowly the molecules move through the column. By selectively slowing the molecules, by mechanisms based on their properties, they are separated and later detected upon exiting the column.

33. The differences between the boiling temperatures of the components of the crude oil are used to separate them in the fractional distillation process. The boiling temperature will depend on the magnitude of the intermolecular forces. For the nonpolar molecules generally found in crude oil, there are only dispersion forces, which are mainly (but not entirely) dependent on the size of the molecules: The larger and straighter molecules experience greater forces and higher boiling temperatures. Therefore, fractional distillation generally separates the molecules by size.

35. Octane is C_8H_{18}. The combustion of octane: $2\ C_8H_{18} + 25\ O_2 \rightarrow 16\ CO_2 + 18\ H_2O$.

37. We start the problem by assuming 100 g of the material, so there are 84.0 g of carbon and 16.0 g of hydrogen. Converting these values to moles:

$$84.0\,\text{g C} \times \frac{1\,\text{mol C}}{12.01\,\text{g C}} = 6.99\,\text{mol C}$$

$$16.0\,\text{g C} \times \frac{1\,\text{mol H}}{1.008\,\text{g H}} = 15.9\,\text{mol C}$$

The simplest carbon-to-hydrogen is 7:16, so the empirical formula is C_7H_{16}, which has a molar mass of 100.19 g/mol. Therefore, that formula is also the molecular formula.

39. Because isooctane has a rating of 100 and heptane a rating of 0, an octane rating of 87 will correspond with 87% isooctane and 13% heptane.

41. $CH_4 + Br_2 \rightarrow CH_2Br_2 + H_2$

43. The structures and names are

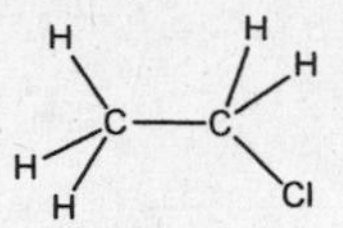

1-chloroethane 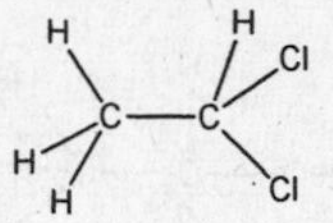 1,1-dichloroethane 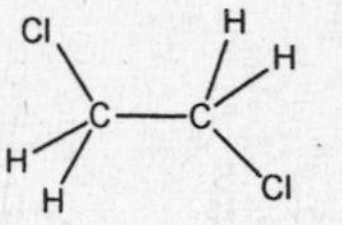 1,2-dichloroethane 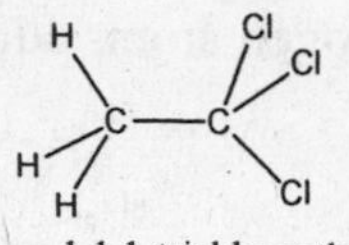 1,1,1-trichloroethane

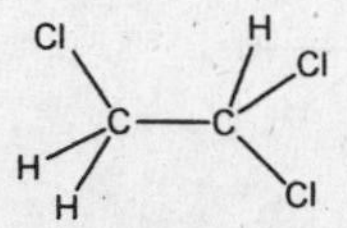

1,1,2-trichloroethane 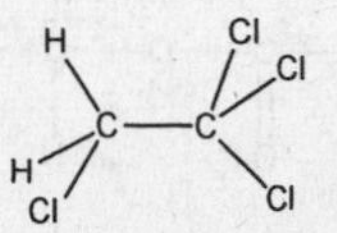 1,1,1,2-tetrachloroethane 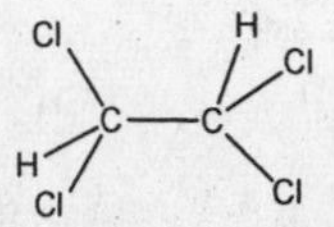 1,1,2,2-tetrachloroethane 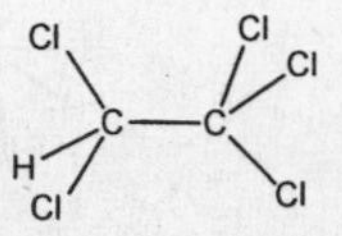 pentachloroethane

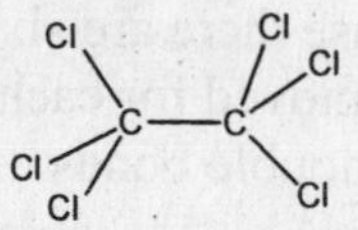

hexachloroethane

45. For the combustion of an alkane, the typical products are CO_2 and H_2O. If there is insufficient oxygen, some CO may also be present.

47. $42.0\,g\,C_3H_8 \times \frac{1\,mol\,C_3H_8}{44.09\,g\,C_3H_8} \times \frac{-2200\,kJ}{1\,mol\,C_3H_8} = -2095\,kJ = -2.10 \times 10^3\,kJ$ (The negative sign indicates that energy is released.)

49. Many answers are possible. We use the shortest possible structure.

a. CH_3OH,

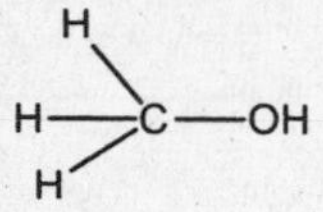

b. CH_2O,

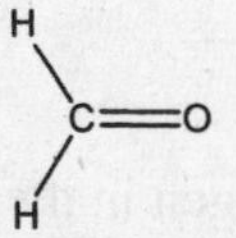

c. C_2H_4,

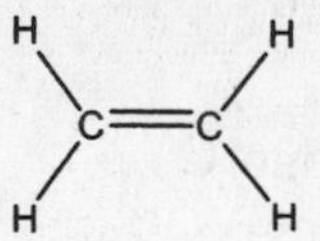

d. CH_3COCH_3,

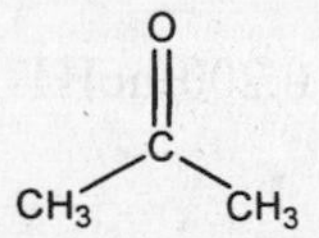

51. The groups are labeled on the structure.

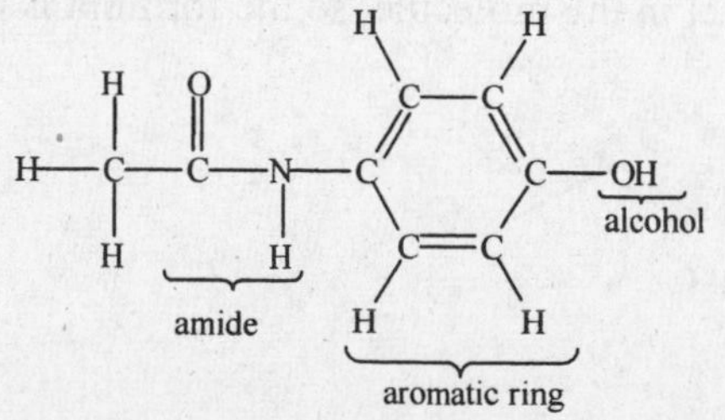

amide aromatic ring alcohol

53. With the hydrogen atoms added:

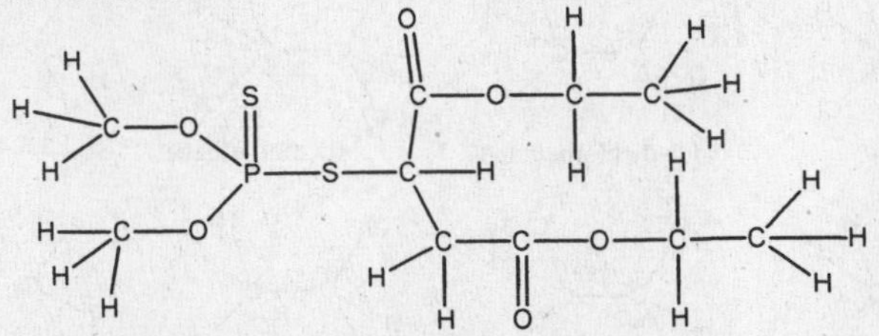

55. a. A saturated three-carbon hydrocarbon would be C_3H_8, so there are no double bonds.
b. With six carbons, the saturated formula would be C_6H_{14}. Because there are six hydrogen atoms missing in this structure, and two hydrogen atoms are removed for each pi bond, there are three pi bonds. (The three pi bonds would be seen as three double bonds or as the combination of one triple bond and one double bond.)
c. With ten carbons, the saturated formula would be $C_{10}H_{22}$. Because there are fourteen hydrogen atoms missing in this structure, and two hydrogen atoms are removed for each multiple bond, there are seven pi bonds.

57. a.

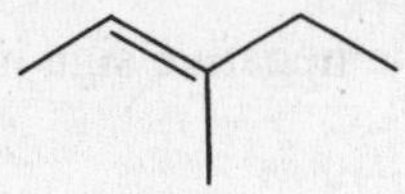

b.

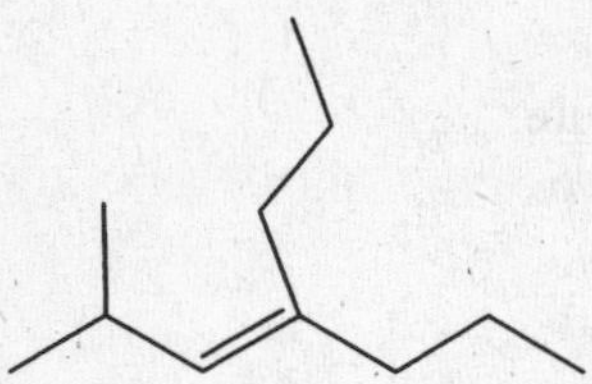

59. To find the formula, we find the number of moles of carbon and hydrogen in the products and then use the ratio to get the empirical formula. By comparing the empirical mass to the molar mass, we can predict the molecular formula:

$$2.93\,g\,CO_2 \times \frac{1\,mol\,CO_2}{44.01\,g\,CO_2} \times \frac{1\,mol\,C}{1\,mol\,CO_2} = 0.0666\,mol\,C$$

$$1.80\,g\,H_2O \times \frac{1\,mol\,H_2O}{18.02\,g\,H_2O} \times \frac{2\,mol\,H}{1\,mol\,H_2O} = 0.200\,mol\,H$$

$$\frac{0.200\,mol\,H}{0.0666\,mol\,C} = 3.00\,H : 1.00\,C$$

The empirical formula is CH_3 and the empirical mass is 15.0 g/mol. There are two empirical units [(30 g/mol) / (15 g/mol) = 2] in the molecule, so the formula is C_2H_6. This molecule is fully saturated.

61. a. Propene is CH_2CHCH_3: 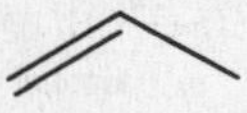
b. $C_3H_6 + H_2O \rightarrow C_3H_7OH$

63. a. The monomer is propene: CH_2CHCH_3,

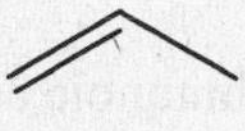

b. A four-unit portion of the polymer:

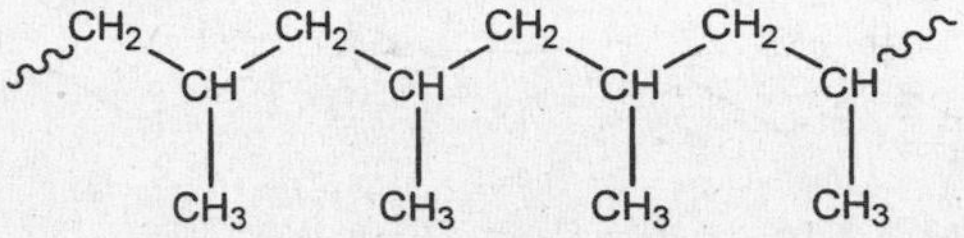

65. The HDPE units are long and unbranched, while the LDPE units are shorter and more branched. Without the branching, the HDPE polymer chains pack more closely together, giving the compound a higher density and strength. Because LDPE units are branched, the polymers cannot pack so closely, yielding a lower density and greater flexibility.

67. This molecule was numbered from the wrong end; it should be 1,4-pentanediol.

69.

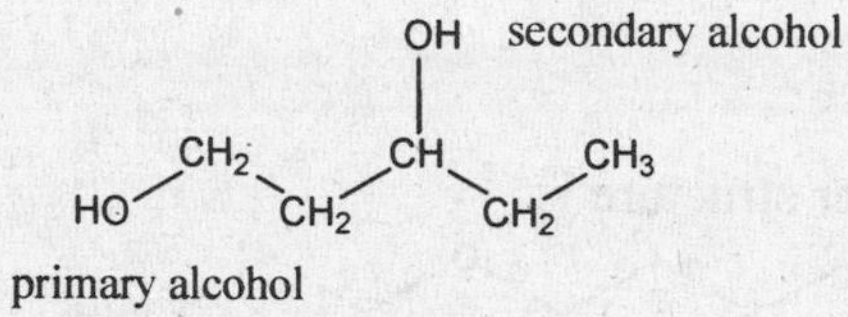

OH secondary alcohol

primary alcohol

71.

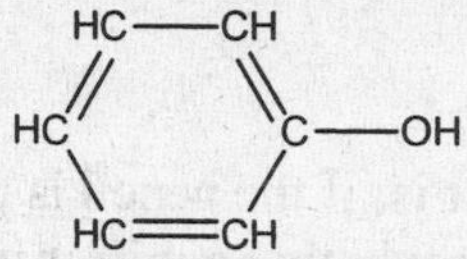

73. Oxidizing an alcohol yields either aldehydes (from primary alcohols) or ketones (from secondary alcohols). Tertiary alcohols do not form carbonyl groups (C=O). Propanol, 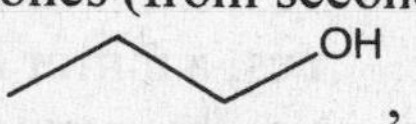 , forms propanal,  .

75. Propanoic acid is 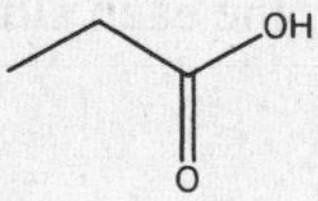 and could be formed from propanal,  .

77. Methyl ethyl ketone is 2-butanone, 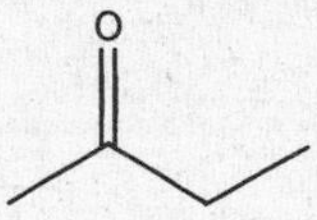 , and is formed from 2-butanol, 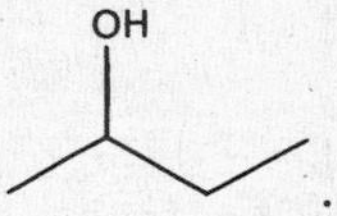 .

79. $CH_3CH_2COOH + H_2O \rightarrow CH_3CH_2COO^- + H_3O^+$

81. 3-Methylbutyl ethanoate is an ester with a 3-methylbutyl group on the alcohol side (attached to the O) and a two-carbon group on the carboxylic acid side (attached to and including the C=O). The balanced reaction between 3-methylbutanol and ethanoic acid to produce 3-methylbutyl ethanoate is

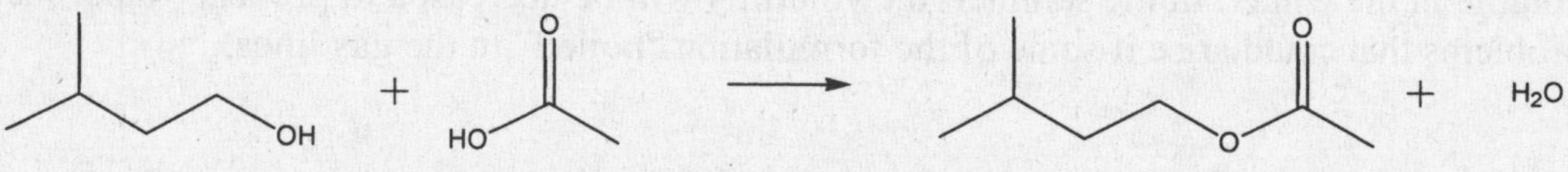

83. This ester is the product of 1-pentanol and butanoic acid:

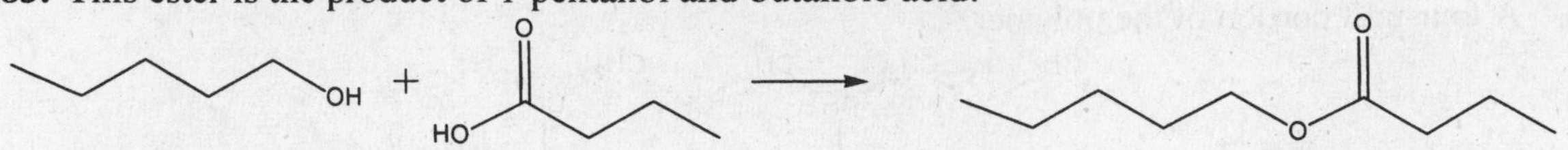

85.

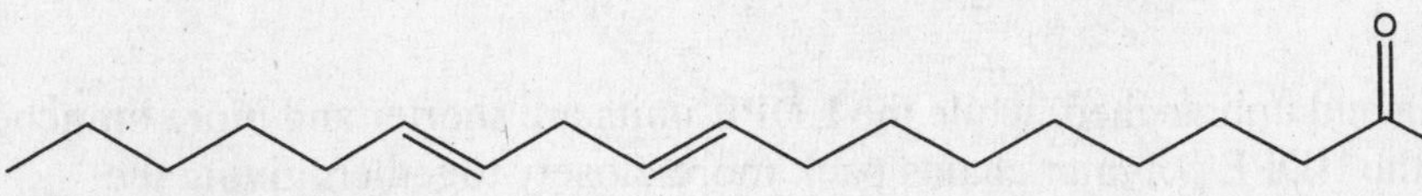

87. As each monomer unit adds to the polymer, H_2O is lost (OH from the acid and H from the amine):

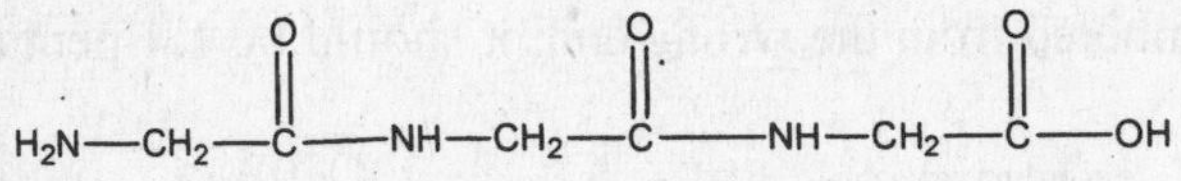

$$H_2N-CH_2-C(=O)-NH-CH_2-C(=O)-NH-CH_2-C(=O)-OH$$

89. NaCl is formed as the result of this reaction. The polymer structure is

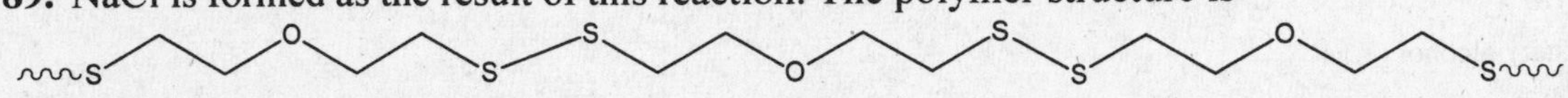

91. The names are dimethyl ether and butyl methyl ether.

93. a. Yes if there is writing or designs on the pencil; otherwise, no. That is, if the pencil is painted a solid color or lacks any design, it is not chiral. The key is to note whether a mirror image of the object is superimposable on the original. If it cannot be, the object displays chirality.

b. Yes; a mirror image of the foot is not the same as the original.

c. Yes; a mirror image of the book is not the same as the original.

d. No; the mirror image is the same for most forks.

95. The labeled carbon is chiral because there are four different groups attached to it. For the left structure, note that even though there is a C=O immediately on the each side, the next group after the C=O is different.

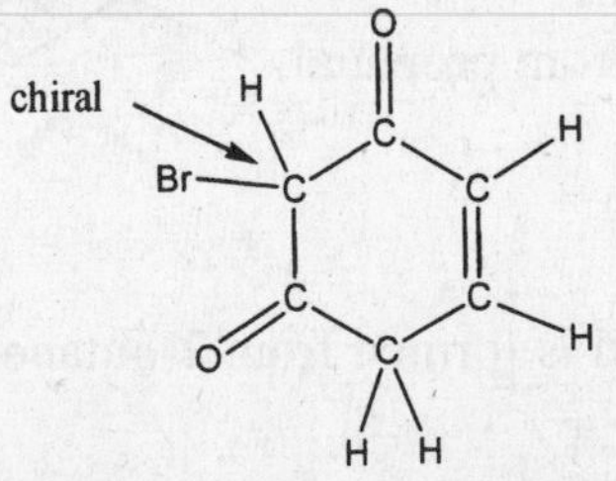

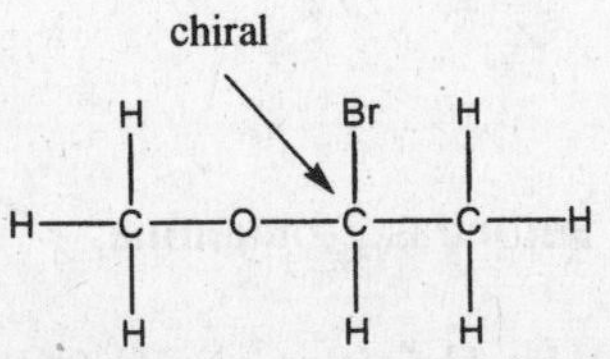

97. The total yield will be the product of the yields of the individual steps: $(0.95)^9 = 0.63 = 63\%$.

99. Because of the different temperatures, the volatility of the different components changes. In order to maintain similar vaporization and combustion characteristics, the components should be more volatile in the winter. In the summer, the volatility will be decreased to prevent "vapor lock" problems that could arise if some of the formulation "boiled" in the gas lines.

101. a. The compound is ethandoic acid (also called oxalic acid), HO–C(=O)–C(=O)–OH, which can be produced by oxidizing 1,2-ethandiol (each alcohol becomes a carboxylic acid) and will react with two molecules of ethanol to form a diester (shown in part c below).

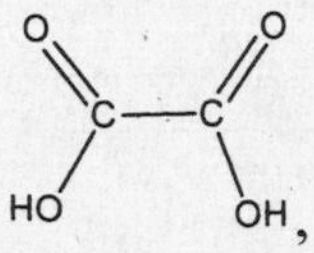

b. With two very polar carboxylic acid units that can also hydrogen-bond, ethandoic acid should be solid at room temperature. It also has all the properties of a weak acid with two acidic protons.

c. The reaction to form the ester is 2 $C_2H_5OH + H_2C_2O_4 \rightarrow C_6H_{10}O_4 + 2\ H_2O$:

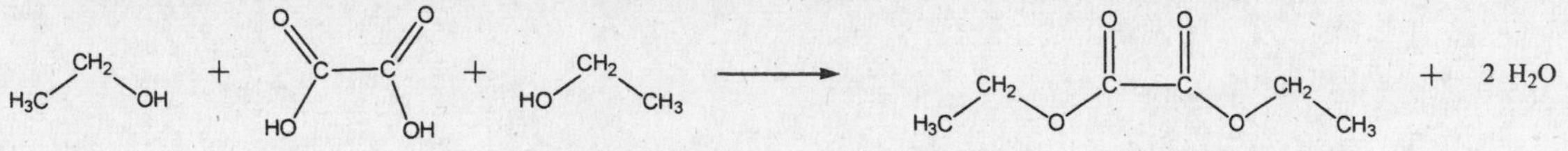

$$10.0\,g\,H_2C_2O_4 \times \frac{1\,mol\,H_2C_2O_4}{90.04\,g\,H_2C_2O_4} \times \frac{1\,mol\,C_6H_{10}O_4}{1\,mol\,H_2C_2O_4} \times \frac{146.14\,g\,C_6H_{10}O_4}{1\,mol\,C_6H_{10}O_4}$$

$$= 16.23\,g\,C_6H_{10}O_4 \text{ theoretical yield}$$

$$\text{Percent yield} = \frac{10.00\,g\,C_6H_{10}O_4}{16.23\,g\,C_6H_{10}O_4} \times 100\% = 61.6\%$$

d. The ester breaks down into its original components (the reverse of the reaction in part c) to regenerate the acid:

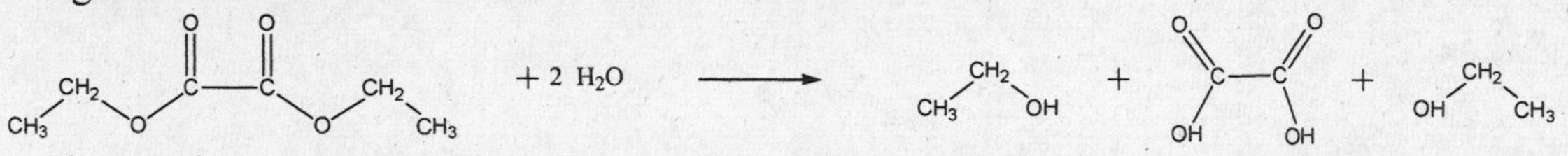

e. The reaction proceeds as the –OH on the glycol and the –OH on the acid condense to form an ester, and the opposite end of the molecule can continue the reaction with another molecule:

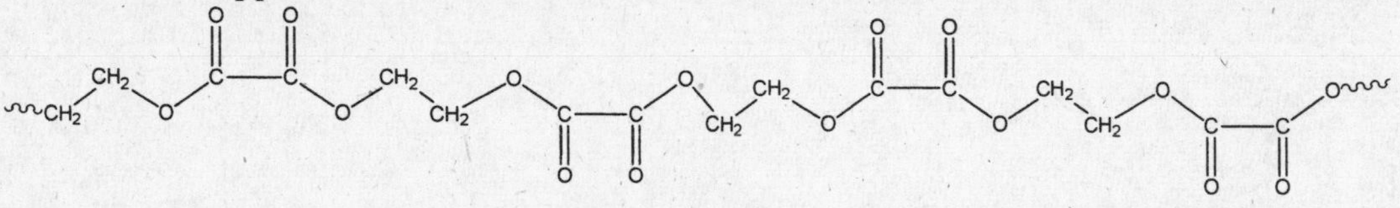

Chapter 13: Modern Materials

The Bottom Line

- In a crystalline solid, the atoms, ions, or molecules are highly ordered in repeating units that make up the lattice of the crystal. Amorphous solids, although their atoms, ions, or molecules inhabit rigid and fixed locations, lack the long-range order in a crystal. (Section 13.1)
- The unit cell is the basic repeating unit that, by simple translation in three dimensions, can be used to represent the entire crystalline lattice. (Section 13.1)
- The simple cubic unit cell, the body-centered cubic unit cell, and the face-centered cubic unit cell make up most of the crystalline lattices of the metallic elements. (Section 13.1)
- Metals often adopt a closest packed structure. These structures include the hexagonal closest packed structure and the cubic closest packed structure. (Section 13.1)
- We can calculate the volume and density of a unit cell by means of simple geometry. In doing so, we assume the atoms are hard spheres. (Section 13.1)
- Band theory describes why metals are electrical and thermal conductors, shiny, malleable, and ductile. (Section 13.2)
- The valence band and the conduction band result from the nearly infinite number of atomic orbitals that overlap to form the molecular orbitals in a metal. The band gap can be used to assess whether a compound is a conductor, a semiconductor, or an insulator. (Section 13.2)
- Alloys are mixtures of two or more metals. They include the interstitial and substitutional alloys. Amalgams are special alloys of mercury. (Section 13.2)
- Ceramics, which include glasses and superconductors, are compounds that contain both ionic and covalent bonds. (Section 13.3)
- Plastics, named after their ability to be molded into shape, are polymeric materials. (Section 13.4)
- Composite materials are made of an intimate combination of two or more materials. (Section 13.5)
- "Green Chemistry" is the practice of chemistry via environmentally benign methods. (Section 13.6)

Solutions to Practice Problems

13.1. a. 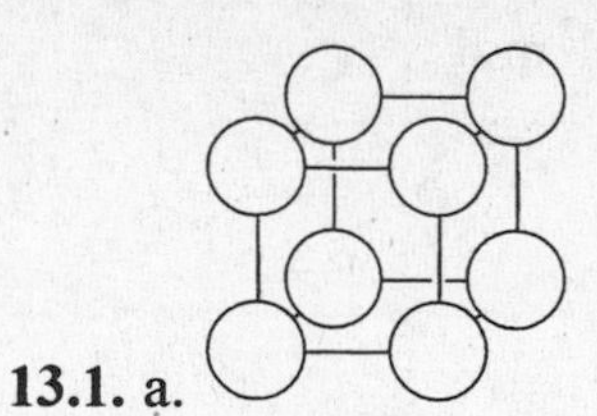 b. 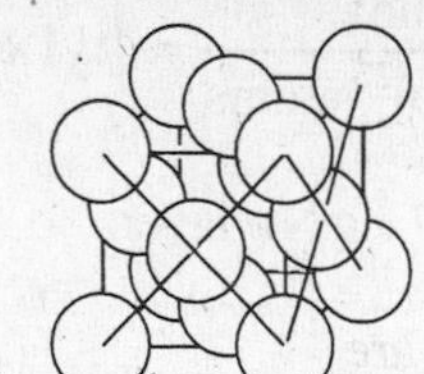  c. 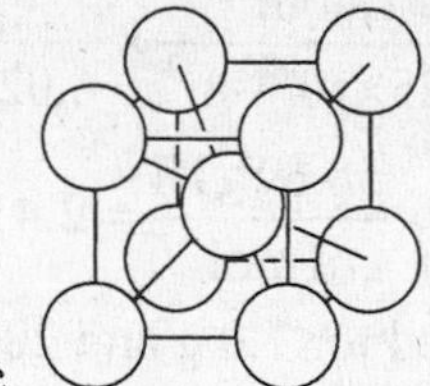

13.2. For a simple cubic cell, the atoms "touch" along the edge such that each edge is 2 atomic radii (r) in length. The volume of the cube is $(2r)^3 = 8r^3$. Only 1/8 of each corner atom actually lies within the unit cell, so there is a total of 1 atom within the cell. The volume of that 1 atom (as a sphere) is $4/3\pi r^3$. The percent of the volume occupied is given by

$$\text{Fraction ocuppied} = \frac{\text{volume of atoms}}{\text{volume of unit cell}} = \frac{4/3\pi r^3}{8r^3} = \frac{4/3\pi}{8} = \frac{\pi}{6} = 0.524$$

Hence only 52.4% of the unit cell is occupied, and 47.6% is empty space.

13.3. A fcc cell has 4 atoms per unit cell and has a volume of $V = \left(\sqrt{8}r\right)^3 = 22.63r^3$. The density is

$$\text{Density} = \frac{\text{mass}}{V} = \frac{4\,\text{atoms} \times \dfrac{207.2\,\text{g Pb}}{1\,\text{mol Pb}} \times \dfrac{1\,\text{mol Pb}}{6.022\times10^{23}\ \text{atoms Pb}}}{22.63\left(\left(175\times10^{-12}\ \text{m}\right)\times\dfrac{100\,\text{cm}}{1\,\text{m}}\right)^3} = 11.35\,\text{g/cm}^3$$

. (The actual density of lead is 11.34 g/cm^3!)

13.4. Recall that 1 eV = 1.602 × 10^{-19} J. Using $E = \dfrac{hc}{\lambda}$, we can solve for the wavelength of light that just has enough energy to promote the electron to the conduction band:

$$\lambda = \frac{hc}{E} = \frac{(6.626\times10^{-34}\,\text{J}\cdot\text{s})(3.00\times10^{8}\,\text{m/s})}{\left(0.95\,\text{eV} \times 1.602\times10^{-19}\,\text{J/eV}\right)} = 1.306\times10^{-6}\,\text{m} = 1306\,\text{nm}$$

. This wavelength falls in the infrared portion of the EM spectrum.

13.5. The rims on the car are metal; a casserole dish is probably glass or ceramic; the toilet is porcelain, which is a ceramic; and the wedding band is metal (or a metal alloy).

13.6. A floppy disk is made from a magnetic material (semiconductors and others) on a plastic disk, so it is a composite material. The nonstick coating on most pans is a fluorinated polymer—a plastic. The windshield on most cars is a layered structure of glass and plastics (sometimes antenna materials)—a composite material.

13.7. The area of one atom is a square the is 2 radii long on each side:

$$\left(2\left(125\times10^{-12}\,\text{m}\right)\right)^2 = 6.25\times10^{-20}\ \text{m}^2/\text{atom}$$

The area of the photograph is $4\text{in}\times6\text{in} = 24\text{in}^2 \times\left(\dfrac{2.54\,\text{cm}}{1\,\text{in}}\times\dfrac{1\,\text{m}}{100\,\text{cm}}\right)^2 = 0.01548\,\text{m}^2$.

The moles and mass of nickel required are as follows:

$$0.01548\,\text{m}^2 \times \frac{1\,\text{atom}}{6.25\times10^{-20}\,\text{m}^2} \times \frac{1\,\text{mol}}{6.022\times10^{23}\ \text{atoms}} = 4.11\text{x}10^{-7}\ \text{mol}$$

$$\left(4.11\text{x}10^{-7}\,\text{mol}\right)\times\frac{58.69\,\text{g Ni}}{1\,\text{mol}} = 2.41\times10^{-5}\ \text{g Ni/layer}$$

The numbers of layers required for 1.0 g and 20.0 g are

$$1.00\,\text{g}\times\frac{1\,\text{layer}}{2.41\times10^{-5}\,\text{g}} = 4.15\times10^{4}\ \text{layers}$$

$$20.0\,\text{g}\times\frac{1\,\text{layer}}{2.41\times10^{-5}\,\text{g}} = 8.30\times10^{5}\ \text{layers}$$

The area of one gold atom is $\left(2\left(144\times10^{-12}\,\text{m}\right)\right)^2 = 8.29\times10^{-20}\ \text{m}^2/\text{atom}$.

The mass of gold per layer is

$$0.01548\,\text{m}^2 \times \frac{1\,\text{atom}}{8.29\times10^{-20}\,\text{m}^2} \times \frac{1\,\text{mol}}{6.022\times10^{23}\,\text{atoms}} = 3.10\text{x}10^{-7}\,\text{mol}$$

$$\left(3.10\times10^{-7}\,\text{mol}\right) \times \frac{197.97\,\text{g Au}}{1\,\text{mol}} = 6.14\times10^{-5}\,\text{g Au / layer}$$

No, it is not possible to coat the photograph using 0.050 mg = 5.0×10^{-5} g of gold.

Solutions to Student Problems

1. A crystalline solid has a regular arrangement of the atoms or molecules, and an amorphous material has the atoms or molecules in a random arrangement.

3. A simple unit cell consists of atoms arranged within a parallelepiped (often at the corners, centered inside or on the faces and edges). The crystal lattice consists of rows and columns of the parallelepiped (unit cells) connected at the corners.

5. Choice a has the bcc unit cell. The repeating unit in the structure is

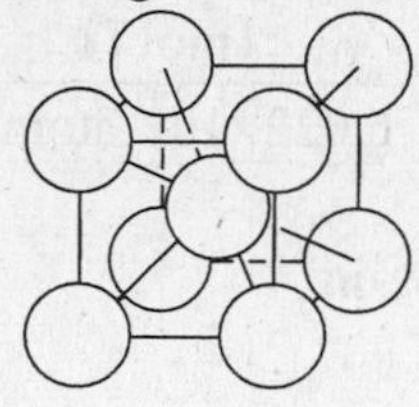

7. Each edge atom (dark circles in diagram) would be 1/4 inside the cell, and each corner atom (white circles in diagram) is 1/8 inside the cell. There are 8 corners and 12 edges, so 1/8(8) + 1/4(12) = 4 atoms.

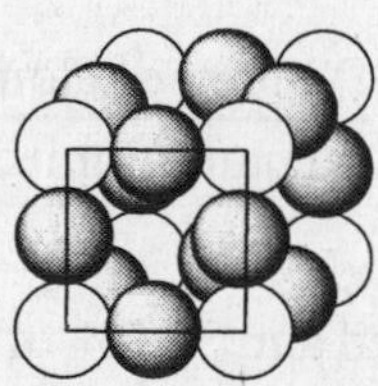

9. A (conductive as a solid) is a metal, B (low melting point) is a molecular solid, and C (electrolyte) is an ionic solid.

11. For fcc cells, the edge of the cube is $e = \sqrt{8} \times r$, so the volume is $V = e^3 = \left(\sqrt{8}r\right)^3 = 22.63r^3$.

a. For $r = 125$ pm, $V = \left(\sqrt{8}r\right)^3 = 22.63r^3 = 22.63(125\times10^{-12}\,\text{m})^3 = 4.42\times10^{-29}\,\text{m}^3$

b. For $r = 210$ pm, $V = \left(\sqrt{8}r\right)^3 = 22.63r^3 = 22.63(210\times10^{-12}\,\text{m})^3 = 2.10\times10^{-28}\,\text{m}^3$

c. For $r = 302$ pm, $V = \left(\sqrt{8}r\right)^3 = 22.63r^3 = 22.63(302\times10^{-12}\,\text{m})^3 = 6.23\times10^{-28}\,\text{m}^3$

13. $0.220\,\text{g Na} \times \frac{1\,\text{mol Na}}{22.99\,\text{g Na}} \times \frac{1\,\text{mol NaCl}}{1\,\text{mol Na}} \times \frac{58.44\,\text{g NaCl}}{1\,\text{mol NaCl}} = 0.559\,\text{g NaCl}$

15. For a fcc cell, the volume is $V = e^3 = \left(\sqrt{8}r\right)^3 = 22.63r^3$ and there are 4 atoms within the unit cell. We can use the density to find the volume per cell and then apply the above formula to get the radius:

$$\frac{1\,\text{cm}^3}{8.90\,\text{g Ni}} \times \left(\frac{1\,\text{m}}{100\,\text{cm}}\right)^3 \times \frac{58.69\,\text{g Ni}}{1\,\text{mol Ni}} \times \frac{1\,\text{mol Ni}}{6.022\times10^{23}\ \text{atom Ni}} \times \frac{4\,\text{atoms}}{\text{cell}} = 4.38\times10^{-29}\ \text{m}^3/\text{cell}$$

$$V = \left(\sqrt{8}r\right)^3 = 22.63r^3 = 4.38\times10^{-29}\ \text{m}^3$$

$$r = \left(\frac{4.38\times10^{-29}\ \text{m}^3}{22.63}\right)^{1/3} = 1.246\times10^{-10}\ \text{m} = 125\,\text{pm}$$

17. For a bcc cell, the volume is $V = e^3 = \left(\frac{4}{\sqrt{3}}r\right)^3 = 12.32r^3$ and there are 2 atoms within the unit cell. We can use the density to find the volume per cell and then apply the above formula to get the radius:

$$\frac{1\,\text{cm}^3}{4.5\,\text{g Ti}} \times \left(\frac{1\,\text{m}}{100\,\text{cm}}\right)^3 \times \frac{47.88\,\text{g Ti}}{1\,\text{mol Ti}} \times \frac{1\,\text{mol Ti}}{6.022\times10^{23}\ \text{atom Ti}} \times \frac{2\,\text{atoms}}{\text{cell}} = 3.53\times10^{-29}\ \text{m}^3/\text{cell}$$

$$V = \left(\frac{4}{\sqrt{3}}r\right)^3 = 12.32r^3 = 3.53\times10^{-29}\ \text{m}^3$$

$$r = \left(\frac{3.53\times10^{-29}\ \text{m}^3}{12.32}\right)^{1/3} = 1.42\times10^{-10}\ \text{m} = 142\,\text{pm}$$

The ratio of the volume of two atoms to the volume of the cell is the fraction occupied:

$$\text{Fraction ocuppied} = \frac{\text{volume of atoms}}{\text{volume of unit cell}} = \frac{2\left(4/3\pi r^3\right)}{\left(\frac{4}{\sqrt{3}}r\right)^3} = 0.680$$

Hence 68.0% of the cell is occupied and 32.0% of the cell is vacant!

19. Using the given density we can solve for the density in units of mol/cm^3, and using the cell data we can find the density in units of atoms/cm^3. The ratio of the atom density to the mole density should be Avogadro's number:

$$\frac{8.9\,\text{g Cu}}{1\,\text{cm}^3} \times \frac{1\,\text{mol Cu}}{63.55\,\text{g Cu}} = 0.140\,\text{mol}/\text{cm}^3$$

$$\text{fcc density} = \frac{\text{atoms}}{V} = \frac{4\,\text{atoms}}{\left(3.6\times10^{-10}\ \text{m}\times\frac{100\,\text{cm}}{1\,\text{m}}\right)^3} = 8.57\times10^{22}\ \text{atoms}/\text{cm}^3$$

$$\frac{8.57\times10^{22}\ \text{atoms}/\text{cm}^3}{0.140\,\text{mol}/\text{cm}^3} = 6.1\times10^{23}\ \text{atoms}/\text{mol}$$

Given that the edge data have only 2 significant figures, this is an acceptably close value.

21. A fcc cell has 4 atoms per unit cell and has a volume of $V = \left(\sqrt{8}r\right)^3 = 22.63r^3$. The density is

For Co: $\text{Density} = \dfrac{\text{mass}}{V} = \dfrac{4\,\text{atoms} \times \dfrac{58.93\,\text{g Co}}{1\,\text{mol Co}} \times \dfrac{1\,\text{mol Co}}{6.022 \times 10^{23}\,\text{atoms Co}}}{22.63\left(\left(125 \times 10^{-12}\,\text{m}\right) \times \dfrac{100\,\text{cm}}{1\,\text{m}}\right)^3} = 8.86\,\text{g/cm}^3$.

For Rh: $\text{Density} = \dfrac{\text{mass}}{V} = \dfrac{4\,\text{atoms} \times \dfrac{102.91\,\text{g Rh}}{1\,\text{mol Rh}} \times \dfrac{1\,\text{mol Rh}}{6.022 \times 10^{23}\,\text{atoms Rh}}}{22.63\left(\left(135 \times 10^{-12}\,\text{m}\right) \times \dfrac{100\,\text{cm}}{1\,\text{m}}\right)^3} = 12.28\,\text{g/cm}^3$

Rh, then, has the higher density.

23. Na, Al, Ca, Cr, and Bi are metals. (B and Sb are metalloids, and Se is a nonmetal.)

25. Each potassium atom has one valence electron ($4s^1$). The total number of molecular orbitals is always the same as the number of atomic orbitals used to create them; therefore, there are three molecular orbitals that result from combining the three potassium orbitals.

27. The size of the band gap increases as you go from a metal to a semiconductor to an insulator; therefore, the band gaps are metal (2.5 kJ/mol), semiconductor (85 kJ/mol), and insulator (450 kJ/mol).

29. Because the valence electrons from the metals are free to travel in the delocalized orbitals spanning the entire metal (the electron gas), the remaining positive atomic cores are all attracted to the sea of electrons that surrounds them. All cores are attracted, so it doesn't matter what the specific identity of the metal is; all metals can be accommodated in the metal alloy crystal lattice. (How the packing works out for different sizes is another matter.)

31. Consider the analogy of a display of apples that are originally all one color (green, for example). Alloys can be created in two ways: (1) Remove some of the green apples and replace each with a red apple in that same position. This alloy is a substitutional alloy. (2) Place some smaller red crabapples in the open spaces left between the green apples, without rearranging the apples or removing any apples. This represents an interstitial alloy.

33. a. Sodium has one valence electron ($3s^1$).

b. Sodium donates one orbital to the metal "molecular" orbitals. The number of molecular orbitals that are created must equal the number of atomic orbitals that were used. For an even number of orbitals, half are bonding and the other half are antibonding. (For odd numbers, one is nonbonding with the remaining orbitals equally split as before). Because each orbital can hold two electrons and sodium donates only one per atom, only half of the orbitals are filled. In other words, the lower half, all bonding, are filled. Any transition from the highest occupied molecular orbital to the lowest unoccupied molecular orbital will be a transition from a bonding orbital to an antibonding orbital.

35. For one 820-nm photon: $E(\text{J}) = \dfrac{hc}{\lambda} = \dfrac{(6.626 \times 10^{-34}\,\text{J}\cdot\text{s})(3.00 \times 10^{8}\,\text{m/s})}{\left(820\,\text{nm} \times \dfrac{10^{-9}\,\text{m}}{1\,\text{nm}}\right)} \times \dfrac{1\,\text{kJ}}{1000\,\text{J}} = 2.42 \times 10^{-19}\,\text{J}$

37. Indium, from Group IIIA, has one fewer valence electron than does silicon. For each indium in the semiconductor, there is one fewer electron (one more *positive* hole), making the semiconductor p-type.

39. Adding boron (Group IIIA) to carbon (Group IVA) results in fewer electrons than pure carbon would have, resulting in openings in the valence band and a larger gap to the conduction band than in pure carbon. There are now lower energy transitions available within the valence band. Because of these lower energy transitions, the electron absorb some visible light, leaving behind the blue color we see.

41. Ceramics are characterized by ionic bonding, which has a very large gap between the valence and conduction bands. A large amount of energy is required to bridge the gap and very few (if any) electrons are promoted into the conduction band. Without electrons in the conduction band, no (or very little) current can be conducted through the ceramic.

43. The bonding in glass is very disordered and random, so different regions within the glass experience different localized bonding strengths. Different amounts of energy will be required to overcome these interactions. Because the bonding strengths and disrupting energies required are different, the melting occurs over a range of temperatures rather than at a sharp temperature characteristic of purely crystalline compounds.

45. $$\lambda = \frac{c}{\nu} = \frac{3.00\text{x}10^8\ \text{m/s}}{250\text{x}10^6\ /\text{s}} = 1.2\,\text{m}$$

$$E = h\nu = (6.626\text{x}10^{-34}\ \text{J}\cdot\text{s})(250\text{x}10^6\ /\text{s}) = 1.66\text{x}10^{-25}\ \text{J}$$

47. The tetrahedral structure is

49. $$10.0\,\text{g HF} \times \frac{1\,\text{mol HF}}{20.01\,\text{g HF}} \times \frac{1\,\text{mol SiO}_2}{4\,\text{mol HF}} \times \frac{60.09\,\text{SiO}_2}{1\,\text{mol SiO}_2} = 7.51\,\text{g SiO}_2$$

51. (a) plastic or metal, (b) plastic, (c) metal or plastic

53. All plastics are formed by the linking of several smaller monomer units, which are then thermally molded and hardened in a particular shape. DNA is a polymer, but because it is not thermally molded and hardened, it is not a plastic.

55. (a) plastic or metal, (b) glass and metal, (c) plastic or metal, (d) plastic, glass, ceramic, or metal

57. A composite material is both strong, providing protection in crashes, and lightweight, increasing the performance or economy by lowering the weight of the car. Composites tend to be expensive to produce.

59. Most of the bonds in most polymers are sigma bonds and are strongly localized. The electrons are not free to conduct electricity. (Pi bonds, such as those in graphite, and the *p* orbitals from which they came can be delocalized much like the valence electrons in metals and can conduct electricity if they can be delocalized along the entire molecule.)

61. We can make a guess that the visor would be 1 ft^2 in area or 144 in^2. The total volume of the gold on the visor is

$$144\,\text{in}^2 \times \left(\frac{2.54\,\text{cm}}{1\,\text{in}}\right)^2 \times 215\,\text{nm} \times \frac{10^{-9}\,\text{m}}{1\,\text{nm}} \times \frac{100\,\text{cm}}{1\,\text{m}} = 0.01997\,\text{cm}^3$$

$$= 0.01997\,\text{cm}^3 \times \left(\frac{1\,\text{m}}{100\,\text{cm}}\right)^3 = 1.997 \times 10^{-8}\,\text{m}^3$$

Assuming that the gold atom is a cube with a side equal to the diameter of the atom (288 pm), the volume of each atom is $(288 \times 10^{-12}\,\text{m})^3 = 2.389 \times 10^{-29}\,\text{m}^3$. Therefore, the number of gold atoms is

$$\frac{1.997 \times 10^{-8}\,\text{m}^3}{2.389 \times 10^{-29}\,\text{m}^3/\text{atom}} = 8.4 \times 10^{20}\,\text{atoms}$$

A better approximation can be found by using the density of gold (19.4 g/cm^3, from the problem in the chapter. We can find the number of gold atoms:

$$0.01997\,\text{cm}^3 \times \frac{19.4\,\text{g Au}}{1\,\text{cm}^3} \times \frac{1\,\text{mol Au}}{196.97\,\text{g Au}} \times \frac{6.022 \times 10^{23}\,\text{atom Au}}{1\,\text{mol Au}} = 1.18 \times 10^{21}\,\text{atoms Au}$$

63. If we assumed that the silver formed a layer by arranging in simple columns and rows of squares whose sides are the silver diameter (2 × radius = 290 pm), each atom would have an area of $(290 \times 10^{-12}\,\text{m})^2 = 8.41 \times 10^{-20}\,\text{m}^2/\text{atom}$. The area covered is (2.54 cm)^2 = 6.45 cm^2. The number of atoms needed is $\dfrac{6.45\,\text{cm}^2 \times \left(\dfrac{1\,\text{m}}{100\,\text{cm}}\right)^2}{8.41 \times 10^{-20}\,\text{m}^2/\text{atom}} = 7.67 \times 10^{15}\,\text{atoms}$.

65. The mass of silver in one layer is

$$7.67 \times 10^{15}\,\text{atoms} \times \frac{1\,\text{mol}}{6.022 \times 10^{23}\,\text{atoms}} \times \frac{107.87\,\text{g Ag}}{\text{mol}} = 1.37 \times 10^{-6}\,\text{g Ag}$$

The mass in three layers is 3 $(1.37 \times 10^{-6}$ g Ag$) = 4.12 \times 10^{-6}$ g Ag.

67. Depending how the TiO_2 is obtained in the gas phase, the process could be physical deposition (if the solid TiO_2 is heated until it sublimes and deposits on the glass) or chemical-vapor deposition (if a reaction forming TiO_2 from Ti and O_2 gases deposits the film on the glass).

69. The use of the zeolites depends largely on the size of the holes or pores and cavities within the zeolite (see Chapter 8).

71. TVs are composed of many plastics, metals, and glass, and there is sometimes a large portion of lead on the tube. The plastics, metals, and lead are often recycled.

73. One type of biopolymer is the nucleic acids (DNA and RNA); a second type is proteins, consisting mostly of enzymes; the third type is some carbohydrates, including cellulose, chitin, and starches.

75. Solid zeolites used in liquid and gaseous hydrocarbons are an example of a heterogeneous (different phases) catalyst. The use of a catalyst (reusable!), reduces the use of other chemicals and the resulting waste by-products.

77. Aerogels are very light and are more heat-resistant and insulating than most ceramics. Furthermore, if the aerogel is flexible and therefore crack-resistant, it will be very easy to use and apply this highly insulating material. For example, the ceramic tiles on the NASA space shuttle are very brittle. Aerogels would be perfect for aeronautic applications.

79. Neither molecular (excluding graphite-like compounds) nor ionic solids conduct electricity, because movement of the charges is not possible in the solid state. Even when molten or dissolved, the molecular species cannot carry charge and therefore is nonconductive. The ionic solid, when molten or dissolved, does have separate positive and negative charges that are free to move, and therefore the ionic species can carry current.

81. Because so many metal atoms are in the bulk form of a metal, and each donates an orbital to the overall molecular orbital, there are an extremely large number of bonding and antibonding orbitals produced. Because the molecular orbitals are so plentiful and close together, electrons can absorb and release almost any wavelength of light, from the infrared to the ultraviolet. The absorption promotes the electron into the conduction band; the release of a photon returns the electron to the valence band. The result is the luster that we associate with metals.

83. An interesting site to visit is http://www.dnr.state.oh.us/recycling/awareness/facts/tires/goodyear.htm

85. A persistent pesticide will often work its way into the groundwater and contaminate the local water supply. It then can have deleterious affects on people either directly or through bioaccumulation in the food supply.

86. a. The structures are

H_3C, O, C, O, H, CH_2 ⟷ H_3C, O, C, $O^{\ominus}$, C, H, $\overset{\oplus}{C}H_2$

b. The polymer is

CH_3 CH_3 CH_3 CH_3 CH_3

~CH—CH₂—CH—CH₂—CH—CH₂—CH—CH₂—CH~ .

It forms from the formation of a bond between the carbons in the C=C double bonds.

c. Because each monomer unit within the polymer can react with one water to revert to the acid and alcohol, this polymer should react with water in the environment. It may find use as a biodegradable polymer.

d. The reaction is $C_4H_6O_2 + H_2O \rightarrow C_3H_4O_2 + CH_4O$. The mass of ethanol is

$$1.50\,g\,C_4H_6O_2 \times \frac{1\,mol\,C_4H_6O_2}{86.09\,g\,C_4H_6O_2} \times \frac{1\,mol\,CH_3OH}{1\,mol\,C_4H_6O_2} \times \frac{32.04\,g\,CH_3OH}{1\,mol\,CH_3OH} = 0.558\,g\,CH_3OH$$

e. The structure of the polymer formed just from acrylic acid is

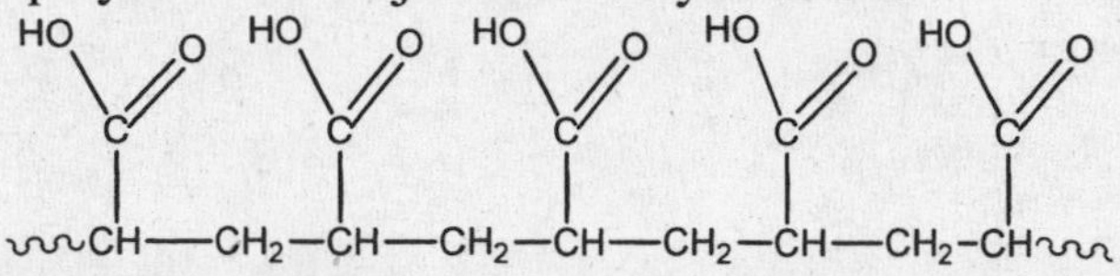

HO O HO O HO O HO O HO O (C=O on each CH)

~CH—CH₂—CH—CH₂—CH—CH₂—CH—CH₂—CH~ .

Unlike the polymer above, this polymer has acidic properties. Also, it does not react with water in an ester cleavage reaction.

Chapter 14: Thermodynamics: A Look at Why Reactions Happen

The Bottom Line

- The multiplicity of a system can be used to determine the behavior of the system. The most probable macrostate is the one that has the most contributing microstates. (Section 14.1)
- Spontaneous processes occur without outside assistance. The reverse of the spontaneous process is nonspontaneous. (Section 14.2)
- Entropy is a measure of how energy and matter can be distributed in a chemical system. Entropy is *not* disorder. (Section 14.2)
- The second law of thermodynamics says that the entropy of the universe (as we defined the term *universe* in Chapter 5) continues to increase.Any process that is spontaneous must correspond to an increase in the entropy of the universe (ΔSuniverse > 0). (Section 14.2)
- The third law of thermodynamics says that the entropy of a pure perfect crystalline material is zero. This law enables us to calculate the entropy of any compound in any state. (Section 14.4)
- The free energy, ΔG, can be calculated via the Gibbs equation ($\Delta G = \Delta H - T\Delta S$) to determine the spontaneity of a process. (Section 14.5)
- When $\Delta G = 0$, the system is at equilibrium. The forward and reverse reactions still proceed, but their rates are equal. (Section 14.6)
- Coupled reactions are used by the body tomake nonspontaneous processes become spontaneous. The use of ATP in this manner assists in the overall production of energy from glucose. (Section 14.6)

Solutions to Practice Problems

14.1. We could fill more than one page with drawings showing the different possible combinations. Instead, if we "number" the atoms 1 through 8, we can list the possible combinations of only three numbers representing the three atoms in the left jar. The number 123 would represent atoms 1, 2, and 3 in the left jar, with the other five atoms (4, 5, 6, 7, and 8) in the right jar. The order of writing the three numbers is unimportant, but we cannot use the same number twice. The possible combinations with atoms 1 and 2 in the left jar are 123, 124, 125, 126, 127, and 128. Then we exchange atom 2 for atom 3 in the jar. The possible combinations with atoms 1 and 3 in the left jar are 134, 135, 136, 137, and 138. We do not include the combination 132 because it is the same as 123, which we used earlier. Completing this process gives 56 unique combinations. Look for the pattern in the first set.

Combinations containing atom 1: (21 total)
123, 124, 125, 126, 127, 128
134, 135, 136, 137, 138
145, 146, 147, 148
156, 157, 158
167, 168
178

Combinations containing atom 2 but not atom 1: (15 total)
234, 235, 236, 237, 238, 245, 246, 247, 248, 256, 257, 258, 267, 268, 278

Combinations with atom 3, but not atom 1 or atom 2: (10 total)
345, 346, 347, 348, 356, 357, 358, 367, 368, 378

Combinations with atom 4, but not atom 1, 2, or 3: 456, 457, 458, 467, 468, 478 (6 total)

Combinations with atom 5, but not atom 1, 2, 3, or 4: 567, 568, 578 (3 total)
Combinations with atom 6, but not atom 1, 2, 3, 4, or 5: 678
There are no other unique combinations. The total number of combinations shown is 56.

We have accounted for only the combinations having atom 3 in the left jar and atom 5 in the right. We could repeat the process and find the combinations allowing three atoms in the right. These combinations are exactly the same as above. Therefore, the total number of ways to have a 3 atoms / 5 atom split is twice the above value, $2 \times 56 = 112$, to account for having 5 atoms in the left jar and 3 in the right, or vice versa. There are $2^8 = 256$ different unique combinations possible, so $112 / 256 \times 100\% = 43.8\ \%$ of the possible microstates have 3 atoms in one jar and 5 atoms in the second jar.

14.2. Reaction of potassium in water is vigorous without outside intervention: spontaneous. Leaves falling from a tree require no outside intervention: spontaneous. A puddle evaporates from the sidewalk occurs without outside intervention: spontaneous. Photosynthesis will not occur without the continuous input of solar energy: not spontaneous.

14.3. If the total entropy change is positive, the process will be spontaneous.

a. $\Delta S = \Delta S_{system} + \Delta S_{surroundings} = (-23\ \text{J/mol·K}) + (-55\ \text{J/mol·K}) = -78\ \text{J/mol·K}$; not spontaneous
b. $\Delta S = \Delta S_{system} + \Delta S_{surroundings} = (38\ \text{J/mol·K}) + (59\ \text{J/mol·K}) = 97\ \text{J/mol·K}$; spontaneous
c. $\Delta S = \Delta S_{system} + \Delta S_{surroundings} = (-84\ \text{J/mol·K}) + (132\ \text{J/mol·K}) = 48\ \text{J/mol·K}$; spontaneous

14.4. Because the process is reversible, we *can* use the enthalpy change to calculate the entropy change. (Under nonreversible conditions this is not possible.)

$$\text{For } I_2(g) \rightarrow I_2(s):\ \Delta S_{system} = \frac{\Delta H}{T} = \frac{+62.4\,\text{kJ} \times \frac{1000\,\text{J}}{1\,\text{kJ}}}{298\,\text{K}} = +209\,\text{J/K}$$

$$\text{For } H_2O(l) \rightarrow H_2O(g):\ \Delta S_{system} = \frac{\Delta H}{T} = \frac{40.7\,\text{kJ} \times \frac{1000\,\text{J}}{1\,\text{kJ}}}{298\,\text{K}} = 136\,\text{J/K}$$

14.5. The conversion of liquid water to gaseous form greatly increases the possible microstates of the system, resulting in an increase in the system entropy; ΔS = "+". In the second reaction, there are fewer moles of gas as products, so the entropy of the system decreases during the reaction; ΔS = "–".

14.6. For $CH_3OH(l) + HCl(g) \rightarrow CH_3Cl(g) + H_2O(g)$, there are more moles of gas on the product side, indicating an increase in entropy of the system:

$$1 \text{ mol } CH_3Cl(g) \times 234 \text{ J/mol·K} = 234 \text{ J/K}$$
$$1 \text{ mol } H_2O(g) \times 189 \text{ J/mol·K} = \underline{189 \text{ J/K}}$$
$$\text{Total } S^\circ_{\text{products}} = 423 \text{ J/K}$$

$$1 \text{ mol } CH_3OH(l) \times 127 \text{ J/mol·K} = 127 \text{ J/K}$$
$$1 \text{ mol } HCl(g) \times 187 \text{ J/mol·K} = \underline{187 \text{ J/K}}$$
$$\text{Total } S^\circ_{\text{reactants}} = 314 \text{ J/K}$$

$$\text{Total } S^\circ_{\text{products}} = 423 \text{ J/K}$$
$$-\text{Total } S^\circ_{\text{reactants}} = \underline{-314 \text{ J/K}}$$
$$\Delta S^\circ = 109 \text{ J/K}$$

Or, more simply, take products minus reactants:

$$\Delta S = \left(1(234\,\text{J/K}\cdot\text{mol}) + 1(189\,\text{J/K}\cdot\text{mol})\right) - \left(1(127\,\text{J/K}\cdot\text{mol}) + 1(187\,\text{J/K}\cdot\text{mol})\right)$$
$$\Delta S = 109\ \text{J/K}$$

These values agree well with the prediction.

For $3\ O_2(g) \rightarrow 2\ O_3(g)$, there is a loss of one mole of gas, so the value should be large and negative:

$$\Delta S = \left(2(239\,\text{J/K}\cdot\text{mol})\right) - \left(3(205\,\text{J/K}\cdot\text{mol})\right) = -137\ \text{J/K}$$

Again, this value is in good agreement with the prediction.

14.7. The melting of ice should not be spontaneous below the freezing point (0°C). Using the formula for the free energy change:
At −10°C :

$$\Delta G = \Delta H - T\Delta S = 6.01\text{kJ/mol} - (263\text{K})(22.0\text{J/mol}\cdot\text{K})\left(\frac{1\text{kJ}}{1000\text{J}}\right) = 0.224\text{kJ/mol}$$

At 45°C:

$$\Delta G = \Delta H - T\Delta S = 6.01\text{kJ/mol} - (318\text{K})(22.0\text{J/mol}\cdot\text{K})\left(\frac{1\text{kJ}}{1000\text{J}}\right) = -0.986\text{kJ/mol}$$

Only the reaction at 45°C has a negative, and therefore spontaneous, value for the free energy change.

14.8. For $C_6H_{12}O_6(s) + 6\ O_2(g) \rightarrow 6\ CO_2(g) + 6\ H_2O(g)$:

$$6 \text{ mol } CO_2(g) \times -394 \text{ kJ/mol} = -2364 \text{ kJ}$$
$$6 \text{ mol } H_2O(g) \times -229 \text{ kJ/mol} = \underline{-1374 \text{ kJ}}$$
$$\text{Total } \Delta G^\circ_{\text{products}} = -3738 \text{ kJ}$$

$$1 \text{ mol } C_6H_{12}O_6(s) \times -911 \text{ kJ/mol} = -911 \text{ kJ}$$
$$6 \text{ mol } O_2(g) \times 0 \text{ kJ/mol} = \underline{0 \text{ kJ}}$$
$$\text{Total } \Delta G^\circ_{\text{reac tan ts}} = -911 \text{ kJ}$$

$$\text{Total } \Delta G^\circ_{\text{products}} = -3738 \text{ kJ}$$
$$-\text{Total } \Delta G^\circ_{\text{reac tan ts}} = \underline{+911 \text{ kJ}}$$
$$\Delta G^\circ = -2827 \text{ kJ}$$

Again, more simply, we could take:

$$\Delta G^\circ = \left(6(-394\text{kJ/mol}) + 6(-229\text{kJ/mol})\right) - \left(1(-911\text{ kJ/mol}) + 6(0)\right)$$
$$\Delta G^\circ = -2827\text{kJ}$$

Because ΔG° is negative, the reaction is spontaneous.

For $CH_3CH_2CH_3(g) + 5\ O_2(g) \rightarrow 3\ CO_2(g) + 4\ H_2O(g)$

$$\Delta G^\circ = \left(3(-394\text{kJ/mol}) + 4(-229\text{kJ/mol})\right) - \left(1(-24\text{kJ/mol}) + 5(0)\right)$$
$$\Delta G^\circ = -2074\text{kJ}$$

Because ΔG° is negative, the reaction is spontaneous.

For $3\ Mg(s) + N_2(g) \rightarrow Mg_3N_2(s)$:

$$\Delta G^\circ = \left(1(-401\text{kJ/mol})\right) - \left(3(0) + 1(0)\right)$$
$$\Delta G^\circ = -401\text{kJ}$$

Because ΔG° is negative, the reaction is spontaneous.

14.9. We can use the equation for the free energy change to solve for the temperature by setting ΔG equal to zero: $\Delta G = \Delta H - T\Delta S = 0$ then $\Delta H = T\Delta S$ and $T = \dfrac{\Delta H}{\Delta S}$

For methanol: $T = \dfrac{\Delta H}{\Delta S} = \dfrac{35.27\text{kJ/mol} \times \dfrac{1000\text{J}}{1\text{kJ}}}{104.6\text{J/K}\cdot\text{mol}} = 337.2\text{K} = 64°\text{C}$

For water: $T = \dfrac{\Delta H}{\Delta S} = \dfrac{40.66\text{kJ/mol} \times \dfrac{1000\text{J}}{1\text{kJ}}}{109.0\text{J/K}\cdot\text{mol}} = 373.0\text{K} = 100°\text{C}$

Solutions to Student Problems

1. Because there are two possible orientations for the coin and there are four coins, the total number of microstates is $2^4 = 16$. Incidentally, there are only 5 macrostates represented in those 16 possible microstates.

3. If we split the four coins into two groups of two, there are 4 separate ways of splitting the coins heads up or heads down. Using numbers to represent the coins, the possible ways to get two coins are: 12, 13, 14, 23, 24, and 34. Therefore, the probability is 6 / 16 = 37.5%.

5. A macrostate represents the total properties of the system. In the previous four problems, the macrostate was the total number each of heads up and heads down. A microstate is a representation of a single way to arrange the parts of a system. In Problem 3, there were 6 individual microstates that had a macrostate of two heads and two tails.

7. There are many, many possible answers here. One possibility would be all students in one room with vacant seats alternating with students. Another possibility would be half the students in each room with vacant seats between.

9. Because there are two possible containers for each atom and there are 1000 atoms, the total number of microstates is $2^{1000} = 1.07 \times 10^{301}$. The number of microstates become very large, so the number of individual microstates that correspond with either exact or very close to exact even distributions is much larger than the number of unequally distributed microstates. Figure 14.6 shows how quickly the distribution of microstates narrows to the center even by 128 atoms.

11. The initial state involves a small volume of solid (creamer) with few microstates, and a cup of liquid (coffee) with a larger number of microstates due to the larger volume and the additional movement allowed in the liquid state. Upon mixing, the number of microstates increases dramatically as a consequence of the possibilities now of having either the liquid or the creamer in each of the available volumes. We know that there have to be many more microstates available after mixing, because the process is spontaneous, indicating a final state with more available microstates.

13. Only the evaporation process results in more microstates, because the confined liquid now has a much larger available volume as a gas. The other two process result in confining a freely moving substance (the water or the ions) into a constrained solid with fewer available microstates.

15. At room temperature combustion (a) needs the intervention of a heater or ignition source, because the reaction does not proceed without intervention at low temperatures. Hard-boiling an egg (b) required the addition of heat. Adding muscle mass (c) does not happen without a lot of work lifting weights.

17. The entropy of the universe decreases in a nonspontaneous process.

19. The greater the number of microstates, the greater the entropy in a system.

21. The number of microstates corresponding with an even dispersal of the aroma-producing molecules is much larger than the number of microstates in which the aroma remains contained in one part of the room. The spontaneous process is then the dispersal of the aroma corresponding with the greater number of microstates.

23. If the total entropy of the universe increases, the process is spontaneous (second law of thermodynamics). The conversion of the gasoline from a liquid to a vapor involves a large increase in entropy, which offsets the small decrease in entropy due to the removal of heat (and therefore available microstates) of your hand.

25. The change in entropy of the carbon dioxide is very large and positive as the molecules go from solid to gas. Because the heat to sublime the CO_2 is taken from the environment, the nearby gases are cooled, lowering their entropy, which is a negative entropy change. Because the process is spontaneous, the entropy of the universe increases. The positive entropy change of the CO_2 must be larger than the negative entropy change of the nearby gases.

27. As the reaction proceeds from left to right, the number of moles of gas is reduced. Removal of a mole of gas corresponds with a large decrease in the entropy; therefore, a reaction from right to left will cause an increase in the entropy of the system.

29. a. Positive, because both are positive
b. Need more information
c. Need more information

31. The more ways a molecule can move, the greater the multiplicity in a system.

33. As heat is added to the surroundings from the system, the molecular motion of the surroundings increases. This results in additional available microstates and an increase in the entropy of the surroundings.

35. a. Negative; solidifying the mercury dramatically reduces the number of available microstates.
b. Positive; adding heat causes the motion of the atoms to increase, increasing the available microstates.
c. Positive; the gaseous atoms have many more available microstates as a gas than as a liquid.

37. For the CO_2 the transition from a dissolved species, confined to the liquid, to a gas with a much larger available volume represents a large increase in entropy (ΔS = "+").

39. $\underline{1}\ C_3H_8(g) + \underline{5}\ O_2(g) \rightarrow \underline{3}\ CO_2(g) + \underline{4}\ H_2O(g)$; reactants: 6 moles of gas; products: 7 moles of gas. Entropy increases because of the increase in the number of moles of gas from reactants to products.

41. $C_3H_8(l) + 5\ O_2(g) \rightarrow 3\ CO_2(g) + 4\ H_2O(l)$; reactants: 5 moles of gas; products: 3 moles of gas. Entropy decreases because of the decrease in the number of moles of gas from reactants to products.

43. Because the number of microstates that become available in a gas rather than a liquid or solid is so large, any conversion to a gas greatly influences the entropy. The change in the number of microstates of liquids and solids is much lower and influences the reaction much less.

45. a. Negative, due to conversion of gases to solid
b. Positive, due to conversion of solids to liquids and gas
c. Negative, due to the loss of moles of gas

46. a. $\Delta S° = \sum nS°_{\text{products}} - \sum nS°_{\text{reactants}}$

$\Delta S° = (1(95\,J/K \cdot mol)) - (1(192\,J/K \cdot mol) + 1(187\,J/K \cdot mol)) = -284\,J/K$

b. $\Delta S° = (2(76\,J/K \cdot mol) + 1(205\,J/K \cdot mol)) - (2(70\,J/K \cdot mol)) = 217\,J/K$

c. $\Delta S° = (1(55\,J/K \cdot mol)) - (1(52\,J/K \cdot mol) + 1/2(205\,J/K \cdot mol))$

$= -99.5\,J/K = -1.00 \times 10^2\,J/K$

49. For $\underline{1}\ C_2H_5OH(l) + \underline{3}\ O_2(g) \rightarrow \underline{2}\ CO_2(g) + \underline{3}\ H_2O(g)$:

$$\Delta S° = (2(214\,J/K \cdot mol) + 3(189\,J/K \cdot mol)) - (1(161\,J/K \cdot mol) + 3(205\,J/K \cdot mol))$$

$$\Delta S° = 219\,J/K$$

51. For C(*s*, diamond) $\rightarrow$ C(*s*, graphite):

$$\Delta S° = \sum nS°_{\text{products}} - \sum nS°_{\text{reactants}}$$

$$\Delta S° = (1(6\,J/K \cdot mol)) - (1(2\,J/K \cdot mol))$$

$$\Delta S° = 4\,J/K$$

53. a. Because $\Delta G = \Delta H - T\Delta S$, we can input the signs into the equation to predict what type of temperature may allow for spontaneous reaction (negative ΔG): $\Delta G = (+) - T(+)$. In order for ΔG to be negative, T must be large so that the overall negative second term is larger than the positive first term.

b. $\Delta G = (+) - T(-)$. In this case, ΔG will always be positive; there is no temperature where this reaction could be spontaneous.

55. a. Using the equation $\Delta G = \Delta H - T\Delta S$, we can solve for the enthalpy change:

$$\Delta H = \Delta G + T\Delta S = -24.5\,kJ/mol + \left((298K)(287\,J/K \cdot mol)\left(\frac{1\,kJ}{1000\,J}\right)\right) = 61.0\,kJ/mol$$

c. $$\Delta H = \Delta G + T\Delta S = 500.0\,kJ/mol + \left((325K)(-6439\,J/K \cdot mol)\left(\frac{1\,kJ}{1000\,J}\right)\right)$$

$$= -1593\,kJ/mol = 1.59 \times 10^3\,kJ/mol$$

56. a. Using the equation $\Delta G = \Delta H - T\Delta S$, we can solve for the free energy change:

$$\Delta G = \Delta H - T\Delta S = -20.5\,kJ/mol - (298\,K)(259\,J/mol \cdot K)\left(\frac{1\,kJ}{1000\,J}\right) = -97.7\,kJ/mol$$

c. $$\Delta G = \Delta H - T\Delta S = 299\,kJ/mol - (325\,K)(-639\,J/mol \cdot K)\left(\frac{1\,kJ}{1000\,J}\right) = 507\,kJ/mol$$

57. a. Using the equation $\Delta G = \Delta H - T\Delta S$, we can solve for the free energy change:

$$\Delta S = \frac{\Delta H - \Delta G}{T} = \frac{\left((259\,\text{kJ/mol}) - (-20.5\,\text{kJ/mol})\right)\left(\frac{1000\,\text{J}}{1\,\text{kJ}}\right)}{298\,\text{K}} = 938\,\text{J/K}\cdot\text{mol}$$

c. $$\Delta S = \frac{\Delta H - \Delta G}{T} = \frac{\left((-639\,\text{kJ/mol}) - (209\,\text{kJ/mol})\right)\left(\frac{1000\,\text{J}}{1\,\text{kJ}}\right)}{35.0\,\text{K}} = -2.42\times10^{4}\,\text{J/K}\cdot\text{mol}$$

58. a. Using the equation $\Delta G = \Delta H - T\Delta S$, we can solve for the temperature:

$$T = \frac{\Delta H - \Delta G}{\Delta S} = \frac{\left((259\,\text{kJ/mol}) - (-20.5\,\text{kJ/mol})\right)\left(\frac{1000\,\text{J}}{1\,\text{kJ}}\right)}{260\,\text{J/K}\cdot\text{mol}} = 1.1\times10^{3}\,\text{K}$$

c. $$T = \frac{\Delta H - \Delta G}{\Delta S} = \frac{\left((639\,\text{kJ/mol}) - (209\,\text{kJ/mol})\right)\left(\frac{1000\,\text{J}}{1\,\text{kJ}}\right)}{560\,\text{J/K}\cdot\text{mol}} = 7.7\times10^{2}\,\text{K}$$

59. a. $\Delta G^\circ = \sum n\Delta_f G^\circ_{\text{(products)}} - \sum n\Delta_f G^\circ_{\text{(reactants)}}$

$\Delta G^\circ = \left(1(-203\,\text{kJ/mol})\right) - \left(1(-16\,\text{kJ/mol}) + 1(-95\,\text{kJ/mol})\right)$

$\Delta G^\circ = -92\,\text{kJ}$, spontaneous

b. $\Delta G^\circ = \left(2(0\,\text{kJ/mol}) + 1(0\,\text{kJ/mol})\right) - \left(2(-59\,\text{kJ/mol})\right) = 118\,\text{kJ}$, nonspontaneous

c. $\Delta G^\circ = \left(1(-228\,\text{kJ/mol})\right) - \left(1(0\,\text{kJ/mol}) + 1/2(0\,\text{kJ/mol})\right) = -228\,\text{kJ}$, spontaneous

61. $\Delta G^\circ = \sum n\Delta_f G^\circ_{\text{(products)}} - \sum n\Delta_f G^\circ_{\text{(reactants)}}$

$\Delta G^\circ = \left(2(-740\,\text{kJ/mol})\right) - \left(3(0\,\text{kJ/mol}) + 4(0\,\text{kJ/mol})\right)$

$\Delta G^\circ = -1480\,\text{kJ} = -1.48\times10^{3}\,\text{kJ}$

63. We can rearrange the two equations and add the resulting free energies to obtain the value for the reaction:

$S(s) + 3/2O_2(g) \rightarrow SO_3(g)$	$\Delta G^\circ = -371$ kJ/mol
$SO_2(g) \rightarrow S(s) + O_2(g)$	$\Delta G^\circ = +300$ kJ/mol
$SO_2(g) + ½\,O_2(g) \rightarrow SO_3(g)$	$\Delta G^\circ = -71$ kJ/mol

65. For the reaction $H_2S(aq) + O_2(g) \rightarrow 2\ H_2O(l) + 2\ S(s)$:

$$\Delta H° = \sum n\Delta_f H°_{(\text{products})} - \sum n\Delta_f H°_{(\text{reactants})}$$

$$\Delta H° = \big(2(-286\,\text{kJ/mol}) + 2(0\,\text{kJ/mol})\big) - \big(1(-40\,\text{kJ/mol}) + 1(0\,\text{kJ/mol})\big)$$

$$\Delta H° = -532\ \text{kJ/mol}$$

$$\Delta S° = \sum nS°_{\text{products}} - \sum nS°_{\text{reactants}}$$

$$\Delta S° = \big(2(70\,\text{J/K}\cdot\text{mol}) + 2(32\,\text{J/K}\cdot\text{mol})\big) - \big(1(121\,\text{J/K}\cdot\text{mol}) + 1(205\,\text{J/K}\cdot\text{mol})\big)$$

$$\Delta S° = -122\ \text{J/K}$$

$$\Delta G° = \sum n\Delta_f G°_{(\text{products})} - \sum n\Delta_f G°_{(\text{reactants})}$$

$$\Delta G° = \big(2(-237\,\text{kJ/mol}) + 2(0\,\text{kJ/mol})\big) - \big(1(-28\,\text{kJ/mol}) + 1(0\,\text{kJ/mol})\big)$$

$$\Delta G° = -446\ \text{kJ/mol}$$

67. a. We can use the equation for the free energy change to solve for the temperature by setting ΔG equal to zero: $\Delta G = \Delta H - T\Delta S = 0$ then $\Delta H = T\Delta S$ and $T = \dfrac{\Delta H}{\Delta S}$. Therefore,

$$T = \frac{\Delta H}{\Delta S} = \frac{46.4\,\text{kJ/mol} \times \dfrac{1000\,\text{J}}{1\,\text{kJ}}}{27.6\,\text{J/K}\cdot\text{mol}} = 1681\,\text{K}$$

b. $$T = \frac{\Delta H}{\Delta S} = \frac{10.6\,\text{kJ/mol} \times \dfrac{1000\,\text{J}}{1\,\text{kJ}}}{77\,\text{J/K}\cdot\text{mol}} = 138\,\text{K}$$

c. $$T = \frac{\Delta H}{\Delta S} = \frac{124\,\text{kJ/mol} \times \dfrac{1000\,\text{J}}{1\,\text{kJ}}}{295.5\,\text{J/K}\cdot\text{mol}} = 420\,\text{K}$$

69. For this reaction, $Q_p = P_{CO_2}$, so $\Delta G = \Delta G° + RT \ln Q_p = \Delta G° + RT \ln Q_p$. At equilibrium, $\Delta G = 0$, so

$$0 = \Delta G° + RT \ln P_{CO_2}$$

$$\Delta G° = -RT \ln P_{CO_2}$$

$$\ln P_{CO_2} = -\frac{\Delta G°}{RT}$$

$$P_{CO_2} = e^{-\frac{\Delta G°}{RT}} = e^{-\left(\frac{131\,\text{kJ/mol} \times \frac{1000\,\text{J}}{1\,\text{kJ}}}{(8.3145\,\text{J/mol}\cdot\text{K})(298\,\text{K})}\right)} = 1.09 \times 10^{-23}\ \text{atm}$$

71. At equilibrium, $\Delta G = \Delta H - T\Delta S = 0$ then $\Delta H = T\Delta S$ and $\Delta S = \dfrac{\Delta H}{T}$, so

$$\Delta S = \frac{40.7\,\text{kJ/mol} \times \dfrac{1000\,\text{J}}{1\,\text{kJ}}}{373\,\text{K}} = 109\,\text{J/K}\cdot\text{mol}$$

The total entropy change will be

$$1000.0\,\text{g}\,H_2O \times \frac{1\,\text{mol}\,H_2O}{18.02\,\text{g}\,H_2O} \times \frac{109\,\text{J/K}}{1\,\text{mol}\,H_2O} = 6049\,\text{J/K} = 6.05\times10^3\,\text{J/K}$$

73. For a reversible process or one at equilibrium:

$$\Delta G = \Delta H - T\Delta S = 0 \text{ then } \Delta H = T\Delta S \text{ and } T = \frac{\Delta H}{\Delta S} = \frac{10.50\,\text{kJ/mol} \times \dfrac{1000\,\text{J}}{1\,\text{kJ}}}{9.6\,\text{J/K}\cdot\text{mol}} = 1094\,\text{K}$$

75. a. $\Delta G° = \sum n\Delta_f G^o_{(\text{products})} - \sum n\Delta_f G^o_{(\text{reactants})}$

$$\Delta G° = \big(1(-394\,\text{kJ/mol}) + 2(-229\,\text{kJ/mol})\big) - \big(1(-50\,\text{kJ/mol}) + 2(0\,\text{kJ/mol})\big)$$

$$\Delta G° = -802\text{ kJ/mol}$$

b. For this reaction, $Q_p = \dfrac{P_{CO_2}\left(P_{H_2O}\right)^2}{P_{CH_4}\left(P_{O_2}\right)^2} = \dfrac{(1)(1)^2}{(1)(0.2)^2} = 25$; therefore, the free energy change is

$$\Delta G = \Delta G° + RT\ln Q_p = -802\,\text{kJ} + \left(8.3145\,\text{J/mol}\cdot\text{K} \times \frac{1\,\text{kJ}}{1000\,\text{J}}\right)(298\,\text{K})\ln 25$$

$$\Delta G = -794\,\text{kJ}$$

c. The negative value shows that the reaction is spontaneous at room temperature but says nothing about the rate at which the reaction proceeds. The additional heat from the spark or flame is necessary to get the reaction to happen at an appreciable rate.

77. The balanced reaction is 2 $C_4H_{10}(g)$ + 13 $O_2(g) \rightarrow$ 8 $CO_2(g)$ + 10 $H_2O(g)$. As the reaction proceeds, 15 moles of gas are converted to 18 moles of gas. The increase of 3 moles of gas will cause an increase in entropy.

79. In order to predict spontaneity, we must determine whether the entropy of the universe increases. To complete this calculation, we must consider both the system and the surroundings. In other words, we need to account for the entire universe. Using Gibb's equation, we need only consider variables related to the system.

81. If $\Delta G = 0$ at 25°C, then the value of Q can be obtained from the value of $\Delta G°$. Having $\Delta G = 0$ means that the reaction is neither spontaneous nor nonspontaneous (that is, it is under equilibrium conditions). Under this condition, Q becomes an equilibrium constant.

83. a. The balanced reaction is 1 $C_6H_{12}O_6(s)$ + 6 $O_2(g) \rightarrow$ 6 $CO_2(g)$ + 6 $H_2O(l)$. There are equal numbers of moles of gas on each side, although CO_2 will have a higher entropy than diatomic O_2. Additionally, there are 6 moles of liquid H_2O versus a single mole of solid. There should be an increase of entropy on the order of a few hundred J/K·mol.

b. Assuming that the values for the fructose and glucose are the same:

$$\Delta H° = \big(6(-394\,\text{kJ/mol}) + 6(-286\,\text{kJ/mol})\big) - \big(1(-1268\,\text{kJ/mol}) + 6(0)\big)$$

$$\Delta H° = -2812\,\text{kJ}$$

$$\Delta S° = (6(214\,\text{kJ/mol}) + 6(70\,\text{kJ/mol})) - (1(212\,\text{kJ/mol}) + 6(205))$$
$$\Delta S° = 262\,\text{kJ}$$
$$\Delta G° = (6(-394\,\text{kJ/mol}) + 6(-237\,\text{kJ/mol})) - (1(-911\,\text{kJ/mol}) + 6(0))$$
$$\Delta G° = -2875\,\text{kJ}$$

c. With water as a gas:

$$\Delta H° = (6(-394\,\text{kJ/mol}) + 6(-242\,\text{kJ/mol})) - (1(-1268\,\text{kJ/mol}) + 6(0))$$
$$\Delta H° = -2548\,\text{kJ}$$
$$\Delta S° = (6(214\,\text{kJ/mol}) + 6(189\,\text{kJ/mol})) - (1(212\,\text{kJ/mol}) + 6(205))$$
$$\Delta S° = 976\,\text{kJ}$$
$$\Delta G° = (6(-394\,\text{kJ/mol}) + 6(-229\,\text{kJ/mol})) - (1(-911\,\text{kJ/mol}) + 6(0))$$
$$\Delta G° = -2827\,\text{kJ}$$

d. To solve for the volume of CO_2, we first use the reaction stoichiometry to solve for the moles and then use the ideal gas law to find the volume:

$$5.0\,\text{g}\,C_6H_{12}O_6 \times \frac{1\,\text{mol}\,C_6H_{12}O_6}{180.16\,\text{g}\,C_6H_{12}O_6} \times \frac{6\,\text{mol}\,CO_2}{1\,\text{mol}\,C_6H_{12}O_6} = 0.1665\,\text{mol}\,CO_2$$

$$V = \frac{nRT}{P} = \frac{(0.1665\,\text{mol})(0.08206\,\text{L}\cdot\text{atm/mol}\cdot\text{K})(298\,\text{K})}{1\,\text{atm}} = 4.07\,\text{L}\,CO_2$$

A living organism removes the gas via respiration.

e. $5.0\,\text{g}\,C_6H_{12}O_6 \times \frac{1\,\text{mol}\,C_6H_{12}O_6}{180.16\,\text{g}\,C_6H_{12}O_6} \times \frac{1\,\text{mol reaction}}{1\,\text{mol}\,C_6H_{12}O_6} \times \frac{-2548\,\text{kJ}}{1\,\text{mol reaction}} = -71\,\text{kJ}$. The negative sign indicates that heat is released. (At room temperature, the water will condense to a liquid, so we used the combustion reaction that includes liquid water. If you assume the water remains as a gas, the answer becomes −71 kJ.)

f. Combining the two reactions, after reversing the second, yields

$C_6H_{12}O_6(aq) + ATP(aq) \rightarrow ADP(aq) + C_6H_{11}O_9P^{2-}(aq)$	$\Delta G = -16.7$ kJ/mol.
$ADP(aq) + HPO_4^{2-}(aq) \rightarrow ATP(aq) + H_2O(l)$	$\Delta G = +30.5$ kJ/mol
$C_6H_{12}O_6(aq) + HPO_4^{2-}(aq) \rightarrow C_6H_{11}O_9P^{2-}(aq) + H_2O(l)$	$\Delta G = 13.8$ kJ/mol

g. If there were no glucose directly available to the organism, it would have to rely on other sources of energy, such as fructose. Fructose is naturally available in many fruits and vegetables; it also is a product in the hydrolysis of table sugar (sucrose).

Chapter 15: Chemical Kinetics

The Bottom Line

- Reaction rates determine the speed at which a reaction progresses but do not reveal anything about the extent to which they produce a product. (Section 15.1)
- The average rate can be measured if you know the initial and final concentrations over a particular time period. (Section 15.1)
- The instantaneous rate is the rate at a given point in the reaction. It can be determined by measuring the concentrations at points when the time difference approaches zero, or it can be measured by determining the slope of a line tangent to a plot of the rate versus time. (Section 15.1)
- The initial rate of a reaction is the instantaneous rate as the reaction starts. (Section 15.1)
- The half-life of a reaction is the time required for the concentration of a reaction to reach 50% of the initial value. (Section 15.3)
- Complex reaction orders can often be reduced to pseudo-first-order reactions by keeping the concentration of one of the reactants large. (Section 15.3)
- The rate law is experimentally determined using the method of initial rates or the method of graphical analysis. (Section 15.4)
- Transition state theory, which is based on collision theory, describes the energy of a reaction during the course of a reaction. A plot of the reaction coordinate versus the energy can give meaningful information, such as the activation energy, the presence of any intermediates, the effect of a catalyst, and the enthalpy for the forward and reverse process. (Section 15.6)
- Catalysts speed up a reaction without being consumed in the reaction. (Section 15.7)

Solutions to Practice Problems

15.1. When writing an equation relating the individual rates of change of the products or reactants, we must multiply reactants by a negative sign to ensure that the rate of reaction remains positive. Each species's rate is also reduced by the inverse of its stoichiometric coefficient:

$$\text{Rate} = -\frac{1}{2}\frac{\Delta[\text{HI}]}{\Delta t} = \frac{\Delta[\text{H}_2]}{\Delta t} = \frac{\Delta[\text{I}_2]}{\Delta t}$$

15.2. For the Haber process, we can write a rate expression using H_2:

$$\text{Rate} = -\frac{1}{3}\frac{\Delta[\text{H}_2]}{\Delta t} = -\frac{1}{3}\frac{(0.258\,M - 0.355\,M)}{(83.3\,\text{s} - 12.5\,\text{s})} = 4.6\times10^{-4}\ M/\text{s}$$

15.3. This reaction has only a single reactant (HI), and the power on [HI] in the equation is 2. Therefore, the reaction is second-order with respect to the HI concentration and second-order overall.

15.4. The decay of radioactive materials is a first-order process. In order to relate time to amount, we must use the integrated form of the first-order rate law. Using 875 kg for the initial amount and 87.5 kg for the final amount, the time required is

$$\ln\left(\frac{A_t}{A_0}\right) = -kt$$

$$t = -\frac{\ln\left(\frac{A_t}{A_0}\right)}{k} = -\frac{\ln\left(\frac{87.5}{875}\right)}{2.874\times10^{-5}\,/\,\text{yr}} = 8.01\times10^{4}\ \text{yr}$$

Half of the original 875 kg is 437.5 kg. The time required for plutonium-239 to decay to half the initial amount is

$$t = -\frac{\ln\left(\frac{A_t}{A_0}\right)}{k} = -\frac{\ln\left(\frac{437.5}{875}\right)}{2.874\times10^{-5}\,/\,\text{yr}} = 2.41\times10^{4}\ \text{yr}$$

15.5. Alachlor has a soil half-life of 15 days. A drop to 25% of the initial value is two half-lives: 30 days. A drop to 6.25% is four half-lives: 60 days. A value of 0.001% is not a simple multiple of 1/2, so we must use the integrated first-order rate law to solve. It is necessary to use the half-life to find the rate constant for the equation:

$$k = \frac{0.693}{t_{1/2}} = \frac{0.693}{15\,\text{day}} = 0.046_2\,/\,\text{day}$$

$$t = -\frac{\ln\left(\frac{A_t}{A_0}\right)}{k} = -\frac{\ln\left(\frac{0.001\%}{100\%}\right)}{0.046_2\,/\,\text{day}} = 249\,\text{days} = 2.5\times10^{2}\,\text{days}$$

15.6. The general rate law for the reaction of carbon monoxide with hemoglobin is

$$\text{Rate} = k[\text{CO}]^m[\text{Hb}]^n$$

We can use ratios of this equation for separate experiments to find the exponents in the equation, along with the rate constant. In experiments 1 and 2, the concentration of the CO remains constant, so that part of the equation will cancel and allow us to solve for n:

$$\frac{\text{Rate}\#2}{\text{Rate}\#1} = \frac{1.24\times10^{-6}\,M/\text{s}}{0.619\times10^{-6}\,M/\text{s}} = \frac{k[\text{CO}]_2^{\,m}[\text{Hb}]_2^{\,n}}{k[\text{CO}]_1^{\,m}[\text{Hb}]_1^{\,n}} = \frac{[\text{Hb}]_2^{\,n}}{[\text{Hb}]_1^{\,n}} = \left(\frac{[\text{Hb}]_2}{[\text{Hb}]_1}\right)^n = \left(\frac{4.42\times10^{-6}\,\text{M}}{2.21\times10^{-6}\,\text{M}}\right)^n$$

$$2 = 2^n$$

$$n = 1$$

We can use experiments 3 and 2 to find m:

$$\frac{\text{Rate}\#3}{\text{Rate}\#2} = \frac{3.71\times10^{-6}\,M/\text{s}}{1.24\times10^{-6}\,M/\text{s}} = \frac{k[\text{CO}]_3^{\,m}[\text{Hb}]_3^{\,n}}{k[\text{CO}]_2^{\,m}[\text{Hb}]_2^{\,n}} = \frac{[\text{CO}]_3^{\,m}}{[\text{CO}]_2^{\,m}} = \left(\frac{[\text{CO}]_3}{[\text{CO}]_2}\right)^m = \left(\frac{3.00\times10^{-6}\,\text{M}}{1.00\times10^{-6}\,\text{M}}\right)^m$$

$$3 = 3^m$$

$$m = 1$$

Using experiment 1 and the exponents, we can solve for k:

$$\text{Rate} = k[\text{CO}]_1^1[\text{Hb}]_1^1$$

$$k = \frac{\text{Rate}}{[\text{CO}]_1^1[\text{Hb}]_1^1} = \frac{0.619\times10^{-6}\ M/\text{s}}{(1.00\times10^{-6}\ M)(2.21\times10^{-6}\ M)} = 2.80\times10^5\ M^{-1}\text{s}^{-1}$$

Therefore, the overall rate law is $\text{Rate} = (2.80\times10^5\ M^{-1}\text{s}^{-1})[\text{CO}]^1[\text{Hb}]^1$, which is second-order overall.

15.7. We create zero-, first-, and second-order reaction plots with the data. Included in the plots are the equations of the best-fit lines. R is called the coefficient of determination, which indicates a closer fit to a line as the value of R^2 approaches 1. The second-order plot has the best fit to a line both visually and with an R^2 value closest to 1. Therefore, the reaction is second-order.

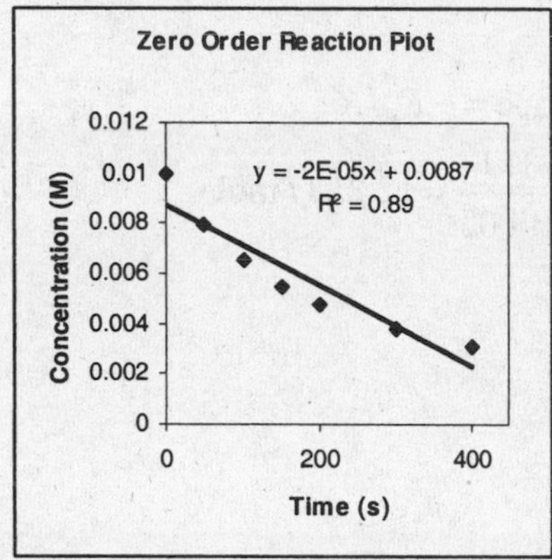

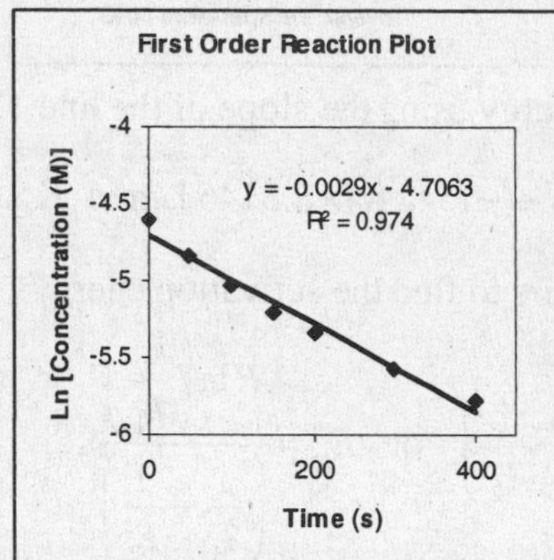

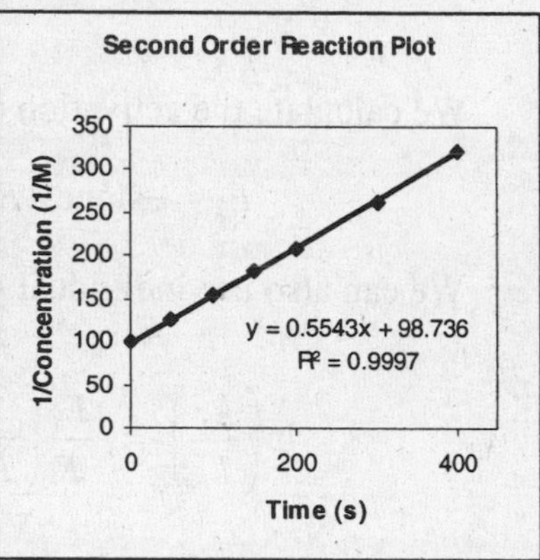

15.8. The overall reaction is the sum of the two steps: $2\ \text{NO}(g) + \text{O}_2(g) \rightarrow 2\ \text{NO}_2(g)$. An intermediate appears in the mechanism, but not in the net stoichiometric reaction. Within the mechanistic steps, it is produced during one step and destroyed in a later step. The intermediate in this reaction is $\text{NO}_3(g)$. We use the rate law for the slow step as the rate law of the mechanism: Rate = k_2 [NO_3][NO]. However, $\text{NO}_3(g)$ is an intermediate and cannot remain in the rate law. The first step will reach an equilibrium state where the forward and reverse rates are equal. We can then set the rate equations for the forward reaction (Rate = k_1 [NO][O_2]) and the reverse reaction (Rate = k_{-1} [NO_3]) equal to one another to solve for the concentration of NO_3:

$$k_1[\text{NO}][\text{O}_2] = k_{-1}[\text{NO}_3]$$

$$[\text{NO}_3] = \frac{k_1}{k_{-1}}[\text{NO}][\text{O}_2]$$

This solution can be placed into the original rate law:

$$\text{Rate} = k_2[\text{NO}_3][\text{NO}]$$

$$\text{Rate} = k_2\left(\frac{k_1}{k_{-1}}[\text{NO}][\text{O}_2]\right)[\text{NO}]$$

$$\text{Rate} = \frac{k_1}{k_{-1}}k_2[\text{NO}]^2[\text{O}_2] = k[\text{NO}]^2[\text{O}_2]$$

15.9. To find the activation energy, we create an Arrhenius plot (ln k versus 1/T). Included in the plot is the equation of the best-fit line. R is called the coefficient of determination, which indicates a closer fit to a line as the value of R^2 approaches 1.

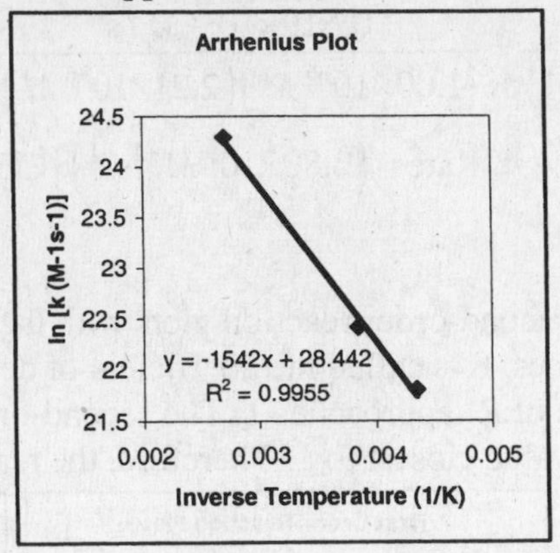

We calculate the activation energy using the slope of the line: slope $= -E_a/R$.

$$E_a = -\text{slope} \times R = -(-1542\,\text{K}) \times 8.3145\,\text{J/mol}\cdot\text{K} \times \frac{1\,\text{kJ}}{1000\,\text{J}} = 12.8\,\text{kJ/mol}$$

We can also use individual pairs to find the activation energy:

$$\ln\left(\frac{k_2}{k_1}\right) = \frac{E_a}{R}\left(\frac{1}{T_1} - \frac{1}{T_2}\right) \quad \text{or} \quad E_a = \frac{R\ln\left(\frac{k_2}{k_1}\right)}{\left(\frac{1}{T_1} - \frac{1}{T_2}\right)}$$

$$E_a = \frac{(8.3145\ \text{J}\cdot\text{mol}^{-1}\text{K}^{-1})\ln\left(\frac{2.95\times10^9}{5.42\times10^9}\right)}{\left(\frac{1}{260\ \text{K}} - \frac{1}{230\ \text{K}}\right)} = 10082\,\text{J/mol} = 10.1\,\text{kJ/mol}$$

$$\text{or } E_a = \frac{(8.3145\ \text{J}\cdot\text{mol}^{-1}\text{K}^{-1})\ln\left(\frac{2.95\times10^9}{35.5\times10^9}\right)}{\left(\frac{1}{369\ \text{K}} - \frac{1}{230\ \text{K}}\right)} = 12629\,\text{J/mol} = 12.6\,\text{kJ/mol}$$

$$\text{or } E_a = \frac{(8.3145\ \text{J}\cdot\text{mol}^{-1}\text{K}^{-1})\ln\left(\frac{5.42\times10^9}{35.5\times10^9}\right)}{\left(\frac{1}{369\ \text{K}} - \frac{1}{260\ \text{K}}\right)} = 13754\,\text{J/mol} = 13.8\,\text{kJ/mol}$$

Small variations and rounding errors can lead to slightly different values for the activation energy. When more than two data pairs are available, the preferred method is to create a graph as shown above.

Solutions to Student Problems

1. a. The total distance run by the runner during the race would represent the extent of reaction.
b. The average speed of the runner (total distance divided by total time) would be the same as the average rate of reaction.
c. The exact speed the runner is moving at a particular moment (perhaps determined using a radar gun) would be the same as the instantaneous rate.
d. How quickly the runner started the race (also determined with the radar gun) would be the initial rate.

3. The concentration consumed is 1.00×10^{-4} mol / 4.00 L = 2.50×10^{-5} M. The elapsed time is 17 minutes. Because C_4H_8 is a reactant and has a stoichiometric coefficient of 1, the average rate of the reaction can be written $\text{Rate} = -\frac{\Delta[C_4H_8]}{\Delta t} = \frac{2.50\times10^{-5}\,M}{17\,\text{min}\times\frac{60\,\text{s}}{1\,\text{min}}} = 2.5\times10^{-8}\,M/\text{s}$.

For many reactions, the initial rate of the reaction is fastest and the rate of the reaction slows as the reaction proceeds. In this case, the initial (and instantaneous) rate and other rates recorded near the start of the reaction would be larger than the average rate.

5. Because C_2H_4 is produced at twice the rate (according to the reaction stoichiometry), we must use a factor of ½ to equate the two rates of change. Additionally, because C_4H_8 is consumed, we must include a negative sign with that rate to ensure that the overall reaction rate is positive.

$$\text{Rate} = -\frac{\Delta[C_4H_8]}{\Delta t} = \frac{1}{2}\frac{\Delta[C_2H_4]}{\Delta t}$$

7. a. For the reaction given, we would write the rate of the reaction as

$$\text{Rate} = -\frac{\Delta[A]}{\Delta t} = -\frac{[A]_2 - [A]_1}{t_2 - t_1} = -\frac{0.300\,M - 0.350\,M}{100\,\text{s} - 0\,\text{s}} = 5\times10^{-4}\,M/\text{s}$$

b. $$\text{Rate} = -\frac{0.350\,M - 1.522\,M}{(15\,\text{min} - 0\,\text{min})\times\frac{60\,\text{s}}{1\,\text{min}}} = 1.3\times10^{-3}\,M/\text{s}$$

c. $$\text{Rate} = -\frac{0.010\,M - 0.050\,M}{(399\,\text{days} - 0\,\text{days})\times\frac{24\,\text{h}}{1\,\text{day}}\times\frac{60\,\text{min}}{1\,\text{h}}\times\frac{60\,\text{s}}{1\,\text{min}}} = 1.2\times10^{-9}\,M/\text{s}$$

d. $$\text{Rate} = -\frac{0.140\,M - 0.280\,M}{\left(2.5\,\text{h}\times\frac{60\,\text{min}}{1\,\text{h}} - 35\,\text{min}\right)\times\frac{60\,\text{s}}{1\,\text{min}}} = 2.0\times10^{-5}\,M/\text{s}$$

9. a. For the reaction, we can write the rate of the reaction in terms of each reactant and product:

$$\text{Rate} = -\frac{1}{2}\frac{\Delta[A]}{\Delta t} = -\frac{1}{3}\frac{\Delta[B]}{\Delta t} = \frac{\Delta[C]}{\Delta t} = \frac{1}{2}\frac{\Delta[D]}{\Delta t}$$

Solving for B in terms of A:

$$-\frac{1}{2}\frac{\Delta[A]}{\Delta t} = -\frac{1}{3}\frac{\Delta[B]}{\Delta t}$$

$$\frac{\Delta[B]}{\Delta t} = \frac{3}{2}\frac{\Delta[A]}{\Delta t} = \frac{3}{2}\left(-4.5\times10^{-2}\,M/\text{s}\right) = -6.8\times10^{-2}\,M/\text{s}$$

b. Solving for C in terms of A:

$$-\frac{1}{2}\frac{\Delta[A]}{\Delta t} = \frac{\Delta[C]}{\Delta t}$$

$$\frac{\Delta[C]}{\Delta t} = -\frac{1}{2}\frac{\Delta[A]}{\Delta t} = -\frac{1}{2}\left(-4.5\times10^{-2}\,M/\text{s}\right) = 2.2\times10^{-2}\,M/\text{s}$$

c. Solving for D in terms of A:

$$-\frac{1}{2}\frac{\Delta[A]}{\Delta t} = \frac{1}{2}\frac{\Delta[D]}{\Delta t}$$

$$\frac{\Delta[D]}{\Delta t} = -\frac{\Delta[A]}{\Delta t} = -\left(-4.5\times10^{-2}\,M/\text{s}\right) = 4.5\times10^{-2}\,M/\text{s}$$

11. The plot of the data is shown in the figure at the right.

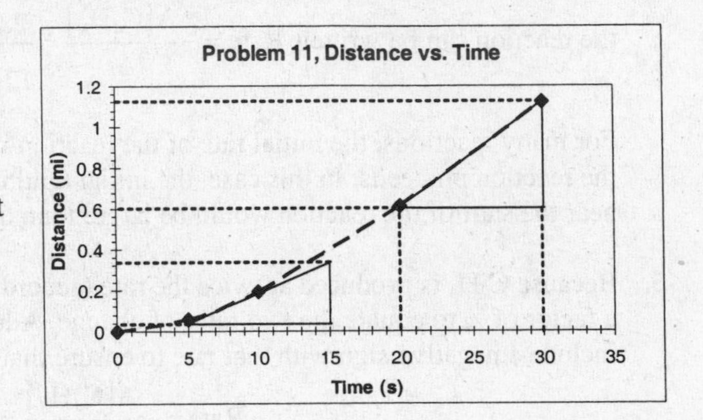

a. The initial speed (at 0 s) is 0 mph.

b. To find the speed from the graph, we must determine the slope of the line at that point. By drawing a tangent line at each point, we can estimate the speed. At 8 s, the speed is about

$$\text{Speed} = \frac{0.35\,\text{mi} - 0.04\,\text{mi}}{\left(10\,\text{s}\times\frac{1\,\text{h}}{3600\,\text{s}}\right)} = 112\,\text{mph}$$

c. At 25 s, the speed is about

$$\text{Speed} = \frac{1.12\,\text{mi} - 0.60\,\text{mi}}{\left(10\,\text{s}\times\frac{1\,\text{h}}{3600\,\text{s}}\right)} = 187\,\text{mph}$$

d. The average speed is calculated from the overall distance over the total time:

$$\text{Average speed} = \frac{1.12\,\text{mi} - 0.00\,\text{mi}}{\left(30\,\text{s}\times\frac{1\,\text{h}}{3600\,\text{s}}\right)} = 134\,\text{mph}$$

e. This plot is the opposite of most reactions, wherein the product is produced rapidly at first and then levels off as the reaction slows down. Here the distance covered is lowest initially and gets larger as the racer speeds up.

13. a. Because the stoichiometry of the reaction shows that there are two H_2O produced for each O_2, the rate of appearance of water is twice the rate of appearance of oxygen.

b. $\text{Rate} = -\frac{1}{2}\frac{\Delta[H_2O_2]}{\Delta t} = \frac{1}{1}\frac{\Delta[O_2]}{\Delta t} = \frac{1}{2}\frac{\Delta[H_2O]}{\Delta t}$

15. Because the stoichiometry has O_2 and O_3 in a 1:1 ratio, the ozone curve will be a mirror image of the oxygen curve:

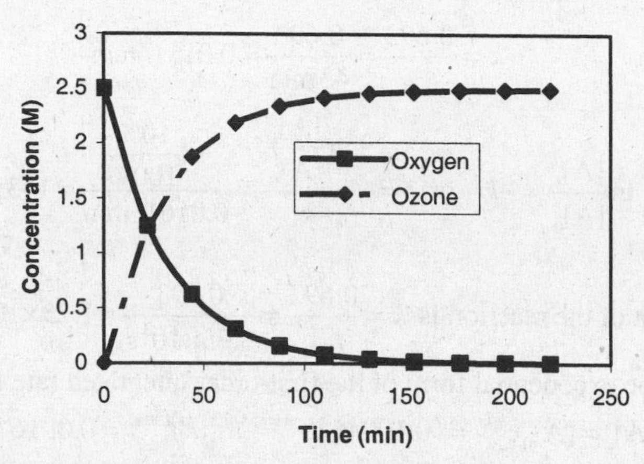

17. a. Increasing the temperature will increase the rate of reaction because (1) it increases the number of collisions and (2) it increases the speed of the molecules, which increases the energy in those collisions, making reaction more likely.

b. Will not increase the rate: see part a above.

c. Increasing the initial concentration of the reactants will make collisions between those reactants more likely, thereby increasing the rate of the reaction.

d. Will not increase the rate; lowering the concentration of the reactants will make collisions between those reactants less likely, thereby decreasing the rate of the reaction.

19. a. The reaction is third order overall, first-order in $[H^+]$, and second-order in [Pest].

b. k is the rate constant of the reaction.

c. Doubling the reaction will quadruple the rate of the reaction.

d. Changing any of the concentrations has no effect on the rate constant. It is a constant (at a given temperature)!

21. The rate law is Rate = k [A], so Rate = $(3.95\times10^{-4}\ /\text{s})\ (0.509\ M) = 2.01\times10^{-4}\ M/\text{s}$.

23. Rate laws are experimentally determined, so there is no way (without additional information) to know what order the reaction is. Further, we cannot merely look at a reaction to determine the order of the reaction.

25. a. We can use the special relation between half-life and rate constant for a first-order reaction to solve these problems: $k = \dfrac{0.693}{t_{1/2}} = \dfrac{0.693}{1920\,\text{s}} = 3.61\times10^{-4}\ /\text{s}$

b. $k = \dfrac{0.693}{t_{1/2}} = \dfrac{0.693}{525\,\text{min}} = 1.32\times10^{-3}\ /\text{min}$

c. $k = \dfrac{0.693}{t_{1/2}} = \dfrac{0.693}{1.25\times10^{9}\ \text{yr}} = 5.55\times10^{-10}\ /\text{yr}$

d. $k = \dfrac{0.693}{t_{1/2}} = \dfrac{0.693}{5730\,\text{yr}} = 1.21\times10^{-4}\ /\text{yr}$

27. a. A drop to 50% is one half life: 43 min.

b. A drop to 25% is two half-lives: 86 min.

c. A drop to 10% is not a simple multiple of half-lives. To solve this problem, we first convert the half-life to the rate constant and then use the integrated first-order rate to solve for the time.

$$k = \frac{0.693}{t_{1/2}} = \frac{0.693}{43\,\text{min}} = 0.0161/\text{min}$$

$$\ln\frac{[\text{A}]_t}{[\text{A}]_0} = -kt \text{ or } t = -\frac{\ln\frac{[\text{A}]_t}{[\text{A}]_0}}{k} = \frac{\ln\frac{10\%}{100\%}}{0.0161/\text{min}} = 143\,\text{min}$$

29. a. The rate constant of the reaction is $k = \frac{0.693}{t_{1/2}} = \frac{0.693}{3.95\times10^3\,\text{s}} = 1.75\times10^{-4}/\text{s}$. The concentration is found using the exponential form of the first-order integrated rate law.

$$[\text{A}]_t = [\text{A}]_0 e^{-kt} = (0.100M)e^{-(1.754\times10^{-3}/\text{s})(50\text{s})} = 0.0916M$$

b. $[\text{A}]_t = [\text{A}]_0 e^{-kt} = (0.100M)e^{-(1.754\times10^{-3}/\text{s})(100\text{s})} = 0.0839M$

c. $[\text{A}]_t = [\text{A}]_0 e^{-kt} = (0.200M)e^{-(1.754\times10^{-4}/\text{s})(50\text{s})} = 0.183M$

d. $[\text{A}]_t = [\text{A}]_0 e^{-kt} = (0.200M)e^{-(1.754\times10^{-4}/\text{s})(20\text{s})} = 0.193M$

31. The rate constant of the reaction is $k = \frac{0.693}{t_{1/2}} = \frac{0.693}{1.18\,\text{day}} = 0.587/\text{day}$. Using the exponential form of the first-order integrated rate law, we can find the initial concentration:

$$[\text{A}]_t = [\text{A}]_0 e^{-kt} \quad \text{or} \quad [\text{A}]_0 = \frac{[\text{A}]_t}{e^{-kt}} = \frac{0.0388M}{e^{-(0.587/\text{day})(2.75\text{day})}} = 0.195M$$

33. a. $[\text{A}]_0 = 0.100\ M$ cannot be correct because the initial concentration cannot be lower than the later concentration. (Alternatively, one could say that $[\text{A}]_t$ is incorrect because it should be lower than the initial concentration.)

b. The time values here are reversed; the initial time should be lower than the later time.

35. Using the zero-order integrated rate law, $[\text{A}]_t = [\text{A}]_0 - \text{kt}$, we can solve for the rate constant and half-life: $k = -\frac{[\text{A}]_t - [\text{A}]_0}{t} = -\frac{0.0333M - 0.0150M}{450.0\text{s}} = 4.07\times10^{-5}\ M\cdot\text{s}^{-1}$

$$t_{1/2} = \frac{[\text{A}]_0}{2k} = \frac{0.0333M}{2(4.07\times10^{-5}\ M\cdot\text{s}^{-1})} = 409\text{s for } [\text{A}]_0 = 0.0333M$$

37. The zero-order rate constant can be found using the half-life formula:

$$t_{1/2} = \frac{[\text{A}]_0}{2k} \quad \text{or} \quad k = \frac{[\text{A}]_0}{2t_{1/2}} = \frac{0.100M}{2(3.55\text{h})} = 0.0141M\cdot\text{h}^{-1}$$

39. Using the second-order integrated rate law, we can find the rate constant:

$$\frac{1}{[\text{A}]_t} = \frac{1}{[\text{A}]_0} + kt \quad \text{so} \quad k = \frac{\frac{1}{[\text{A}]_t} - \frac{1}{[\text{A}]_0}}{t} = \frac{\frac{1}{2.00\times10^{-4}\ M} - \frac{1}{2.00\times10^{-3}\ M}}{520\,\text{min}} = 8.65\,M^{-1}\,\text{min}^{-1}$$

41. a. Solving directly for the concentration in the second-order integrated rate law yields

$$\frac{1}{[\mathrm{A}]_t} = \frac{1}{[\mathrm{A}]_0} + kt \qquad [\mathrm{A}]_t = \frac{1}{\dfrac{1}{[\mathrm{A}]_0} + kt}$$

$$[\mathrm{A}]_t = \frac{1}{\dfrac{1}{0.100\,M} + \left(0.0312\,M^{-1}\,\mathrm{min}^{-1}\right)\left(10.0\,\mathrm{s} \times \dfrac{1\,\mathrm{min}}{60\,\mathrm{s}}\right)} = 0.0999\,M$$

b. $$[\mathrm{A}]_t = \frac{1}{\dfrac{1}{0.500\,M} + \left(0.0312\,M^{-1}\,\mathrm{min}^{-1}\right)\left(10.0\,\mathrm{s} \times \dfrac{1\,\mathrm{min}}{60\,\mathrm{s}}\right)} = 0.499\,M$$

c. $$[\mathrm{A}]_t = \frac{1}{\dfrac{1}{0.339\,M} + \left(0.0312\,M^{-1}\,\mathrm{min}^{-1}\right)\left(200.0\,\mathrm{s} \times \dfrac{1\,\mathrm{min}}{60\,\mathrm{s}}\right)} = 0.327\,M$$

d. $$[\mathrm{A}]_t = \frac{1}{\dfrac{1}{0.0050\,M} + \left(0.0312\,M^{-1}\,\mathrm{min}^{-1}\right)\left(24\,\mathrm{d} \times \dfrac{24\,\mathrm{h}}{1\,\mathrm{day}} \times \dfrac{60\,\mathrm{min}}{1\,\mathrm{h}}\right)} = 7.82\times10^{-4}\,M$$

43. Because collisions between the reactants must occur for a reaction to take place, and the number of reactants is decreasing with time, the rate of collisions between reactants declines with time. A lower rate of collision results in a lower reaction rate.

45. The half-life is constant for a first-order reaction and does not depend on the initial concentration. Pablo's experimental measure should be consistent. The half-life of a second-order reaction is inversely proportional to the initial concentration. Peter's experiment will be different because the initial concentration, and therefore the half-life, will be different.

47. a. Using the integrated first-order rate law to determine the rate constant:

$$\ln\frac{[\mathrm{A}]_t}{[\mathrm{A}]_0} = -kt \text{ or } k = -\frac{\ln\dfrac{[\mathrm{A}]_t}{[\mathrm{A}]_0}}{t} = -\frac{\ln\left(\dfrac{75\%}{100\%}\right)}{168\,\mathrm{h}} = 1.71\times10^{-3}\,\mathrm{h}^{-1}$$

b. $$\frac{1}{[\mathrm{A}]_t} = \frac{1}{[\mathrm{A}]_0} + kt \text{ so } k = \frac{\dfrac{1}{[\mathrm{A}]_t} - \dfrac{1}{[\mathrm{A}]_0}}{t} = \frac{\dfrac{1}{0.750\times10^{-3}\,M} - \dfrac{1}{1.00\times10^{-3}\,M}}{168\,\mathrm{h}} = 1.98\,M^{-1}\cdot\mathrm{h}^{-1}$$

c. $$[\mathrm{A}]_t = [\mathrm{A}]_0 - kt \qquad k = -\frac{[\mathrm{A}]_t - [\mathrm{A}]_0}{t} = -\frac{0.750\times10^{-3}\,M - 1.00\times10^{-3}\,M}{168\,\mathrm{h}} = 1.49\times10^{-6}\,M\cdot\mathrm{h}^{-1}$$

49. Using the exponential form of the integrated first-order rate law:

$$[\mathrm{A}]_t = [\mathrm{A}]_0 e^{-kt} = (0.980\,M)e^{-(0.00205/\mathrm{h})(244\mathrm{h})} = 0.594\,M$$

51. Using the half-life formula and the logarithmic form of the integrated first-order rate law:

$$k = \frac{0.693}{t_{1/2}} = \frac{0.693}{241\,\text{yr}} = 0.00288/\,\text{yr}$$

$$\ln\frac{[\text{A}]_t}{[\text{A}]_0} = -kt \quad \text{or} \quad t = -\frac{\ln\frac{[\text{A}]_t}{[\text{A}]_0}}{k} = \frac{-\ln\left(\frac{66.6\%}{100.0\%}\right)}{0.00288/\,\text{yr}} = 141\,\text{yr}$$

53. We create plots for zero-, first-, and second-order reactions. Included in the plots are the equations of the best-fit lines. R is called the coefficient of determination, which indicates a closer fit to a line as the value of R^2 approaches 1. The second-order plot is most linear (that is it has the R^2 value closest to 1), and the reaction is second-order.

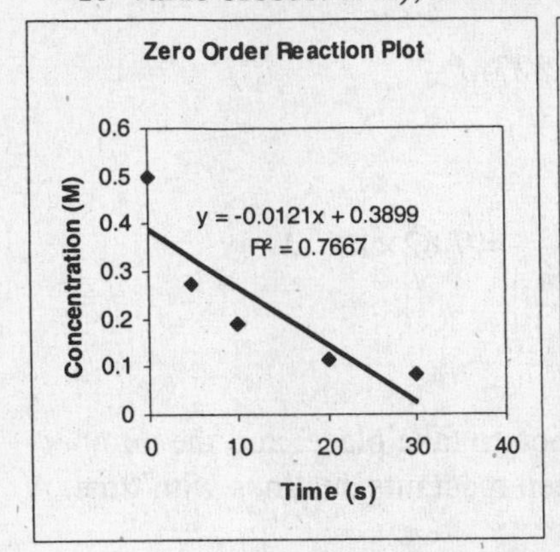

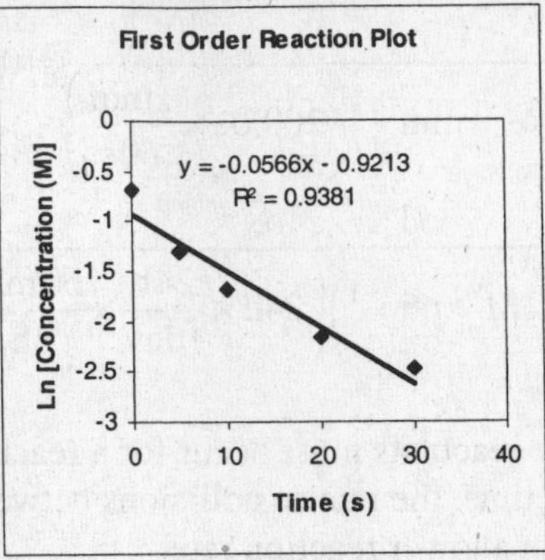

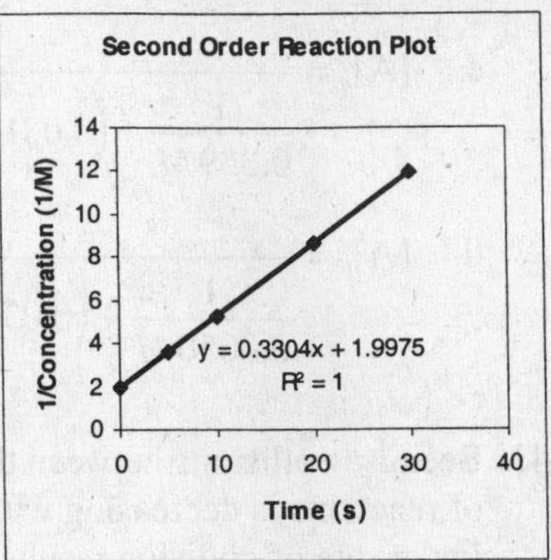

55. a. Doubling [A] will double the rate of the reaction because the reaction is first-order in [A].
b. Doubling [B] will double the rate of the reaction because the reaction is first-order in [B].
c. Doubling both [A] and [B], will increase the rate of the reaction fourfold: double due to [A] and double due to [B].

57. The general rate law for the reaction is

$$\text{Rate} = k[\text{A}]^m[\text{B}]^n$$

We can use ratios of this equation for separate experiments to find the exponents in the equation, along with the rate constant. In experiments 1 and 2, the concentration of A remains constant, so that part of the equation will cancel and allow us to solve for n:

$$\frac{\text{Rate}\#2}{\text{Rate}\#1} = \frac{0.444\,M/\text{s}}{0.222\,M/\text{s}} = \frac{k[\text{A}]_2^m[\text{B}]_2^n}{k[\text{A}]_1^m[\text{B}]_1^n} = \frac{[\text{B}]_2^n}{[\text{B}]_1^n} = \left(\frac{[\text{B}]_2}{[\text{B}]_1}\right)^n = \left(\frac{0.20\,M}{0.10\,M}\right)^n$$

$$2 = 2^n$$

$$n = 1$$

We can use experiments 3 and 2 to find m:

$$\frac{\text{Rate}\#3}{\text{Rate}\#2} = \frac{0.444\,M/\text{s}}{0.444\,M/\text{s}} = \frac{k[\text{A}]_3^m[\text{B}]_3^n}{k[\text{A}]_2^m[\text{B}]_2^n} = \frac{[\text{A}]_3^m}{[\text{A}]_2^m} = \left(\frac{[\text{A}]_3}{[\text{A}]_2}\right)^m = \left(\frac{0.20\,M}{0.10\,M}\right)^m$$

$$1 = 2^m$$

$$m = 0$$

Using experiment 1 and the exponents, we can solve for k:

$$\text{Rate} = k[\text{A}]^0[\text{B}]^1$$

$$k = \frac{\text{Rate}_1}{[\text{B}]_1^1} = \frac{0.222\,M/\text{s}}{(0.10\,M)} = 2.2\,\text{s}^{-1}$$

Therefore, the overall rate law is $\text{Rate} = 2.2\,\text{s}^{-1}\,[\text{B}]^1$. The reaction is first-order overall.

59. The general rate law for the reaction is

$$\text{Rate} = k[\text{A}]^m[\text{B}]^n$$

We can use ratios of this equation for separate experiments to find the exponents in the equation, along with the rate constant. In experiments 1 and 2, the concentration of A remains constant, so that part of the equation will cancel and allow us to solve for n:

$$\frac{\text{Rate\#2}}{\text{Rate\#1}} = \frac{0.420\,M/\text{s}}{0.105\,M/\text{s}} = \frac{k[\text{A}]_2^m[\text{B}]_2^n}{k[\text{A}]_1^m[\text{B}]_1^n} = \frac{[\text{B}]_2^n}{[\text{B}]_1^n} = \left(\frac{[\text{B}]_2}{[\text{B}]_1}\right)^n = \left(\frac{0.20\text{M}}{0.10\text{M}}\right)^n$$

$$4 = 2^n$$

$$n = 2$$

We can use experiments 3 and 2 to find m:

$$\frac{\text{Rate\#3}}{\text{Rate\#2}} = \frac{0.840\,M/\text{s}}{0.420\,M/\text{s}} = \frac{k[\text{A}]_3^m[\text{B}]_3^n}{k[\text{A}]_2^m[\text{B}]_2^n} = \frac{[\text{A}]_3^m}{[\text{A}]_2^m} = \left(\frac{[\text{A}]_3}{[\text{A}]_2}\right)^m = \left(\frac{0.20\text{M}}{0.10\text{M}}\right)^m$$

$$2 = 2^m$$

$$m = 1$$

Using experiment 1 and the exponents, we can solve for k:

$$\text{Rate} = k[\text{A}]^1[\text{B}]^2$$

$$k = \frac{\text{Rate}_1}{[\text{A}]_1^1[\text{B}]_1^2} = \frac{0.105\,M/\text{s}}{(0.10\,M)(0.10\,M)^2} = 1.0\times10^2\,M^{-2}\,\text{s}^{-1}$$

Therefore, the overall rate law is $\text{Rate} = \left(1.0\times10^2\,M^{-2}\,\text{s}^{-1}\right)[\text{A}]^1[\text{B}]^2$. The reaction is third-order overall.

61. We create plots for zero-, first-, and second-order reactions. Included in the plots are the equations of the best-fit lines. R is called the coefficient of determination, which indicates a closer fit to a line as the value of R^2 approaches 1. The zero-order plot is most linear (that is, it has the R^2 value closest to 1), and the reaction is zero-order. The slope of the zero-order graph is equal to $-k$. Therefore, the rate constant is +0.047 M/min.

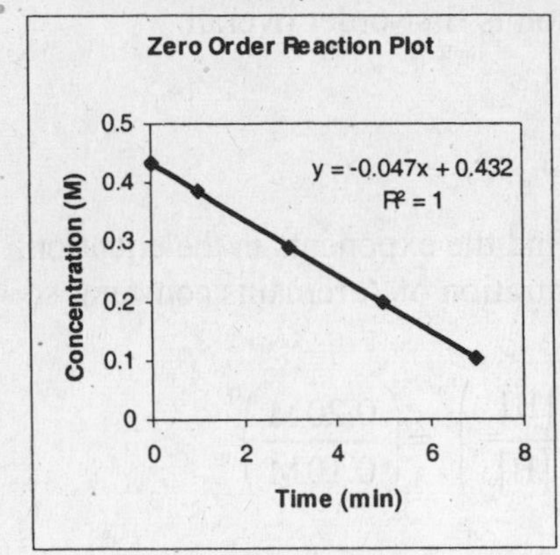

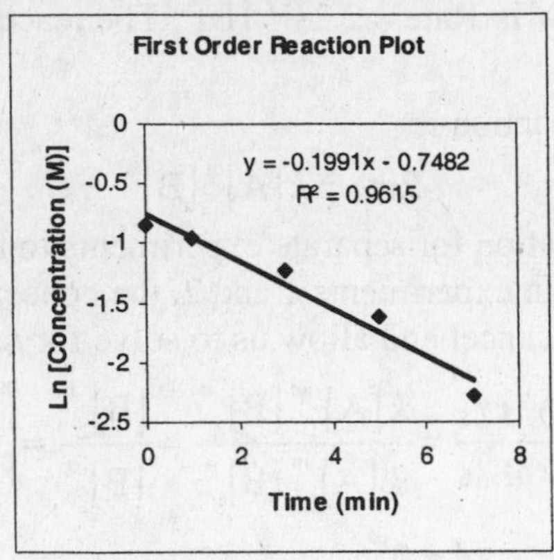

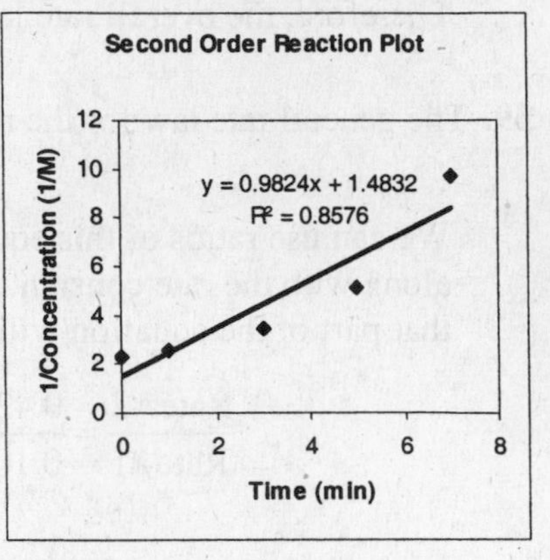

63. a. A first-order plot would be needed: natural log (large dark atoms) versus time.

b. Because there are twice as many large dark atoms in the left diagram than in the right diagram and the rate law is Rate = k [large dark atoms], the rate on the left will be twice that on the right.

c. A first-order rate constant has units of inverse time: s^{-1}.

65. We create plots for zero-, first-, and second-order reactions. Included in the plots are the equations of the best-fit lines. R is called the coefficient of determination, which indicates a closer fit to a line as the value of R^2 approaches 1. The first-order plot is most linear (that is, it has the R^2 value closest to 1), and the reaction is first-order. The rate constant is the negative of the slope and equals 0.2286 / min, or 0.229 / min to three significant figures.

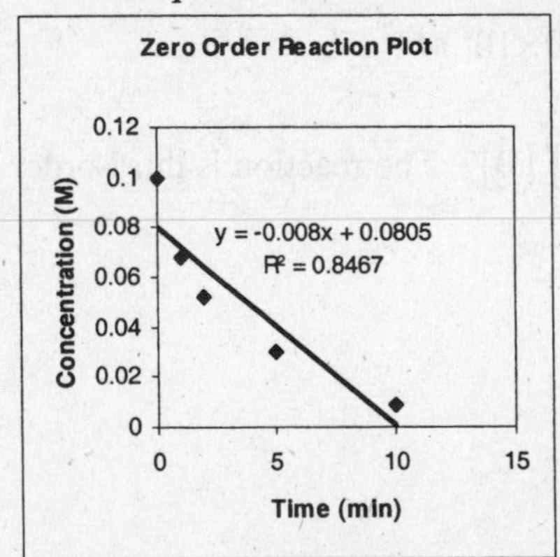

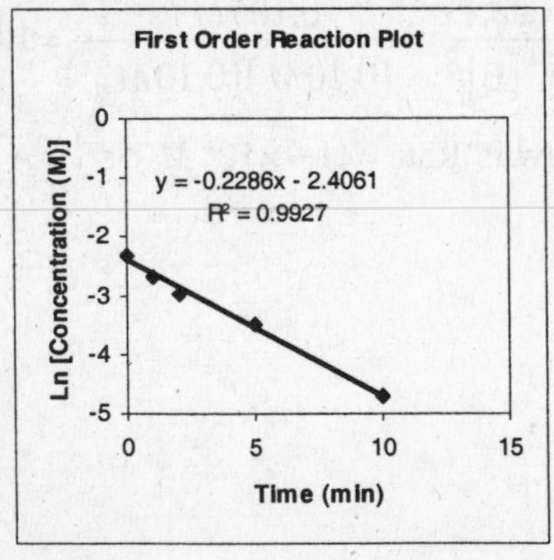

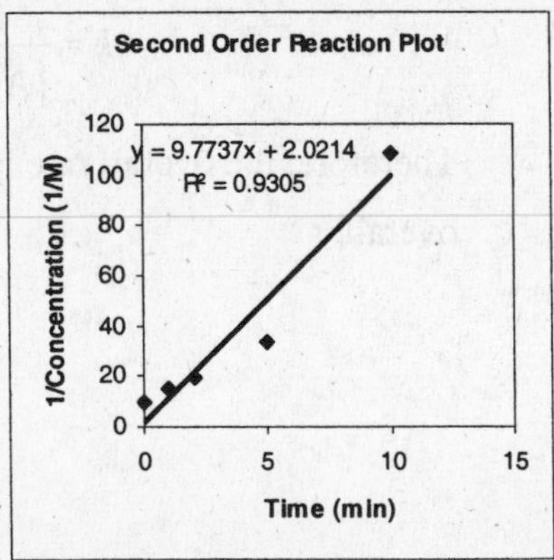

67. a. The general rate law for the reaction is

$$\text{Rate} = k[\text{Stabilizer}]^m[\text{H}^+]^n[\text{O}_2]^p$$

We can use ratios of this equation for separate experiments to find the exponents in the equation, along with the rate constant. In experiments 3 and 2, the concentrations of the Stablilizer and O_2 remain constant, so that part of the equation will cancel and allow us to solve for n:

$$\frac{\text{Rate\#2}}{\text{Rate\#3}} = \frac{4.55\times10^{-5}\ M/\text{s}}{4.55\times10^{-5}\ M/\text{s}} = \frac{k[\text{Stabilizer}]_2^m[\text{H}^+]_2^n[\text{O}_2]_2^p}{k[\text{Stabilizer}]_3^m[\text{H}^+]_3^n[\text{O}_2]_3^p} = \frac{[\text{H}^+]_2^n}{[\text{H}^+]_3^n} = \left(\frac{[\text{H}^+]_2}{[\text{H}^+]_3}\right)^n = \left(\frac{0.500\,M}{0.100\,M}\right)^n$$

$$1 = 5^n$$

$$n = 0$$

Therefore, the rate law does not depend on $[\text{H}^+]$. We can use experiments 1 and 5 to find m:

$$\frac{\text{Rate\#1}}{\text{Rate\#5}} = \frac{7.14\times10^{-4}\ M/\text{s}}{3.57\times10^{-4}\ M/\text{s}} = \frac{k[\text{Stabilizer}]_1^m[\text{O}_2]_1^p}{k[\text{Stabilizer}]_5^m[\text{O}_2]_5^p} = \frac{[\text{Stabilizer}]_1^m}{[\text{Stabilizer}]_5^m} = \left(\frac{[\text{Stabilizer}]_1}{[\text{Stabilizer}]_5}\right)^n = \left(\frac{0.400\,M}{0.100\,M}\right)^n$$

$$2 = 4^n$$

$$n = 0.5$$

We can use experiments 1 and 4 to find p:

$$\frac{\text{Rate\#4}}{\text{Rate\#1}} = \frac{1.28\times10^{-3}\ M/\text{s}}{7.14\times10^{-4}\ M/\text{s}} = \frac{k[\text{Stabilizer}]_4^m[\text{O}_2]_4^p}{k[\text{Stabilizer}]_1^m[\text{O}_2]_1^p} = \frac{[\text{O}_2]_4^p}{[\text{O}_2]_1^p} = \left(\frac{[\text{O}_2]_4}{[\text{O}_2]_1}\right)^p = \left(\frac{0.750\text{M}}{0.560\text{M}}\right)^p$$

$$1.79 = 1.34^p$$

$$p = 2$$

The rate law is $\text{Rate} = k[\text{Stabilizer}]^{0.5}[\text{O}_2]^2$

b. Using experiment 1 and the exponents, we can solve for k:

$$\text{Rate} = k[\text{Stabilizer}]^{0.5}[\text{O}_2]^2$$

$$k = \frac{\text{Rate}_1}{[\text{Stabilizer}]^{0.5}[\text{O}_2]^2} = \frac{7.14\times10^{-4}\ M/\text{s}}{(0.400\,M)^{0.5}(0.560\,M)^2} = 3.60\times10^{-3}\ M^{-1.5}\,\text{s}^{-1}$$

Therefore, the overall rate law is $\text{Rate} = (3.60\times10^{-3}\ M^{-1.5}\text{s}^{-1})[\text{Stabilizer}]^{0.5}[\text{O}_2]^2$. The reaction is 2.5-order overall.

c. $\text{Rate} = (3.60\times10^{-3}\,M^{-1.5}\,\text{s}^{-1})[\text{Stabilizer}]^{0.5}[\text{O}_2]^2$

$\text{Rate} = (3.60\times10^{-3}\,M^{-1.5}\,\text{s}^{-1})(0.111\,M)^{0.5}(1.00\,M)^2 = 1.20\times10^{-3}\,M/\text{s}$

d. $\text{Rate} = (3.60\times10^{-3}\,M^{-1.5}\,\text{s}^{-1})[\text{Stabilizer}]^{0.5}[\text{O}_2]^2$

$$[\text{O}_2]^2 = \frac{\text{Rate}}{(3.60\times10^{-3}M^{-1.5}\,\text{s}^{-1})[\text{Stabilizer}]^{0.5}}$$

$$[\text{O}_2] = \sqrt{\frac{\text{Rate}}{(3.60\times10^{-3}M^{-1.5}\,\text{s}^{-1})[\text{Stabilizer}]^{0.5}}} = \sqrt{\frac{2.55\times10^{-4}\,M/\text{s}}{(3.60\times10^{-3}M^{-1.5}\,\text{s}^{-1})(0.300\,M)^{0.5}}}$$

$$[\text{O}_2] = 0.360\,M$$

69. Using the half-life equation and the exponential form of the first-order integrated rate law:

$$k = \frac{0.693}{t_{1/2}} = \frac{0.693}{12\,\text{yr}} = 0.0578/\text{yr}$$

$$\text{and } [\text{A}]_t = [\text{A}]_0 e^{-kt} = (545\,\text{cpm})e^{-(0.0578/\text{yr})(3\,\text{yr})} = 458\,\text{cpm}.$$

The sample remaining would be enough to create 458 cpm, or 84% of the original sample.

71. Because each of those values is the *y*-intercept of the graph, each line would be moved vertically until the intercept was at $y = 0$.

73. The slope is negative.

75. With the same initial concentrations, the higher rate constant indicates a faster reaction: $8.95 \times 10^{-2}/\text{s}$

77. Creating a ratio of the new data to the old, we find

$$\frac{\text{Rate\#2}}{\text{Rate\#1}} = \frac{4}{1} = \frac{k\left[\text{Cr}^{3+}\right]_2^2\left[\text{Ce}^{4+}\right]_2^1\left[\text{CrO}_4^{\,2-}\right]_2^x}{k\left[\text{Cr}^{3+}\right]_1^2\left[\text{Ce}^{4+}\right]_1^1\left[\text{CrO}_4^{\,2-}\right]_1^x} = \frac{(2)^2(2)^1(2)^x}{(1)^2(1)^1(1)^x}$$

$$4 = 8 \times 2^x$$

$$\frac{1}{2} = 2^x$$

$$x = -1$$

79. A *reaction mechanism* is a collection of single simple steps called *elementary steps* that represent what is thought to occur in a reaction. The slowest step in the mechanism usually determines the rate of the reaction and is called the *rate-determining step*.

81. The answers will vary. One possibility: 1. Hear phone ring. 2. Decide to answer. 3. Push chair out from table. 4. Stand up. 5. Walk to phone. 6. Pick up phone. 7. Say Hello, etc. Depending on how agile you are and how far the phone is from the table, you might choose step 3, step 4, or step 5 as the slow step.

83. Using the equation $\ln\left(\frac{k_2}{k_1}\right) = \frac{E_a}{R}\left(\frac{1}{T_1} - \frac{1}{T_2}\right)$, we can solve for the activation energy:

$$E_a = \frac{R\ln\left(\frac{k_2}{k_1}\right)}{\left(\frac{1}{T_1} - \frac{1}{T_2}\right)} = \frac{(8.3145\,\text{J/mol}\cdot\text{K})\ln\left(\frac{1.74\times10^{-2}/\text{s}}{4.22\times10^{-2}/\text{s}}\right)}{\left(\frac{1}{400\text{K}} - \frac{1}{300\text{K}}\right)} = 8839\,\text{J/mol} = 8.84\,\text{kJ/mol}$$

85. Mechanism B, with its first step the slow step, has a rate law of Rate = k [H_2O_2][I^-], which does not have the proper hydrogen ion concentration dependence. The rate law for Mechanism A does match and is derived by setting the rates of the forward and reverse reactions in the first step equal, solving for the intermediate concentration, and placing it in the rate law from the second step (the slow step):

The rate law is Rate = k [H_2O_2][HI], but [HI] cannot remain in the rate law because it is an intermediate.

The rate laws for the forward reaction (Rate = $k_1[H^+][I^-]$) and the reverse reaction (Rate = k_{-1}[HI]) are set equal to one another, because the first step will reach equilibrium as a consequence of the following slow step. We can then solve for the intermediate (HI) concentration:

$$k_1\left[H^+\right]\left[I^-\right] = k_{-1}[HI]$$

$$[HI] = \frac{k_1}{k_{-1}}\left[H^+\right]\left[I^-\right]$$

This solution can be placed in the rate law:

$$\text{Rate} = k_2[H_2O_2][HI]$$

$$\text{Rate} = k_2[H_2O_2]\left(\frac{k_1}{k_{-1}}\left[H^+\right]\left[I^-\right]\right)$$

$$\text{Rate} = \frac{k_1}{k_{-1}}k_2[H_2O_2]\left[H^+\right]\left[I^-\right] = k'[H_2O_2]\left[I^-\right]\left[H^+\right]$$

This rate law matches the experimental rate law.

87. a. Using the equation $\ln\left(\frac{k_2}{k_1}\right) = \frac{E_a}{R}\left(\frac{1}{T_1} - \frac{1}{T_2}\right)$, we can solve for the activation energy:

$$E_a = \frac{R\ln\left(\frac{k_2}{k_1}\right)}{\left(\frac{1}{T_1} - \frac{1}{T_2}\right)} = \frac{(8.3145\,\text{J/mol}\cdot\text{K})\ln\left(\frac{0.203\,\text{L/mol}\cdot\text{s}}{6.4\times10^{-5}\,\text{L/mol}\cdot\text{s}}\right)}{\left(\frac{1}{283\text{K}} - \frac{1}{351\text{K}}\right)} = 97{,}920\,\text{J/mol} = 97.9\,\text{kJ/mol}$$

b. At 37°C, the rate constant is

$$k_2 = k_1\,e^{\frac{E_a}{R}\left(\frac{1}{T_1} - \frac{1}{T_2}\right)} = (6.4\times10^{-5}\,\text{L/mol}\cdot\text{s})e^{\left(\frac{97920\,\text{J/mol}}{8.3145\,\text{J/mol}\cdot\text{K}}\left(\frac{1}{283\text{K}} - \frac{1}{310\text{K}}\right)\right)} = 2.4\times10^{-3}\,\text{L/mol}\cdot\text{s}$$

89. a. A reaction intermediate is a compound that is formed and then consumed during the course of a reaction.

b. An activated complex is the species or collection of atoms that exists at the transition state midway between reactants and products in a single elementary step.

c. A homogeneous catalyst is a species that is in the same phase as the rest of the reactants and aids in the reaction by lowering the overall activation energy in the process of converting reactants to products. A homogeneous catalyst is destroyed and then re-created in the course of the reaction.

d. A heterogeneous catalyst is a species or material that is in a different phase from the rest of the reactants and aids in the reaction by lowering the overall activation energy in the process of converting reactants to products.

91. a. The overall reaction is the sum of the individual steps: $2\ N_2O \rightarrow 2\ N_2 + O_2$.
b. O is an intermediate. There are no catalysts.
c. In order for the mechanisms rate law to be Rate = $k\ [N_2O]$, the first step must be the slow step. (If the second step were slow instead, the mechanism would give a rate law of
Rate = $k\ [N_2O]^2[N_2]^{-1}$.

93. a. The overall reaction is the sum of the individual steps: Lactose $\rightarrow$ Glucose + Galactose
b. (Lactase-Lactose) and (Lactase-Glucose-Galactose) are intermediates. Lactase is the catalyst
c. The rate law is Rate = k [Lactase][Lactose].

95. a. The overall reaction is the sum of the individual steps: $N_2(g) + 3\ H_2(g) \rightarrow 2\ NH_3(g)$.
b. FeH_2 and FeN_2 are intermediates.
c. Fe(s) is a catalyst.
d. Because Fe is a solid catalyst and the remaining species are gases, Fe is a heterogeneous catalyst.

97. $$\frac{\Delta[H_2O_2]}{\Delta t} = \frac{0.223M - 0.250M}{8.0\text{s}} = -3.3_8 \times 10^{-3}\ M/\text{s}$$

$$\text{Rate} = -\frac{1}{2}\frac{\Delta[H_2O_2]}{\Delta t} = -\frac{1}{2}\left(-3.3_8 \times 10^{-3}\ M/\text{s}\right) = 1.7\times10^{-3}\ M/\text{s}$$

$$\text{Rate} = 1.7\times10^{-3}\ M/\text{s} \times \frac{60\text{s}}{1\text{h}} = 0.10M/\text{h}$$

99. a. The rate law for the reaction is Rate = k [A], so the rate constant is

$$k = \frac{\text{Rate}}{[A]} = \frac{0.0875M/\text{s}}{0.250M} = 0.350\text{s}^{-1}$$

b. The half-life is $t_{1/2} = \frac{0.693}{k} = \frac{0.693}{0.350/\text{s}} = 1.98\text{s}$.

101. Because $\Delta H = E_a(\text{forward}) - E_a(\text{reverse})$, solving for the forward activation energy yields $E_a(\text{forward}) = \Delta H + E_a(\text{reverse})$=125 kJ + 75 kJ = 200 kJ.

103. a. Adding the two steps gives $C_2H_4(aq) + HCl(aq) \rightarrow C_2H_5Cl(aq)$.
b. The rate law is Rate = $k_1\ [C_2H_4][HCl]$ as written for the first step, which is the slow (and rate-determining) step.

104. See the uncatalyzed profile at the right (solid line).
a. Because the heat of formation for H_2 and N_2 is 0 and for NH_3 is – 46 kJ/mol,

$$\Delta H = \left(2(-46\text{ kJ/mol})\right) - \left(1(0) + 3(0)\right) = -92\text{ kJ} \text{ which is exothermic.}$$

b. To determine the spontaneity of the reaction, we must calculate the free energy change for the reaction:

$$\Delta G = \left(2(-16\text{ kJ/mol})\right) - \left(1(0) + 3(0)\right) = -32\text{kJ}$$

which indicates a spontaneous reaction. To determine the equilibrium temperature, we need the entropy change for the reaction in addition to the enthalpy change:

$$S = [2(192\text{ J/K-mol})] - [1(192\text{ J/K-mol}) + 3(131\text{ J/K-mol})] = -201\text{ J/K}$$

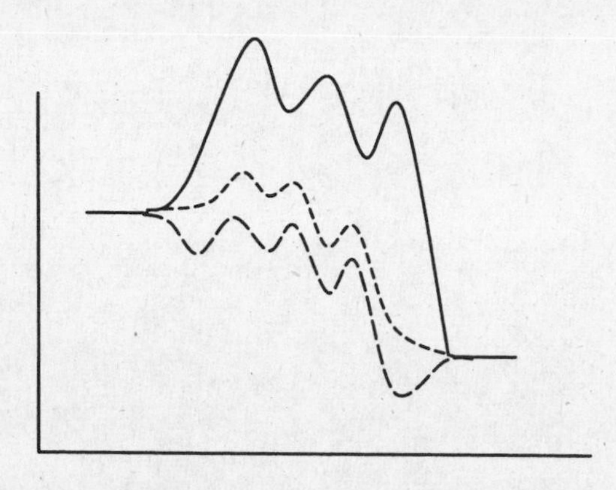

Then the equilibrium temperature (under standard conditions) is

$$T = \frac{\Delta H}{\Delta S} = \frac{-92\,\text{kJ} \times \dfrac{1000\,\text{J}}{1\,\text{kJ}}}{-201\,\text{J/K}} = 458\,\text{K}$$

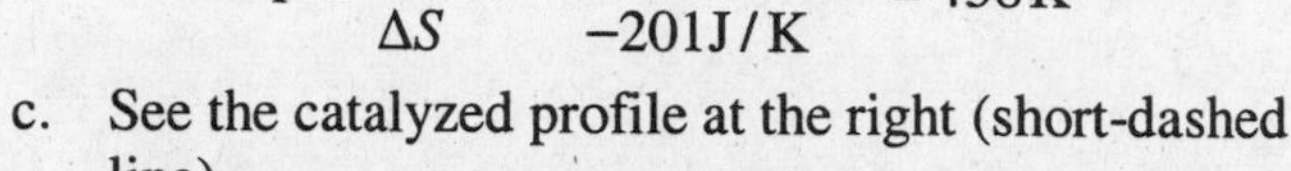

c. See the catalyzed profile at the right (short-dashed line).

d. See the profile at the right (long-dashed line).

e. The two reactants are in stoichiometric proportion, so we need not worry about limiting reagents. The mass of ammonia produced is

$$1\,\text{mol N}_2 \times \frac{2\,\text{mol NH}_3}{1\,\text{mol N}_2} \times \frac{17.03\,\text{g NH}_3\,(\text{theoretical})}{1\,\text{mol NH}_3} \times \frac{45\,\text{g (actual)}}{100\,\text{g (theoretical)}} = 15\,\text{g NH}_3$$

f. These are not stoichiometric ratios, so we need to determine which reagent limits the reaction by solving for the mass of ammonia produced from each:

$$10.0\,\text{kg N}_2 \times \frac{1000\,\text{g}}{1\,\text{kg}} \times \frac{1\,\text{mol N}_2}{28.02\,\text{g N}_2} \times \frac{2\,\text{mol NH}_3}{1\,\text{mol N}_2} \times \frac{17.03\,\text{g NH}_3}{1\,\text{mol NH}_3} \times \frac{1\,\text{kg}}{1000\,\text{g}} = 12.2\,\text{kg NH}_3$$

$$30.0\,\text{kg H}_2 \times \frac{1000\,\text{g}}{1\,\text{kg}} \times \frac{1\,\text{mol H}_2}{2.02\,\text{g H}_2} \times \frac{2\,\text{mol NH}_3}{3\,\text{mol H}_2} \times \frac{17.03\,\text{g NH}_3}{1\,\text{mol NH}_3} \times \frac{1\,\text{kg}}{1000\,\text{g}} = 169\,\text{kg NH}_3$$

The nitrogen limits the reaction, and the yield of ammonia is 12.2 kg.

Chapter 16: Chemical Equilibrium

The Bottom Line

- Reactions can proceed reversibly toward the products or back toward the reactants. (Section 16.1)
- The point in a reaction at which there is no net change in the concentration of reactants or products is known as chemical equilibrium—or, often, simply as equilibrium. (Section 16.1)
- The free energy, G, of a reaction is at a minimum at equilibrium. (Section 16.1)
- The free energy change, ΔG, is equal to 0 at equilibrium. (Section 16.1)
- The rates of the forward and reverse reactions are equal at equilibrium. (Section 16.1)
- The mass-action expression relates the equilibrium concentrations of reactants and products in a reaction. (Section 16.1)
- The equilibrium constant is temperature dependent. (Sections 16.1, 16.6)
- The size of the equilibrium constant gives us information about the extent of a reaction. (Section 16.3)
- Modifying the coefficients of a reaction modifies the value of its equilibrium constant. (Section 16.4)
- The equilibrium constant can be converted for use with partial pressures or molarities. (Section 16.4)
- The equilibrium constant for the sum of chemical reactions is the mathematical product of the individual K values. (Section 16.4)
- We can use the equilibrium constant and massaction expression to calculate the equilibrium concentration of substances in a reaction. (Section 16.5)
- Solving problems relating to reaction equilibria involves asking and answering a series of systematic
- questions. (Section 16.5)
- We can use the reaction quotient, Q, to assess which way a reaction will proceed to reach equilibrium. (Section 16.5)
- Le Châtelier's principle concerns the impact of changing the pressure, temperature, and concentration conditions of a reaction at equilibrium. (Section 16.6)
- A catalyst does not affect the equilibrium position. It changes the reaction mechanism in such a way as to speed up the reaction. (Section 16.6)
- The free energy change of a reaction can be determined from the equilibrium constant for that reaction. (Section 16.7)

Solutions to Practice Problems

16.1. We find that the concentration of I_2 has changed by computing

$$\Delta[I_2] = [I_2] - [I_2]_0 = 0.041\,\text{M} - 0.050\,M = -0.009\,M$$

Using the reaction stoichiometry, the concentration of HI must increase by twice the amount that I_2 decreases (the negative is used because HI *increases* while I_2 *decreases*):

$$[\text{HI}] = [\text{HI}]_0 + \Delta[\text{HI}] = [\text{HI}]_0 - 2\Delta[I_2] = 0.206\,M - 2(-0.009\,M) = 0.224\,M$$

16.2. The mass-action expression for the reaction is $K = \dfrac{[CO][H_2][H_2][H_2]}{[CH_4][H_2O]} = \dfrac{[CO][H_2]^3}{[CH_4][H_2O]}$.

16.3. a. $K = \dfrac{[NO_2][O_2]}{[NO][O_3]}$

b. $K = \dfrac{[NH_4Cl]}{[HCl][NH_3]}$

c. $K = \dfrac{[C_2H_6]}{[C_2H_2][H_2]^2}$

16.4. a. Because the equilibrium constant is much smaller than 1, this reaction provides mostly reactants.

b. Because the equilibrium constant is much smaller than 1, this reaction provides mostly reactants.

c. Because the equilibrium constant is much smaller than 1, this reaction provides mostly reactants.

d. Because the equilibrium constant is much larger than 1, this reaction provides mostly products.

16.5. In creating the new reaction from the original, the original was reversed, so we must invert the constant; and the reaction was multiplied by a factor of ½, so we must raise the new equilibrium constant to the ½ power:

$$K_{\text{new}} = \left(\frac{1}{K_{\text{original}}}\right)^{1/2} = \left(\frac{1}{8.6\times10^{-7}}\right)^{1/2} = 1.1\times10^{3}$$

16.6. This reaction has 2 moles of gaseous reactants and only 1 mole of gaseous product, so $\Delta n = 1 - 2 = -1$. Using this value, we can find the pressure equilibrium constant at 298 K:

$$K_p = K(RT)^{\Delta n} = (3.7\times10^{9})(0.08206\,\text{L}\cdot\text{atm/mol}\cdot\text{K}\times298\,\text{K})^{-1} = 1.5\times10^{8}$$

16.7. a. Because $Q > K$, there are too many products (and too few reactants,) so the reaction shifts left.

b. We first calculate Q: $Q = \dfrac{[NO][O_3]}{[NO_2][O_2]} = \dfrac{(0.5)(1)}{(0.25)(1)} = 2$. Because $Q < K$, there are too many reactants (and too few products), so the reaction shifts right.

c. We first calculate Q: $Q = \dfrac{[NO][O_3]}{[NO_2][O_2]} = \dfrac{(1.78\times10^{-6})(0.0010)}{(5.4\times10^{-3})(2.55)} = 1.29\times10^{-7}$. Because $Q < K$, there are too many reactants (and too few products), so the reaction shifts right.

d. We first calculate Q: $Q = \dfrac{[NO][O_3]}{[NO_2][O_2]} = \dfrac{(9.3)(0.033)}{(0.044)(0.019)} = 367$. Because $Q > K$, there are too many products (and too few reactants), so the reaction shifts left.

16.8. We first create the ICEA table to solve the problem:

	$PbCl_2(s) \rightleftarrows$	$Pb^{2+}(aq)$ +	$2\ Cl^-(aq)$
initial	–	0 *M*	0 *M*
change	–	+ *x*	+ 2*x*
equilibrium	–	*x*	2*x*

$$K=\left[Pb^{2+}\right]\left[Cl^-\right]^2=(x)(2x)^2=4x^3=1.6\times10^{-5}$$

Then

$$x=\left[Pb^{2+}\right]=\sqrt[3]{\frac{1.6\times10^{-5}}{4}}=1.6\times10^{-2}\ M$$

16.9. We notice that because the equilibrium constant is large, the reaction will go nearly to completion. Further, the hydrogen ion will be the species that is nearly depleted and will control how much the others change. We create the ICEA table:

	$C_2H_3O_2^-(aq)$ +	$H^+(aq) \rightleftarrows$	$C_2H_4O_2(aq)$
initial	0.10	0.050 *M*	0 *M*
change	–0.050 + *x*	–0.050 + *x*	+0.050 – *x*
equilibrium	0.050 + *x*	*x*	0.050 – *x*
assumption	0.050	*x*	0.050

$$K=\frac{[C_2H_4O_2]}{[C_2H_3O_2^-][H^+]}=\frac{(0.050)}{(0.050)(x)}=5.6\times10^4$$

$$x=\left[H^+\right]=\frac{(0.050)}{(0.050)(5.6\times10^4)}=1.8\times10^{-5}\ M$$

$$[C_2H_3O_2^-]=0.050M+1.8\times10^{-5}\ M=0.050M$$

$$[C_2H_4O_2]=0.050M-1.8\times10^{-5}\ M=0.050M$$

Because the value for *x* was so low, we are safe in using the approximation to solve the problem.

16.10. a. This reaction is endothermic; raising the temperature will result in a shift toward the products (right).

b. Removing a reactant will cause a shift to restore the concentration—to the left.

c. Catalysts do not affect the equilibrium position—no change.

d. Adding a product will cause a shift to lower the concentration of the product—to the left.

16.11. The free energies of formation ($\Delta_f G°$)for $H_2SO_3(aq)$, $H^+(aq)$, and $HSO_3(aq)$ are –537.9 kJ/mol, 0, and –527.8 kJ/mol, respectively. The reaction free energy change is

$$\Delta_{rxn}G°=\left[1\times(-527.8\text{kJ/mol})+1(0)\right]-\left[1\times(-537.9\text{kJ/mol})\right]=10.1\text{kJ}$$

The thermodynamic equilibrium constant is

$$K_{eq}=e^{-\Delta G/RT}=e^{-10100\text{J/mol}/(8.3145\text{J/mol·K}\times298\text{K})}=0.017$$

The difference between the two values may lie in the difference between the activities and concentrations. Nevertheless, the two values are very close.

Solutions to Student Problems

1. Although the concentrations are not changing, the reactants and products are continuously reacting to form each other, making the equilibrium state dynamic.

3. A reaction that does not go to completion stops before all of the reactants have been converted to products. In other words, the reaction reached equilibrium, and a zero Gibbs energy change was achieved, before all of the reactants were converted.

5. a. We write the equilibrium constant expression by placing the product concentrations over the reactant concentrations, with each concentration raised to the power equal to the stoichiometric coefficient in the balanced reaction. Pure liquids and solids do not appear in the expression.

$$K = \frac{[C_2H_5Cl]}{[HCl][C_2H_4]}$$

b. $K = \frac{[CO_2][H_2O]^2}{[CH_4][O_2]^2}$

c. $K = \frac{1}{[H_2]^2[O_2]}$

7. a. The forward and reverse reaction rates are equal at equilibrium. If $rate_1$ is the forward rate and $rate_2$ is the reverse rate, then $rate_1 = rate_2$.

b. The concentrations at equilibrium will be related by the equilibrium constant expression:

$$K = \frac{[CaSO_4]^3[H_3PO_4]^2}{[H_2SO_4]^3}$$

c. At equilibrium, the Gibbs energy change for the reaction is zero: $\Delta G = 0$.

9. If we rearrange the equation to place all the concentrations on one side (with the products in the numerator), we find that the other side contains only constants, which together create the equilibrium constant:

$$k_1[H_2][Cl_2] = k_{-1}[HCl]^2$$

$$\frac{k_1}{k_{-1}} = \frac{[HCl]^2}{[H_2][Cl_2]} = K$$

11. If equilibria could not be controlled, the amounts of products that are produced could not be maximized for profit. By controlling equilibria, the industry is able to make the products that the consumers want and to make them in the most efficient way possible. Without control of the equilibria, these industries might not even be profitable unless the products were much more expensive for the consumer.

13. a. Chemicals that prefer the stationary phase and move up the plate more slowly have a large K_D. Therefore, the ink with the largest K_D moves the least: the blue ink.

b. Chemicals that prefer the mobile phase and move up the plate more rapidly have a low K_D. Therefore, the ink with the smallest K_D moves the most: the yellow ink.

15. a. Pure solids are not included in the expression: $K = \dfrac{1}{[O_2]}$

b. Aqueous species must be included in the expression: $K = \dfrac{[H_2SO_4]}{[SO_3]}$

17. a. The equilibrium position for this reaction falls short of halfway between reactants and products and therefore would be less than 1 in size. A reasonable estimate for the value of K would be a number that is less than 1 but is not too small.

b. There will be relatively more reactants than products at equilibrium, so there will be more CO at equilibrium than CO_2.

19. The balanced reaction is $C_3H_8(g) + 5\ O_2(g) \rightleftarrows 3\ CO_2(g) + 4\ H_2O(g)$. The equilibrium constant expression is $K = \dfrac{[CO_2]^3[H_2O]^4}{[C_3H_8][O_2]^5}$.

21. The change for O_2 will be half as large as the change for the other two species because it has a coefficient of 1 rather than the coefficient 2 on the other species. H_2O increases by 0.050 *M*; therefore, the change in O_2 will be + 0.025 *M*, giving it a final concentration of 0.050 *M*.

23. Because solids are not included in the expression, only the gases appear: $K = \dfrac{[SO]^4}{[O_2]^5}$

25. Values of $K \gg 1$ are product-favored; values of $K \ll 1$ are reactant-favored; and values of K near 1 produce mixtures. The ammonia synthesis produces a mixture, the ozone depletion is product favored, and the acid ionization is reactant-favored.

27. a. In order to get a quantitative 1:1 reaction between EDTA and the metal, the reaction needs to go to completion. If the reaction does not go to completion, we will not have a firm idea of how much metal is in the solution.

b. A $K = 10^{10}$ ensures complete reaction, while a $K = 10$ produces a mixture. As stated in part a, we need the complete reaction to get a quantitative measurement of the metal in the solution.

29. a. $K = \dfrac{[H_2O]^2}{[H_2]^2[O_2]}$

b. Placing the values in the expression: $K = \dfrac{[H_2O]^2}{[H_2]^2[O_2]} = \dfrac{(1.00)^2}{(0.134)^2(0.673)} = 82.8$

c. If we reverse the reaction, the new equilibrium constant is the inverse of the original:

$K_{reverse} = \dfrac{1}{K} = \dfrac{1}{82.8} = 0.0121$

d. When the reaction is multiplied by a constant, the new equilibrium constant is raised to the power of that constant: $K_{new} = (K)^{1/2} = (82.8)^{1/2} = 9.10$

31. The equilibrium constant is equal to the ratio of the forward rate constant to the reverse rate constant, $K = \frac{k_{\text{forward}}}{k_{\text{reverse}}}$, so solving for the forward rate constant, we find

$$k_{\text{forward}} = K \times k_{\text{reverse}} = \left(1.7\times10^{12}\right)\times\left(6.6\times10^{-12}\right) = 11$$

33. a. $K = \frac{[C_2H_2]^{1/2}[H_2]^{3/2}}{[CH_4]}$

b. For the reaction 2 $CH_4(g) \rightleftarrows C_2H_2(g) + 3\ H_2(g)$, $K = \frac{[C_2H_2][H_2]^3}{[CH_4]^2}$.

c. Because the reaction coefficients were doubled from part a to part b, the constant in part b will be the square of the constant in part a: $K_b = (K_a)^2$.

35. If the reactions are added one to one, the new equilibrium constant is simply the product of the two original constants: $K = K_1 \times K_2 = \left(1.4\times10^5\right)\times\left(4.7\times10^{-3}\right) = 6.6\times10^2$. This is an example where a reactant-favored reaction (phosphate addition to glucose) can be made product-favored by coupling it with a very product-favored reaction (ATP to ADP).

37. If we add the two reactions, we get $C_6H_5OH(aq) + OH^-(aq) \rightleftarrows H_2O(l) + C_6H_5O^-(aq)$. Because we added the two reactions directly, we simply multiply the constants:

$K = K_1 \times K_2 = \left(1.3\times10^{-10}\right)\times\left(1.0\times10^{14}\right) = 1.3\times10^4$. This is a large value, so the reaction should proceed well toward the products.

39. We add the reactions from Problems 37 and 38. (We also need the water formation reaction.)

$$C_6H_5OH \rightleftarrows H^+ + C_6H_5O^- \qquad K_1 = 1.3\times10^{-10}$$

$$C_{13}H_{20}N_2O_2 + H_2O \rightleftarrows OH^- + C_{13}H_{20}N_2O_2H^+ \qquad K_2 = 7.1\times10^{-6}$$

$$H^+ + OH^- \rightleftarrows H_2O \qquad K_3 = 1.0\times10^{14}$$

$$C_{13}H_{20}N_2O_2 + C_6H_5OH \rightleftarrows C_6H_5O^- + C_{13}H_{20}N_2O_2H^+ \qquad K_{net} = K_1K_2K_3 = 0.092$$

This reaction would be a poor way to analyze the procaine solution, because the reaction does not proceed very far toward completion.

41. This reaction has 4 moles of gaseous reactants and only 2 moles of gaseous product, so $\Delta n = 2 - 4 = -2$. Using this value, we can find the concentration equilibrium constant at 673K:

$$K = K_p(RT)^{-\Delta n} = \left(2.5\times10^{-4}\right)\left(0.08206\,\text{L}\cdot\text{atm/mol}\cdot\text{K}\times673\,\text{K}\right)^{-(-2)} = 0.76$$

43. For each problem, we can set up an ICEA table and then solve the problem:

a. $CH_3COOH(aq) \rightleftarrows CH_3COO^-(aq) + H^+(aq)$

	CH_3COOH	CH_3COO^-	H^+
initial	0.500 M	0 M	0 M
change	$-x$	$+x$	$+x$
equilibrium	$0.500 - x$	x	x
assumption	0.500	x	x

$$K = \frac{[CH_3COO^-][H^+]}{[CH_3COOH]}$$

$$1.8\times10^{-5} = \frac{(x)^2}{0.500}$$

$$x = [H^+] = \sqrt{(1.8\times10^{-5})\times(0.500)} = 3.0\times10^{-3}\ M$$

Checking our assumption, we find that x is less than 1% of 0.5, so we are safe in using the assumption.

b. $CH_3COOH(aq) \rightleftarrows CH_3COO^-(aq) + H^+(aq)$

	CH_3COOH	CH_3COO^-	H^+
initial	0.100 M	0.100 M	0 M
change	$-x$	$+x$	$+x$
equilibrium	$0.100 - x$	$0.100 + x$	x
assumption	0.100	0.100	x

$$K = \frac{[CH_3COO^-][H^+]}{[CH_3COOH]}$$

$$1.8\times10^{-5} = \frac{(0.100)(x)}{0.100}$$

$$x = [H^+] = \frac{(1.8\times10^{-5})(0.100)}{(0.100)} = 1.8\times10^{-5}\ M$$

Checking our assumption, we find that x is much less than 1% of 0.1, so we are safe in using the assumption.

c. $CH_3COOH(aq) \rightleftarrows CH_3COO^-(aq) + H^+(aq)$

	CH_3COOH	CH_3COO^-	H^+
initial	0.010 M	0 M	0 M
change	$-x$	$+x$	$+x$
equilibrium	$0.010 - x$	x	x
assumption	0.010	x	x

$$K = \frac{[CH_3COO^-][H^+]}{[CH_3COOH]}$$

$$1.8\times10^{-5} = \frac{(x)^2}{0.010}$$

$$x = [H^+] = \sqrt{(1.8\times10^{-5})\times(0.010)} = 4.2\times10^{-4}\ M$$

Checking our assumption, we find that x is 4.2% of 0.01, so we are safe in using the assumption.

45. For the reaction $H_2(g) + I_2(g) \rightleftarrows 2\,HI(g)$ with $K = 617$, we can calculate the reaction quotient, Q, and compare its value to K to predict the reaction direction:

a. $Q = \frac{[HI]^2}{[H_2][I_2]} = \frac{(0.20)^2}{(0.240)(0.080)} = 2.08$

Because $Q < K$, the reaction will proceed to the right (forward).

b. $Q = \frac{[HI]^2}{[H_2][I_2]} = \frac{(1.500)^2}{(0.030)(0.100)} = 750$

Because $Q > K$, the reaction will proceed to the left (reverse).

47. For the reaction $Mb + O_2 \rightleftarrows MbO_2$ with $K = 8.6 \times 10^{-5}$, we can calculate the reaction quotient, Q, and compare its value to K to predict the reaction direction:

a. $Q = \frac{[MbO_2]}{[Mb][O_2]} = 0$. Because $Q < K$, the reaction will proceed to the right (forward).

b. $Q = \frac{[MbO_2]}{[Mb][O_2]} = \frac{(8.6\times10^{-13})}{(1.0\times10^{-4})(1.0\times10^{-4})} = 8.6\times10^{-5}$. Because $Q = K$, this reaction is at equilibrium. There will be no change.

c. $Q = \frac{[MbO_2]}{[Mb][O_2]} = \frac{(2.6\times10^{-11})}{(2.0\times10^{-4})(1.5\times10^{-4})} = 8.7\times10^{-4}$. Because $Q > K$, the reaction will proceed to the left (reverse).

d. Because one of the reactants is missing, the reaction will proceed to the left (reverse). ($Q = \infty > K$)

49. We begin the problem by setting up the ICE table and then solve for the concentration of Ba^{2+}:

	$BaSO_4(s) \rightleftarrows$	$Ba^{2+}(aq)$ +	$SO_4^{2-}(aq)$	$K = 1.1 \times 10^{-10}$
initial	–	$0\ M$	$0\ M$	
change	–	$+x$	$+x$	
equilibrium	–	x	x	

$$K = [Ba^{2+}][SO_4^{2-}]$$
$$1.1\times10^{-10} = x^2$$
$$x = [Ba^{2+}] = \sqrt{1.1\times10^{-10}} = 1.05\times10^{-5}\ M$$

51. From the dissolution reaction, there are three sulfide ions for every two antimony ions, so the concentration of sulfide in the saturated solution is

$$[S^{2-}] = \frac{3}{2}[Sb^{3+}] = \frac{3}{2}(2.2\times10^{-19}\ M) = 3.3\times10^{-19}\ M$$

We write the mass-action expression for this reaction and then solve for the equilibrium constant:

$$K = [Sb^{3+}]^2[S^{2-}]^3 = (2.2\times10^{-19})^2(3.3\times10^{-19})^3 = 1.7\times10^{-93}$$

53. a. Using the same expression as above:

$$K=\frac{[NH_3]^2}{[N_2][H_2]^3}$$

$$[NH_3]=\sqrt{K[N_2][H_2]^3}=\sqrt{(0.060)(0.0010)(0.010)^3}=7.8\times10^{-6}\,M$$

b. $$[H_2]=\sqrt[3]{\frac{[NH_3]^2}{[N_2]K}}=\sqrt[3]{\frac{(0.020)^2}{(0.015)(0.060)}}=0.76\,M$$

55. a. Because the reaction is taking place in aqueous solution, there is also the self-ionization of water ($H_2O \rightarrow H^+ + OH^-$) in addition to the codeine reaction.

b. Because the equilibrium constant for the water self-ionization is only 1.0×10^{-14}, that reaction is not important in the generation of OH^- relative to the codeine reaction with $K = 1.6 \times 10^{-6}$.

c. The reaction quotient for this reaction is $Q=\frac{[C_{18}H_{21}NO_3H^+][OH^-]}{[C_{18}H_{21}NO_3]}=\frac{(0)(0)}{(0.10)}=0$; because $Q < K$, the reaction will proceed to the right.

d. Using the ICEA table, we can solve for the equilibrium concentrations:

$$C_{18}H_{21}NO_3(aq) + H_2O(l) \rightleftarrows OH^-(aq) + C_{18}H_{21}NO_3H^+(aq)$$

	$C_{18}H_{21}NO_3(aq)$	$OH^-(aq)$	$C_{18}H_{21}NO_3H^+(aq)$
initial	0.10	0 M	0 M
change	$-x$	$+x$	$+x$
equilibrium	$0.10-x$	x	x
assumption	0.10	x	x

$$K=\frac{[C_{18}H_{21}NO_3H^+][OH^-]}{[C_{18}H_{21}NO_3]}$$

$$1.6\times10^{-6}=\frac{(x)^2}{0.10}$$

$$x=[OH^-]=[C_{18}H_{21}NO_3H^+]=\sqrt{(1.6\times10^{-6})(0.10)}=4.0\times10^{-4}\,M$$

$$[C_{18}H_{21}NO_3]=0.10-x=0.10\,M$$

57. We can calculate the reaction quotient for the reaction (which has the same form as the mass-action expression): $Q=\frac{[Ag_{(Zn)}]}{[Ag_{(Pb)}]}=\frac{(0.0010)}{(0.000011)}=91$. The value of the equilibrium constant is 300 (because the silver is 300 times more soluble in zinc than in lead), so we have $Q < K$, and the reaction can proceed to increase the amount of silver in the zinc.

59. a. Using the mass-action expression for the reaction, we can solve for the concentration of CO directly:

$$K = \frac{1}{[\mathrm{CO}][\mathrm{H_2}]^2}$$

$$[\mathrm{H_2}] = \sqrt{\frac{1}{K[\mathrm{CO}]}} = \sqrt{\frac{1}{(13.5)(0.010)}} = 2.72\,M$$

b. The equilibrium line chart should show a final position slightly past center, indicating that the reaction slightly favors the formation of products.

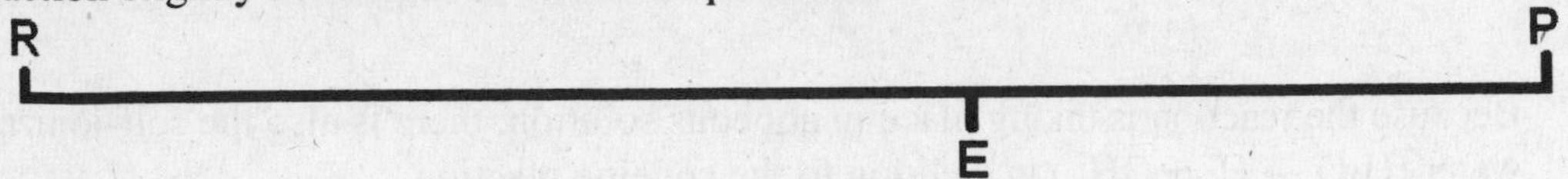

61. Using the mass-action expression for the reaction, we can solve for the concentration of H_2 directly:

$$K = \frac{[\mathrm{H_2}]^3[\mathrm{CO}]}{[\mathrm{CH_4}][\mathrm{H_2O}]}$$

$$[\mathrm{H_2}] = \sqrt[3]{\frac{K[\mathrm{CH_4}][\mathrm{H_2O}]}{[\mathrm{CO}]}} = \sqrt[3]{\frac{(0.25)(0.11)(0.28)}{(0.75)}} = 0.22\,M$$

63. Using the mass-action expression for the reaction, we can solve for the concentration of H_2 directly:

$$K = \frac{[\mathrm{C_2H_4}]}{[\mathrm{C_2H_2}][\mathrm{H_2}]}$$

$$[\mathrm{H_2}] = \frac{[\mathrm{C_2H_4}]}{[\mathrm{C_2H_2}]K} = \frac{(0.025)}{(1.2\times10^{-5})(4.2\times10^{15})} = 5.0\times10^{-13}\,M$$

65. Using the ICE table, we can solve for the equilibrium concentrations. Because the equilibrium constant is in excess of 1, the reaction goes too far for us to apply a simplifying assumption. Because 0.10 M H^+ would require only 0.05 M VO^+ to fully react, the H^+ limits the forward reaction.

	$VO^+(aq)$ +	$2\,H^+(aq)$ ⇄	$V^{3+}(aq)$ +	$H_2O(l)$
initial	0.15	0.10 M	0 M	
change	$-x$	$-2x$	$+x$	
equilibrium	$0.15-x$	$0.10-2x$	x	

$$K = \frac{[\mathrm{V^{3+}}]}{[\mathrm{VO^+}][\mathrm{H^+}]^2}$$

$$14 = \frac{(x)}{(0.15-x)(0.10-2x)^2}$$

This expression cannot be solved using the quadratic equation. Instead, we can solve the equation for x and use a series of successive approximations to find the value of x.

We can start with our first guess of $x = 0.01$:

$$14 = \frac{(x)}{(0.15 - x)(0.10 - 2x)^2}$$

$$x = 14(0.15 - x)(0.10 - 2x)^2$$

$$x = 14(0.15 - 0.01)(0.10 - 2(0.01))^2 = 0.0125$$

We then try again with $x = 0.0125$:

$$x = 14(0.15 - 0.0125)(0.10 - 2(0.0125))^2 = 0.0108$$

We use a third trial with the average of the first two ($x = 0.0116$) and then use the answer from that trial in successive trials:

$$x = 14(0.15 - 0.0116)(0.10 - 2(0.0116))^2 = 0.0114$$

$$x = 14(0.15 - 0.0114)(0.10 - 2(0.0114))^2 = 0.01156$$

$$x = 14(0.15 - 0.01156)(0.10 - 2(0.01156))^2 = 0.01145$$

$$x = 14(0.15 - 0.01145)(0.10 - 2(0.01145))^2 = 0.0115$$

$$x = 14(0.15 - 0.0115)(0.10 - 2(0.0115))^2 = 0.0115$$

After several trials, we find that the answer converges on $x = 0.0115$. The equilibrium concentrations are

$\left[V^{3+}\right] = x = 0.0115\,M$ $\quad\left[VO^{+}\right] = 0.15 - x = 0.138\,M$ $\quad\left[H^{+}\right] = 0.10 - 2x = 0.0770\,M$.

67. Using the ICE table, we can solve for the equilibrium concentrations. The reaction must proceed to the left because there is no O_2 present. But because the equilibrium constant is so small, the concentrations will shift very far to the left:

	Mb +	O_2	$\rightleftarrows$	MbO_2
initial	0.00100	0 M		0.000020 M
change	+0.000020 − x	+0.000020 − x		−0.000020 + x
equilibrium	0.00102 − x	0.000020 − x		x
assumption	0.00102	0.000020		x

$$K = \frac{[MbO_2]}{[Mb][O_2]}$$

$$8.6\times10^{-5} = \frac{(x)}{(0.00102)(0.000020)}$$

$$x = 8.6\times10^{-5}(0.00102)(0.000020) = 1.8\times10^{-12}\,M$$

$$[MbO_2] = x = 1.8\times10^{-12}\,M$$

$$[Mb] = 0.00102\,M$$

$$[O_2] = 0.000020\,M$$

The reaction did indeed shift almost completely to the left, leaving very little MbO_2.

69. a. **Removing CO_2 will cause a shift to the right in order to replace the CO_2 that was lost.**
b. **Adding heat will cause a shift to the left, because exothermic reactions shift toward the reactants for an increase in temperature.**

c. Decreasing the volume of the container (increasing the pressure of all gases in the container) will normally cause a shift toward fewer moles of gas. Because there are equal numbers of moles of gas on both sides of the reaction, the volume change has no effect on the reaction.

d. Adding H_2O will cause a shift to the left in order to remove the extra water added.

e. Adding an inert gas will not change the partial pressure of the reactant and product gases and therefore will have no effect on the reaction.

71. a. Because the equilibrium constant is larger for the dissolution of CuCl and the reactions are essentially the same, there will be more ions in solution at equilibrium for the CuCl system.

b. Because solids do not appear in the equilibrium expression, there is no effect on the reaction.

73. Because the reaction shifts toward the blue upon heating, we could assume that the reaction would shift in the opposite direction upon cooling—toward the pink.

75. The catalyst does nothing to affect the equilibrium; it only allows for the final distribution of concentrations to be reached more quickly.

77. For each case we use the modified equation relating the free energy change to the thermodynamic equilibrium constant:

$$\Delta G^\circ = -RT \ln K_{eq}$$

$$\ln K_{eq} = -\frac{\Delta G^\circ}{RT}$$

$$K_{eq} = e^{-\Delta G^\circ/RT}$$

a. $K_{eq} = e^{-\Delta G^\circ/RT} = e^{-(-1.05\,\text{J/mol·K})/(8.3145\,\text{J/mol·K}\times 298\,\text{K})} = 1.00$

b. $K_{eq} = e^{-\Delta G^\circ/RT} = e^{-(0.230\,\text{J/mol·K})/(8.3145\,\text{J/mol·K}\times 298\,\text{K})} = 1.00$

c. $K_{eq} = e^{-\Delta G^\circ/RT} = e^{-(2550\,\text{J/mol·K})/(8.3145\,\text{J/mol·K}\times 298\,\text{K})} = 0.357$

d. $K_{eq} = e^{-\Delta G^\circ/RT} = e^{-(-9800\,\text{J/mol·K})/(8.3145\,\text{J/mol·K}\times 298\,\text{K})} = 52.2$

79. $\Delta G^\circ = -RT \ln K_{eq} = -(8.3145\,\text{J/mol}\cdot\text{K})(298\,\text{K})\ln(13.5) = -6.45\times10^3\,\text{J/mol} = -6.45\,\text{kJ/mol}$

81. With $K = 0.020$, the free energy change for the reaction is

$\Delta G^\circ = -RT \ln K_{eq} = -(8.3145\,\text{J/mol}\cdot\text{K})(298\,\text{K})\ln(0.020) = 9.69\times10^3\,\text{J/mol} = 9.69\,\text{kJ/mol}$

83. Because the form of the equilibrium constant is the same for each reaction, we can compare the equilibrium constants directly. The strongest acid will have the largest K, indicating the largest shift toward products (and H^+) at equilibrium. The strongest acid is formic acid, HCOOH. The weakest acid has the smallest equilibrium constant: acetic acid, CH_3COOH.

85. To present the arguments, the most important calculation would be the total cost per unit of product formed. Method A uses more expensive reactants, but little will be wasted. With a much smaller equilibrium constant, Method B will produce less product per given amount of reactant. Much more reactant will be needed to produce the same amount of product as Method A, offsetting the lower cost of the reactants. The exact balance of reactant price versus amount of product formed will be the deciding factor.

87. The balanced reaction is $3\ O_2 \rightleftarrows 2\ O_3$. To finish with a single mole of ozone, the reaction must be written $3/2\ O_2 \rightleftarrows O_3$. Because we multiplied the first reaction by a factor of 1/2, we calculate the new equilibrium constant by raising the old constant to the 1/2 power:

$$K_{new} = K_{old}^{1/2} = \left(7.0\times10^{-58}\right)^{1/2} = 2.6\times10^{-29}$$

89. An equilibrium condition would exist if the total number of people inside the shop remained the same. The number of people entering the shop would equal the number of people finishing up and leaving. A larger value of K would indicate a greater number of people in the shop, which should be more profitable. If Q were less than K, there would be fewer people in the shop, a condition indicative of a poor business day.

91. Using the ICE table to solve for the equilibrium concentrations:

	$C(s) + H_2O(g)$	$\rightleftarrows CO(g)$	$+ H_2(g)$
initial	52 atm	0 atm	0 atm
change	$-x$	$+x$	$+x$
equilibrium	$52-x$	x	x

$$K = \frac{[CO][H_2]}{[H_2O]}$$

$$21 = \frac{(x)^2}{(52-x)} \qquad x = \frac{-21\pm\sqrt{(21)^2-4(1)(-1092)}}{2(1)}$$

$$21(52-x) = x^2 \qquad x = -10.5 \pm 34.67$$

$$1092 - 21x = x^2 \qquad x = [CO] = [H_2] = 24.2\,\text{atm}$$

$$x^2 + 21x - 1092 = 0 \qquad [H_2O] = 52 - x = 27.8\,\text{atm}$$

93. Starting in the frame at 1 min, the number of each compound remains the same (2 blue-blue, 2 blue-red, 2 red-black, 2 black-blue). Because the concentrations are the same, the system is at equilibrium.

94. a. The free energies of formation for $CO(g)$, $H_2(g)$ and $CH_3OH(g)$ are −137.2 kJ/mol, 0 kJ/mol, and −162.0 kJ/mol, respectively. The free energy change for the reaction is

$$\Delta G^\circ = \left(1\times(-162.0\,\text{kJ/mol})\right) - \left(2\times(0\,\text{kJ/mol}) + 1\times(-137.2\,\text{kJ/mol})\right) = -24.8\,\text{kJ/mol}$$

The reaction is spontaneous.

b. $K_{eq} = e^{-\Delta G^\circ/RT} = e^{-(-24800\,\text{J/mol}\cdot\text{K})/(8.3145\,\text{J/mol}\cdot\text{K}\times298\,\text{K})} = 2.22\times10^4$. Both the negative free energy change and large equilibrium constant show a reaction that is product-favored.

c. For the reverse reaction, the value of K is inverted: $K_{new} = \dfrac{1}{K_{old}} = \dfrac{1}{2.22\times10^4} = 4.5\times10^{-5}$.

d. At equilibrium, the forward and reverse rates of the reaction are the same. The reverse rate is 3.56×10^{-12} M/s.

e. The initial concentrations are

$$[\mathrm{CO}]=\frac{1.5\,\mathrm{mol}}{5\,\mathrm{L}}=0.30\,M$$

$$[\mathrm{H_2}]=\frac{3.5\,\mathrm{mol}}{5\,\mathrm{L}}=0.70\,M$$

Because the value of the equilibrium constant is so large, the reaction will go nearly to completion. The CO concentration is the limiting reagent for the reaction. Using the ICE table to solve for the equilibrium concentration of methanol:

	$CO(g)$ +	$2\,H_2(g)$ $\rightleftarrows$	$CH_3OH(g)$
initial	$0.30\,M$	$0.70\,M$	$0\,M$
change	$-0.30+x$	$-0.60+x$	$+0.30-x$
equilibrium	x	$0.10+x$	$0.30-x$
approximation	x	0.10	0.30

$$K=\frac{[\mathrm{CH_3OH}]}{[\mathrm{CO}][\mathrm{H_2}]^2}$$

$$2.22\times10^4=\frac{(0.30)}{(x)(0.10)^2}$$

$$x=\frac{(0.30)}{(2.22\times10^4)(0.10)^2}=1.35\times10^{-3}$$

$$[\mathrm{CH_3OH}]=0.30-x=0.30\,M$$

Checking the approximation, we see that x is only 1% of the 0.10 value. We are safe in using the approximation.

f. Adding more CO will shift the reaction to the right, producing more methanol.

g. The 1000 mL of methanol is equivalent to $1000\,\mathrm{mL}\times\frac{0.79\,\mathrm{g}}{1\,\mathrm{mL}}\times\frac{1\,\mathrm{mol}}{32.04\,\mathrm{g}}=24.66\,\mathrm{mol}$.

In 1 L of gas at 1 atm and 298 K, there is

$$n=\frac{PV}{RT}=\frac{(1\,\mathrm{atm})(1\,\mathrm{L})}{(0.08206\,\mathrm{L\cdot atm/mol\cdot K})(298\,\mathrm{K})}=0.0409\,\mathrm{mol}$$

The cost per mole of the reactants is

$$\mathrm{H_2}:\frac{\$0.10}{\mathrm{L}}\times\frac{1\mathrm{L}}{0.0409\,\mathrm{mol}}=\$2.44/\mathrm{mol}$$

$$\mathrm{CO}:\frac{\$0.30}{\mathrm{L}}\times\frac{1\mathrm{L}}{0.0409\,\mathrm{mol}}=\$7.33/\mathrm{mol}$$

The cost of each reactant used to produce 24.66 mol of methanol is

$$\mathrm{H_2}:24.66\,\mathrm{mol\,CH_3OH}\times\frac{2\,\mathrm{mol\,H_2}}{1\,\mathrm{mol\,CH_3OH}}\times\frac{\$2.44}{1\,\mathrm{mol\,H_2}}=\$120.32$$

$$\mathrm{CO}:24.66\,\mathrm{mol\,CH_3OH}\times\frac{1\,\mathrm{mol\,CO}}{1\,\mathrm{mol\,CH_3OH}}\times\frac{\$7.33}{1\,\mathrm{mol\,CO}}=\$180.76$$

The total cost is \$120.32 + \$180.76 = \$301.

Chapter 17: Acids and Bases

The Bottom Line

- Acids and bases can be defined using three different models. (Section 17.1)
- Acids and bases come in different strengths. (Section 17.2)
- Both the strength and the initial concentration of the acid affect the acidity in solution at equilibrium. (Section 17.2)
- Acids and bases have conjugates pairs whose behavior is related to that of the acid or base from which they are derived. (Section 17.1)
- We use pH as our common measure of acidity. (Section 17.3)
- We can interconvert among H+, OH−, pH, and pOH for a given acidic or basic solution. (Section 17.3)
- We can calculate the pH of strong and weak acids and bases in aqueous solutions. (Sections 17.4 and 17.5)
- We can solve for the pH of a polyprotic acid or base, including salts. (Sections 17.6 and 17.7)
- *Ka* and *Kb* are related via *Kw*. (Section 17.7)
- The reaction of an acid anhydride with water results in an acid, and the reaction of a basic anhydride with water results in a base. (Section 17.8)

Solutions to Practice Problems

17.1. Ethylenediamine is a base and will accept a proton from the water, leaving OH^-. The acid–base pairs are noted by the connectors.

$$\underset{\text{Base}}{H_2NCH_2CH_2NH_2\,(aq)} + \underset{\text{Acid}}{H_2O\,(l)} \rightleftarrows \underset{\text{Conjugate Acid}}{H_2NCH_2CH_2NH_3^+\,(aq)} + \underset{\text{Conjugate Base}}{OH^-(aq)}$$

17.2. Lauric acid reacts with water, which accepts the proton:

$$\underset{\text{Acid}}{CH_3(CH_2)_{10}COOH\,(aq)} + \underset{\text{Base}}{H_2O\,(l)} \rightleftarrows \underset{\text{Conjugate Base}}{CH_3(CH_2)_{10}COO^-\,(aq)} + \underset{\text{Conjugate Acid}}{H_3O^+(aq)}$$

17.3. Your answers can vary widely; three possible acids weaker than HF ($K_a = 7.2\times10^{-4}$) are acetic acid (CH_3COOH, $K_a = 1.8\times10^{-5}$), phenol (C_6H_5OH, $K_a = 1.6\times10^{-10}$), and hypobromous acid (HBrO, $K_a = 2.8\times10^{-9}$). Three acids that are stronger than HOCl ($K_a = 2.9\times10^{-8}$) are acetic acid (CH_3COOH, $K_a = 1.8\times10^{-5}$), hydrofluoric acid (HF, $K_a = 7.2\times10^{-4}$), and nitrous acid (HNO_2, $K_a = 4.0\times10^{-4}$).

17.4. Because the strength of an acid is inversely proportional to the strength of its conjugate base, acids whose conjugate bases are weaker than acetate will be stronger acids than acetic acid. Any of these are correct: hydrogen sulfate ion, chlorous acid, hydrofluoric acid, nitrous acid, and lactic acid.

17.5. Because bromine is more electronegative than sulfur (2.8 versus 2.5), electrons will be pulled away from the hydrogen end of the bond in HBr more than in H_2S. HBr should be a stronger acid than H_2S. (It is. HBr is a strong acid, and H_2S is a weak acid.)

17.6. Because HNO_3 is a strong acid, it will completely dissociate into H^+ and NO_3^- in aqueous solution. The number of moles of H^+ produced and the concentration are

$$2.38\,\text{g HNO}_3 \times \frac{1\,\text{mol HNO}_3}{63.02\,\text{g HNO}_3} \times \frac{1\,\text{mol H}^+}{1\,\text{mol HNO}_3} = 0.0378\,\text{mol H}^+$$

$$\frac{0.0378\,\text{mol H}^+}{500.0\,\text{L}} = 7.55\times10^{-5}\,M\,\text{H}^+$$

17.7. a. $\text{pH} = -\log(4.61\times10^{-3}) = 2.336$
b. $\text{p}K_a = -\log(2.77\times10^{-9}) = 8.558$
c. $\text{pOH} = -\log(3.22\times10^{-6}) = 5.492$
d. $\left[\text{H}^+\right] = 10^{-\text{pH}} = 10^{-3.92} = 1.2\times10^{-4}M$
e. $\left[\text{H}^+\right] = 10^{-\text{pH}} = 10^{-1.49} = 3.2\times10^{-2}M$
f. $[\text{OH}^-] = 10^{-\text{pOH}} = 10^{-9.93} = 1.2\times10^{-10}M$

17.8. If the hydrogen ion concentration is 500 times higher, the pH of the more acidic solution will be lower. Using the log rule that log(ab) = log a + log b, we can find the log of the 500 factor and subtract it (remember that pH = –log [H^+]) from the given pH: log 500 = 2.70. Therefore, for the more acidic water, pH = 8.84 – 2.70 = 6.14.

17.9. $\text{pH} = 14 - \text{pOH} = 14 - 12.35 = 1.65$; $\left[\text{H}^+\right] = 10^{-\text{pH}} = 10^{-1.65} = 0.022\,M$

$[\text{OH}^-] = 10^{-\text{pOH}} = 10^{-12.35} = 4.5\times10^{-13}\,M$

17.10. The nitric acid completely dissociates, so [H^+] = 0.0010 *M* and pH = –log(0.0010) = 3.00. The perchloric acid also completely dissociates, giving [H^+] = 0.0000250 *M* and pH = –log(0.0000250) = 4.602.

17.11. Using the approach used in the text:

Step 1. Equilibria in solution

The two important equilibria in the solution are the ionization of CH_3COOH and the autoprotolysis of water.

$$CH_3COOH(aq) \rightleftarrows H^+(aq) + CH_3COO^-(aq) \qquad K_a = 1.8\times10^{-5}$$

$$H_2O(l) \rightleftarrows H(aq) + OH^-(aq) \qquad K_w = 1.00\times10^{-14}$$

Step 2. Determine the equilibria that are the most important contributors to [H^+] in the solution.

Only the CH_3COOH ionization is assumed to be an important contributor to [H^+] in solution, because its K_a is far greater than the K_w of water.

Step 3. Write the equilibrium expression for the important contributors to [H^+].

$$K_a = \frac{[CH_3COO^-][H^+]}{[CH_3COOH]}$$

Step 4. Table of concentrations

	$CH_3COOH \rightleftarrows$	H^+ +	CH_3COO^-
Initial	0.250 *M*	0 *M*	0 *M*
Change	$-x$	$+x$	$+x$
Equilibrium	$0.250 - x$	$+x$	$+x$
With assumption	0.250	$+x$	$+x$

Step 5. Solve

$$1.8 \times 10^{-5} = \frac{(x)^2}{0.250}$$

$$x = [H^+] = [CH_3COO^-] = 0.0021\ M$$

Step 6. Check our assumption of negligible ionization.

$$\%\ \text{dissociated} = \frac{[CH_3COO^-]}{[CH_3COOH]} \times 100\% = \frac{0.0021}{0.250} \times 100\% = 0.8\%$$

We can consider the ionization to be negligible, because it falls within the "5% rule."

Step 7. Solve for the pH of the solution.

$$pH = -\log [H^+] = -\log (0.0021) = 2.67$$

17.12. NaOH is a strong base and will dissociate completely to give $[OH^-] = 1.06\ M$. We can find the pH by first calculating pOH and converting: pOH = –log (1.06) = –0.025 and pH = 14 – pOH = 14 – (–0.025) = 14.025.

17.13. Using the approach used in the text:

Step 1. Equilibria in solution

The two important equilibria in the solution are the base hydrolysis of $(CH_3)_2NH$ and the autoprotolysis of water.

$$(CH_3)_2NH(aq) + H_2O(l) \rightleftarrows (CH_3)_2NH_2^+(aq) + OH^-(aq) \qquad K_b = 5.9 \times 10^{-4}$$

$$H_2O(l) \rightleftarrows H^+(aq) + OH^-(aq) \qquad K_w = 1.00 \times 10^{-14}$$

Step 2. Determine the equilibria that are the most important contributors to [OH⁻] in the solution.

Only the $(CH_3)_2NH$ base hydrolysis is assumed to be an important contributor to $[OH^-]$ in solution because its K_b is far greater than the K_w of water.

Step 3. Write the equilibrium expression for the important contributors to [OH⁻].

$$K_b = \frac{[(CH_3)_2NH_2^+][OH^-]}{[(CH_3)_2NH]}$$

Step 4. Table of concentrations

	$H_2O + (CH_3)_2NH \rightleftarrows$	$(CH_3)_2NH_2^+$ +	OH^-
Initial	0.500 *M*	0 *M*	0 *M*
Change	$-x$	$+x$	$+x$
Equilibrium	$0.500 - x$	$+x$	$+x$
With assumption	0.500	$+x$	$+x$

Step 5. Solve

$$5.9 \times 10^{-4} = \frac{(x)^2}{0.500}$$

$$x = [OH^-] = [(CH_3)_2NH_2^+] = 0.017\ M$$

Step 6. Check our assumption of negligible ionization.

$$\% \text{ dissociated} = \frac{[(CH_3)_2NH_2^+]}{[(CH_3)_2NH]} \times 100\% = \frac{0.017}{0.500} \times 100\% = 3.4\%$$

We can consider the ionization to be negligible since it falls within the "5% rule".

Step 7. Solve for the pOH of the solution.

$$pOH = -\log[OH^-] = -\log(0.017) = 1.77$$
$$pH = 14 - pOH = 14 - 1.15 = 12.23$$

17.14. Using the technique from the text:

Step 1. Equilibria in solution

Step 2. Determine the equilibria that are the most important contributors to [H⁺] in the solution.

The most important equilibrium in the solution is the ionization of H_3PO_4. The autoprotolysis of water and the additional ionization of the products have K_w and K_a values that are much lower.

$$H_3PO_4(aq) \rightleftarrows H^+(aq) + H_2PO_4^-(aq) \qquad K_a = 7.5 \times 10^{-3}$$

Step 3. Write the equilibrium expression for the important contributors to [H⁺].

$$K_a = \frac{[H_2PO_4^-][H^+]}{[H_3PO_4]}$$

Step 4. Table of concentrations

	H_3PO_4 $\rightleftarrows$	H^+ +	$H_2PO_4^-$
Initial	0.200 M	0 M	0 M
Change	$-x$	$+x$	$+x$
Equilibrium	$0.200 - x$	$+x$	$+x$
With assumption	0.200	$+x$	$+x$

Step 5. Solve

$$7.5 \times 10^{-3} = \frac{(x)^2}{0.200}$$
$$x = [H^+] = [H_2PO_4^-] = 0.038\ M$$

Step 6. Check our assumption of negligible ionization.

$$\% \text{ dissociated} = \frac{[H_2PO_4^-]}{[H_3PO_4]} \times 100\% = \frac{0.038}{0.200} \times 100\% = 19\%$$

Therefore, our approximation cannot be used, and we must use the quadratic equation to solve for x:

$$7.5\times10^{-3} = \frac{(x)^2}{0.200 - x}$$
$$7.5\times10^{-3}(0.200 - x) = (x)^2$$
$$x^2 + 7.5\times10^{-3}x - 1.5\times10^{-3} = 0$$
$$x = \frac{-7.5\times10^{-3} \pm \sqrt{(7.5\times10^{-3})^2 - 4(1)(-1.5\times10^{-3})}}{2(1)}$$
$$x = -3.75\times10^{-3} \pm 3.89\times10^{-2} = 0.0352\,M \text{ or } -0.0426\,M$$

Only the positive value for x makes sense, because we cannot finish with negative concentrations. Therefore, $[H^+] = 0.035\ M$.

Step 7. ***Solve for the pH of the solution.***

$$pH = -\log [H^+] = -\log (0.035) = 1.46$$

17.15. Acetate is the conjugate base of acetic acid and will undergo base hydrolysis in aqueous solution:

$$CH_3COO^- + H_2O \rightleftarrows CH_3COOH + OH^-$$

We determine the K_b from the K_a of acetic acid (1.8×10^{-10}):

$$K_b = \frac{K_w}{K_a} = \frac{1.0\times10^{-14}}{1.8\times10^{-5}} = 5.5_6\times10^{-10}$$

We then treat the rest of the problem as a straightforward base problem:

Step 1. ***Equilibria in solution***

The two important equilibria in the solution are the base hydrolysis of CH_3COO^- and the autoprotolysis of water.

$$CH_3COO^- + H_2O \rightleftarrows CH_3COOH + OH^- \qquad K_b = 5.5_6 \times 10^{-10}$$

$$H_2O(l) \rightleftarrows H^+(aq) + OH^-(aq) \qquad K_w = 1.00 \times 10^{-14}$$

Step 2. ***Determine the equilibria that are the most important contributors to [OH⁻] in the solution.***

Only the base hydrolysis is assumed to be an important contributor to $[OH^-]$ in solution, because its K_b is far greater than the K_w of water.

Step 3. ***Write the equilibrium expression for the important contributors to [OH⁻].***

$$K_b = \frac{[CH_3COOH][OH^-]}{[CH_3COO^-]}$$

Step 4. Table of concentrations

	CH_3COO^- + H_2O ⇄	CH_3COOH +	OH^-
Initial	0.250 M	0 M	0 M
Change	$-x$	$+x$	$+x$
Equilibrium	$0.250 - x$	$+x$	$+x$
With assumption	0.250	$+x$	$+x$

Step 5. ***Solve***

$$5.5_6 \times 10^{-10} = \frac{(x)^2}{0.250}$$

$$x = [OH^-] = [CH_3COO^-] = 1.2\times10^{-5}\,M$$

Step 6. ***Check our assumption of negligible reaction.***

$$\%\text{ reacted} = \frac{[CH_3COOH]}{[CH_3COO^-]} \times 100\% = \frac{1.2\times10^{-5}}{0.250} \times 100\% = 0.005\%$$

We can consider the ionization to be negligible, because it falls within the "5% rule."

Step 7. ***Solve for the pOH of the solution.***

$$pOH = -\log [OH^-] = -\log(1.2\times10^{-5}\,M) = 4.93$$

$$pH = 14 - pOH = 14 - 4.93 = 9.07$$

17.16. Note that in Exercise 16, the equation does not depend on the concentration of the solution. The pH will still be 6.10.

Solutions to Student Problems

1. In the Arrhenius model, a base produces OH^- in solution, which ammonia does in the reaction below. In the Brønsted–Lowry model, a base accepts a proton from another species (the acid), which NH_3 does from water in the formation of NH_4^+ in the reaction below.

$$NH_3 + H_2O \rightleftarrows NH_4^+ + OH^-$$

3. In naming the conjugate base of an acid, we need only remove a proton from the acid formula:

a. NO_3^-
b. Br^-
c. OH^-
d. ClO_4^-

5. In naming the conjugate acid of a base, we need only add a proton to the base formula:

a. H_2O (we add the proton to the OH^-)
b. NH_4^+
c. H_3O^+
d. HF (we add the proton to the F^-)

7. The balanced reactions and product names are

a. $2\ Al(s) + 6\ HCl(aq) \rightarrow 2\ AlCl_3(aq) + 3\ H_2(g)$; aluminum chloride
b. $Ca(s) + 2\ HNO_3(aq) \rightarrow Ca(NO_3)_2(aq) + H_2(g)$; calcium nitrate
c. $2\ Na(s) + 2\ HCl(aq) \rightarrow 2\ NaCl(aq) + H_2(g)$; sodium chloride
d. $2\ K(s) + 2\ HNO_3(aq) \rightarrow 2\ KNO_3(aq) + H_2(g)$; potassium nitrate

9. A strong acid is one that dissociates completely in solution to form H^+.

11. a.

HCl +	H_2O	$\rightarrow$	Cl^- +	H_3O^+
acid	base		conj. base	conj. acid

b.

$NaOH$ +	CH_3COOH $\rightarrow$	CH_3COONa +	H_2O
base	acid	conj. base	conj. acid

c.

H_2SO_4 +	$Mg(OH)_2$ $\rightarrow$	$MgSO_4$ +	$2\ H_2O$
acid	base	conj. base	conj. acid

13.

$Mg(OH)_2$	+ $2\ HCl$ $\rightarrow$	$MgCl_2$ +	$2\ H_2O$
base	acid	conj. base	conj. acid

15. a. Strong acids have K_a values that are greater than 1.
b. The conjugate base of a strong acid cannot regain an H^+ in aqueous solution to any measurable degree.
c. In 0.10 *M* solution, a strong acid will be 100% dissociated (as is true for many but the highest of concentrations).

17. Even though bromine is slightly more electronegative than iodine, the iodide ion is much larger and has a much more "spread out" electron density than bromide ion. The attraction of H^+ for the larger I^- with its lower electron density is less, so the HI bond breaks more easily, and HI is the stronger acid.

19. Because both acids have the same number of oxygen atoms that contribute to pulling the electrons from the bond to hydrogen, we must consider the electronegativity of the central halogen atom. Chlorine has a higher electronegativity than bromine and weakens the bond to hydrogen more than in $HBrO_4$; therefore, $HClO_4$ must be the stronger acid. (Recall that in these species, the hydrogen is bound to one of the oxygen atoms, so the electron density argument of Cl versus Br is not useful.)

21. a. Conjugate base strength is inversely proportional to the strength of the acid. The weaker acid, acetic acid, will have the stronger conjugate base.
b. Because the K_a value is larger for formic acid, and the initial concentrations are the same, the formic acid will dissociate more.

23. The reaction of the acid can be written Acid $\rightarrow$ Conj. Base + H^+. The equilibrium constant could then be written (and solved) as follows:

$$K_a = \frac{[\text{Conj. Base}][H^+]}{[\text{Acid}]} = \frac{(0.0011)(0.0011)}{0.10} = 1.2\times10^{-5}$$

25. a. $\text{pH} = -\log(4.55\times10^{-3}) = 2.342$
b. $\text{pH} = -\log(3.27\times10^{-6}) = 5.485$
c. $\text{pH} = -\log(8.11\times10^{-9}) = 8.091$

27. The complete table:

	$[H^+](M)$	pH	$[OH^-](M)$	pOH
(a)	**3.8 x 10^{-5}**	4.42	**2.6 x 10^{-10}**	**9.58**
(b)	0.0056	**2.25**	**1.8 x 10^{-12}**	**11.75**
(c)	**1.3 x 10^{-10}**	**9.89**	0.000078	**4.11**
(d)	**1.3 x 10^{-4}**	**3.9**	**7.9 x 10^{-11}**	10.10

The calculations:

a. $[H^+] = 10^{-\text{pH}} = 10^{-4.42} = 3.8\times10^{-5}\ M$, $\text{pOH} = 14 - \text{pH} = 9.58$,
$[OH^-] = 10^{-\text{pOH}} = 10^{-9.58} = 2.6\times10^{-10}\ M$

b. $\text{pH} = -\log[H^+] = -\log(0.0056) = 2.25$, $\text{pOH} = 14 - \text{pH} = 11.75$,
$[OH^-] = 10^{-\text{pOH}} = 10^{-11.75} = 1.8\times10^{-12}\ M$

c. $\text{pOH} = -\log[OH^-] = -\log(0.000078) = 4.11$, $\text{pH} = 14 - \text{pOH} = 9.89$,
$[H^+] = 10^{-\text{pH}} = 10^{-9.89} = 1.3\times10^{-10}\ M$

d. $[OH^-] = 10^{-pOH} = 10^{-10.10} = 7.9\times10^{-11}\ M$, pH = 14 – pOH = 3.90,
$[H^+] = 10^{-pH} = 10^{-3.90} = 1.3\times10^{-4}\ M$

29. Raising the pH value to twice its initial value lowers the hydronium ion concentration (higher pH corresponds to lower $[H^+]$).

31. The hydroxide concentration is $\dfrac{10.0\,g\,OH^- \times \dfrac{1\,mol\,OH^-}{17.01\,g\,OH^-}}{10.0\,L} = 5.88\times10^{-2}\ M$.

Therefore, pOH = –log(0.0588) = 1.231, and pH = 14 – pOH = 12.769.

33. a. The hydrogen ion concentration is $[H^+] = \dfrac{1.55\times10^{-5}\ mol}{0.500\,L} = 3.10\times10^{-5}\ M$. Therefore, pH = –log $[H^+]$ = $-\log(3.10\times10^{-5}) = 4.509$.

b. The hydrogen ion concentration is $[H^+] = \dfrac{7.25\times10^{-6}\ mol}{0.250\,L} = 2.90\times10^{-5}\ M$. Therefore, pH = –log $[H^+]$ = $-\log(2.90\times10^{-5}) = 4.538$.

35. We need to find the difference in hydrogen ion concentration in the two solutions and use that value to find the moles:

$[H^+] = 10^{-pH} = 10^{-4.35} = 4.5\times10^{-5}\ M$ $\quad [H^+] = 10^{-pH} = 10^{-5.85} = 1.4\times10^{-6}\ M$.

The difference is 4.4×10^{-5} mol/L. Because the hydrogen ion concentration is decreasing, we must add 4.4×10^{-5} moles of hydroxide for each liter.

37. a. All strong acids will dissociate 100% in solution, so the hydrogen ion concentration is the same as the given acid concentration. We can calculate the pH directly from that value: pH = –log $[H^+]$ = –log(0.45) = 0.35

b. pH = –log $[H^+]$ = –log(0.045) = 1.35

c. pH = –log $[H^+]$ = –log(0.000487) = 3.312

d. pH = –log $[H^+]$ = –log(0.00026) = 3.59

39. a. In the first two of these cases, the concentration is large enough or the acid is weak enough (though still stronger than the autoionization of water) that we can safely apply the approximation of negligible ionization in solving the problems. For complete step-by-step solutions to this type of equilibrium, see the solutions to the practice problems. For HOCl, $K_a = 3.5\times10^{-8}$.

	$HOCl \rightleftarrows$	$H^+ +$	OCl^-
Initial	0.45 M	0 M	0 M
Change	$-x$	$+x$	$+x$
Equilibrium	$0.45 - x$	$+x$	$+x$
With assumption	0.45	$+x$	$+x$

$$K_a = \frac{[H^+][OCl^-]}{[HOCl]}$$

$$3.5\times10^{-8} = \frac{(x)^2}{0.45}$$

$$x = [H^+] = \sqrt{(0.45)(3.5\times10^{-8})} = 1.2_5\times10^{-4}\ M$$

$$pH = -\log(1.2_5\times10^{-4}\ M) = 3.90$$

b. For 0.0250M CH_3COOH ($K_a = 1.8 \times 10^{-5}$):

	$CH_3COOH \rightleftarrows$	H^+ +	CH_3COO^-
Initial	0.0250 M	0 M	0 M
Change	$-x$	$+x$	$+x$
Equilibrium	$0.0250 - x$	$+x$	$+x$
With assumption	0.0250	$+x$	$+x$

$$K_a = \frac{[H^+][CH_3COO^-]}{[CH_3COOH]}$$

$$1.8\times10^{-5} = \frac{(x)^2}{0.0250}$$

$$x = [H^+] = \sqrt{(0.0250)(1.8\times10^{-5})} = 6.7\times10^{-4}\ M$$

$$pH = -\log(6.7\times10^{-4}\ M) = 3.17$$

c. For 0.18 M HF ($K_a = 6.3\times10^{-4}$):

	HF $\rightleftarrows$	H^+ +	F^-
Initial	0.18 M	0 M	0 M
Change	$-x$	$+x$	$+x$
Equilibrium	$0.18 - x$	$+x$	$+x$
With assumption	0.18	$+x$	$+x$

$$K_a = \frac{[H^+][F^-]}{[HF]}$$

$$6.3\times10^{-4} = \frac{(x)^2}{0.18}$$

$$x = [H^+] = \sqrt{(0.18)(6.3\times10^{-4})} = 1.1\times10^{-2}\ M$$

$$\frac{1.1\times10^{-2}\ M}{0.18}\times100\% = 6\%$$

Our assumption violates the 5% rule; therefore, we must apply the quadratic equation to the problem:

$$6.3\times10^{-4} = \frac{(x)^2}{0.18-x}$$
$$6.3\times10^{-4}(0.18-x)=x^2$$
$$x^2+6.3\times10^{-4}x-1.134\times10^{-4}=0$$
$$x=\frac{-6.3\times10^{-4}\pm\sqrt{(6.3\times10^{-4})^2-4(1)(-1.134\times10^{-4})}}{2(1)}$$
$$x=-3.15\times10^{-4}\pm1.06\times10^{-2}=1.03\times10^{-2}\,M \text{ or } -1.09\times10^{-2}\,M$$
$$\left[H^+\right]=1.0\times10^{-2}\,M$$
$$pH=-\log\left[H^+\right]=-\log(1.0\times10^{-2})=2.00$$

d. For 0.0010 M HCOOH ($K_a = 1.8\times10^{-4}$):

	$HCOOH \rightleftarrows$	$H^+ +$	$HCOO^-$
Initial	0.0010 M	0 M	0 M
Change	$-x$	$+x$	$+x$
Equilibrium	$0.0010-x$	$+x$	$+x$
With assumption	0.0010	$+x$	$+x$

$$K_a=\frac{\left[H^+\right]\left[HCOO^-\right]}{[HCOOH]}$$
$$1.8\times10^{-4}=\frac{(x)^2}{0.0010}$$
$$x=\left[H^+\right]=\sqrt{(0.0010)(1.8\times10^{-4})}=4.24\times10^{-4}\,M$$
$$\frac{4.24\times10^{-4}\,M}{0.0010}\times100\%=42\%$$

Our assumption violates the 5% rule; therefore, we must apply the quadratic equation to the problem:

$$1.8\times10^{-4}=\frac{(x)^2}{0.0010-x}$$
$$1.8\times10^{-4}(0.0010-x)=x^2$$
$$x^2+1.8\times10^{-4}x-1.8\times10^{-7}=0$$
$$x=\frac{-1.8\times10^{-4}\pm\sqrt{(1.8\times10^{-4})^2-4(1)(-1.8\times10^{-7})}}{2(1)}$$
$$x=-9.0\times10^{-5}\pm4.34\times10^{-4}=3.44\times10^{-4}\,M \text{ or } -5.24\times10^{-4}\,M$$
$$\left[H^+\right]=3.4\times10^{-4}\,M$$
$$pH=-\log\left[H^+\right]=-\log(3.4\times10^{-4})=3.47$$

41. a. $pK_a = -\log K_a = -\log(3.75\times10^{-5}) = 4.426$

b. $pK_a = -\log K_a = -\log(1.84\times10^{-2}) = 1.735$

c. $pK_a = -\log K_a = -\log(4.59\times10^{-8}) = 7.338$

43. The base hydrolysis of codeine in water is $C_{18}H_{21}NO_3 + H_2O \rightarrow HC_{18}H_{21}NO_3^+ + OH^-$. To find the value for K_b, we write the expression for the constant and then plug in the values. The concentration for the conjugate acid and the hydroxide will be given by

$$x = \left[HC_{18}H_{21}NO_3^+\right] = [OH^-] = 10^{-pOH}$$

$$pOH = 14 - pH = 14 - 10.19 = 3.81$$

$$x = \left[HC_{18}H_{21}NO_3^+\right] = [OH^-] = 10^{-3.81} = 1.5_5\times10^{-4}\ M$$

	$C_{18}H_{21}NO_3 + H_2O \rightleftarrows$	$HC_{18}H_{21}NO_3^+$	$+ OH^-$
Initial	$0.015\ M$	$0\ M$	$0\ M$
Change	$-x$	$+x$	$+x$
Equilibrium	$0.015 - x$	$+x$	$+x$

$$K_b = \frac{\left[HC_{18}H_{21}NO_3^+\right][OH^-]}{[C_{18}H_{21}NO_3]} = \frac{(x)^2}{0.015 - x} = \frac{(1.5_5\times10^{-4})^2}{0.015 - 1.5_5\times10^{-4}} = 1.6\times10^{-6}$$

45. a. The concentration of benzoic acid ($K_a = 6.5\times10^{-5}$) is $\dfrac{1.00\,\text{g}\times\dfrac{1\,\text{mol}}{122\,\text{g}}}{0.500\,\text{L}} = 0.0164\,M$. The hydrogen ion concentration of this solution is

	$C_6H_5COOH \rightleftarrows$	H^+	$+ C_6H_5COO^-$
Initial	$0.0164\ M$	$0\ M$	$0\ M$
Change	$-x$	$+x$	$+x$
Equilibrium	$0.0164 - x$	$+x$	$+x$
With assumption	0.0164	$+x$	$+x$

$$K_a = \frac{\left[H^+\right]\left[C_6H_5COO^-\right]}{[C_6H_5COOH]}$$

$$6.5\times10^{-5} = \frac{(x)^2}{0.0164}$$

$$x = \left[H^+\right] = \sqrt{(0.0164)(6.5\times10^{-5})} = 1.0\times10^{-3}\ M$$

b. Sulfuric acid has two acidic protons, and the first one acts as a strong acid. The hydrogen ion concentration from the first proton would be

$$[H^+] = 0.0001\ \text{mol} / 0.100\ \text{L} = 0.0010\ M$$

The second proton is weak and will only raise the concentration. It is unnecessary to calculate the exact value, because the hydrogen ion concentration is already in excess of the threshold in the question.

Both solutions have a hydrogen ion concentration in excess of 0.0000010 M.

47. a. $1.00\times10^3\text{ g CaF}_2 \times \frac{1\text{ mol CaF}_2}{78.08\text{ g CaF}_2} \times \frac{2\text{ mol HF}}{1\text{ mol CaF}_2} \times \frac{20.01\text{ g HF}}{1\text{ mol HF}} = 513\text{ g HF}$

b. Using our standard technique to solve first for the hydrogen ion concentration and then for the other ion concentrations in HF:
For 0.25 *M* HF ($K_a = 7.2\times10^{-4}$):

	HF ⇄	H^+ +	F^-
Initial	0.25 *M*	0 *M*	0 *M*
Change	$-x$	$+x$	$+x$
Equilibrium	$0.25 - x$	$+x$	$+x$
With assumption	0.25	$+x$	$+x$

$$K_a = \frac{[H^+][F^-]}{[HF]}$$

$$7.2\times10^{-4} = \frac{(x)^2}{0.25}$$

$$x = \sqrt{(0.25)(7.2\times10^{-4})} = 1.342\times10^{-2}\ M$$

$$\frac{1.342\times10^{-2}\ M}{0.25}\times100\% = 5.4\%$$

Our assumption violates the 5% rule; therefore, we must apply the quadratic equation to the problem:

$$7.2\times10^{-4} = \frac{(x)^2}{0.25 - x}$$

$$7.2\times10^{-4}(0.25 - x) = x^2$$

$$x^2 + 7.2\times10^{-4}x - 1.8\times10^{-4} = 0$$

$$x = \frac{-7.2\times10^{-4} \pm \sqrt{(7.2\times10^{-4})^2 - 4(1)(-1.8\times10^{-4})}}{2(1)}$$

$$x = -3.6\times10^{-4} \pm 1.34\times10^{-2} = 1.30\times10^{-2}\ M \text{ or } -1.38\times10^{-2}\ M$$

$$[H^+] = 1.3\times10^{-2}\ M$$

$$[F^-] = [H^+] = 1.3\times10^{-2}\ M$$

$$[OH^-] = \frac{K_w}{[H^+]} = \frac{1.0\times10^{-14}}{1.3\times10^{-2}} = 7.7\times10^{-13}\ M$$

49. The concentration of the solution is $\frac{0.5000\text{ g C}_6\text{H}_8\text{O}_6 \times \frac{1\text{ mol C}_6\text{H}_8\text{O}_6}{176.12\text{ g C}_6\text{H}_8\text{O}_6}}{0.355\text{ L}} = 0.00800\ M$. Solving for the pH:

	$C_6H_8O_6$ ⇄	H^+ +	$C_6H_7O_6^-$
Initial	0.00800 *M*	0 *M*	0 *M*
Change	$-x$	$+x$	$+x$
Equilibrium	$0.00800 - x$	$+x$	$+x$
With assumption	0.00800	$+x$	$+x$

$$K_a = \frac{[H^+][C_6H_7O_6^-]}{[C_6H_8O_6]}$$

$$8.0\times10^{-5} = \frac{(x)^2}{0.00800}$$

$$x = [H^+] = \sqrt{(0.00800)(8.0\times10^{-5})} = 8.0\times10^{-4}\ M$$

$$\text{Percent ionized} = \frac{8.0\times10^{-4}}{(0.00800)}\times100\% = 10\%$$

Because the initial concentration was so low relative to the K_a (causing a large percent ionization), we cannot use our approximation and must use the quadratic equation instead:

$$8.0\times10^{-5} = \frac{(x)^2}{0.00800 - x}$$

$$8.0\times10^{-5}(0.00800 - x) = x^2$$

$$x^2 + 8.0\times10^{-5}x - 6.4\times10^{-7} = 0$$

$$x = \frac{-8.0\times10^{-5} \pm \sqrt{(8.0\times10^{-5})^2 - 4(1)(-6.4\times10^{-7})}}{2(1)}$$

$$x = -4.0\times10^{-5} \pm 8.01\times10^{-4} = 7.61\times10^{-4}\ M \text{ or } -8.41\times10^{-4}\ M$$

$$[H^+] = 7.6\times10^{-4}\ M$$

$$pH = -\log[H^+] = -\log(7.6\times10^{-4}) = 3.12$$

51. Solving for the hydroxide ion concentration of a base is very similar to finding the hydrogen ion concentration for acids. The exact step-by-step solution can be found in the practice problems. We will assume that the main hydrolysis is only from the base (not water) and that we can make our simplifying approximation.

a. For 0.100 *M* aniline ($C_6H_5NH_2$, $K_b = 3.8\times10^{-8}$):

	$C_6H_5NH_2 + H_2O$	$\rightleftarrows C_6H_5NH_3^+$	$+ OH^-$
Initial	0.100M	0 *M*	0 *M*
Change	$-x$	$+x$	$+x$
Equilibrium	$0.100 - x$	$+x$	$+x$
With assumption	0.100	$+x$	$+x$

$$K_b = \frac{[C_6H_5NH_3^+][OH^-]}{[C_6H_5NH_2]}$$

$$3.8\times10^{-8} = \frac{(x)^2}{0.100}$$

$$x = [OH^-] = \sqrt{(0.100)(3.8\times10^{-8})} = 6.1_6\times10^{-5}\ M$$

$$pOH = -\log(6.1_6\times10^{-5}) = 4.21$$

$$pH = 14 - pOH = 9.79$$

b. NaOH is a strong base and dissociates completely. $[OH^-] = 0.0100\ M$ and pOH = 2.00, which gives a pH = 12.00.

c. For 0.250 M NH_3 ($K_b = 1.8\times10^{-5}$):

	NH_3 +	$H_2O \rightleftarrows$	NH_4^+ +	OH^-
Initial	0.250M		0 M	0 M
Change	$-x$		$+x$	$+x$
Equilibrium	$0.25-x$		$+x$	$+x$
With assumption	0.250		$+x$	$+x$

$$K_b = \frac{[NH_4^+][OH^-]}{[NH_3]}$$

$$1.8\times10^{-5} = \frac{(x)^2}{0.250}$$

$$x = [OH^-] = \sqrt{(0.250)(1.8\times10^{-5})} = 2.1\times10^{-3}\ M$$

$$pOH = -\log(2.1\times10^{-3}) = 2.68$$

$$pH = 14 - pOH = 11.32$$

53. Solving for the hydroxide ion concentration of a base is very similar to finding the hydrogen ion concentration for acids. The exact step-by-step solution can be found in the practice problems. We will assume that the main hydrolysis is only from the base (not water) and that we can make our simplifying approximation.
For 0.0010 M pyridine ($K_b = 1.4\times10^{-9}$):

	Pyr +	$H_2O \rightleftarrows HPyr^+$ +	OH^-
Initial	0.0010 M	0 M	0 M
Change	$-x$	$+x$	$+x$
Equilibrium	$0.0010-x$	$+x$	$+x$
With assumption	0.0010	$+x$	$+x$

$$K_b = \frac{[HPyr^+][OH^-]}{[Pyr]}$$

$$1.4\times10^{-9} = \frac{(x)^2}{0.0010}$$

$$x = [OH^-] = \sqrt{(0.0010)(1.4\times10^{-9})} = 1.1_8\times10^{-6}\ M$$

$$pOH = -\log(1.1_8\times10^{-6}) = 5.93$$

$$pH = 14 - pOH = 8.07$$

55. A pH = 11.87 corresponds to pOH = 2.13, and $[OH^-] = 10^{-pOH} = 10^{-2.13} = 0.0074\,M$. The concentration of ethylamine is $\dfrac{1.90\,g\,CH_3CH_2NH_2 \times \dfrac{1\,mol\,CH_3CH_2NH_2}{45.09\,g\,CH_3CH_2NH_2}}{0.5000\,L} = 0.0843\,M$. The concentration of the conjugate base will be the same as the hydroxide ion concentration, so we can find the base hydrolysis constant:

$$K_b = \frac{[CH_3CH_2NH_3^+][OH^-]}{[CH_3CH_2NH_2]} = \frac{x^2}{0.0843-x} = \frac{(0.0074)(0.0074)}{0.0843-0.0074} = 7.1\times10^{-4}$$

57. The three ionization steps for H_3PO_4 are

$$H_3PO_3 + H_2O \rightleftarrows H_2PO_3^- + H_3O^+$$
$$H_2PO_3^- + H_2O \rightleftarrows HPO_3^{2-} + H_3O^+$$
$$HPO_3^{2-} + H_2O \rightleftarrows PO_3^{3-} + H_3O^+$$

Phosphorus has an oxidation number of +3 in each of the species.

59. All of the hydrogen ion will come solely from the first ionization step of phosphoric acid (K_{a1} = 7.4×10^{-3}). Because phosphoric acid has such a large ionization constant, we need not even try the approximation and may proceed directly to the quadratic equation.

	$H_3PO_4 \rightleftarrows$	$H^+ +$	$H_2PO_4^-$
Initial	1.0 M	0 M	0 M
Change	$-x$	$+x$	$+x$
Equilibrium	$1.0-x$	$+x$	$+x$

$$K_{a1} = \frac{[H_2PO_4^-][H^+]}{[H_3PO_4]}$$

$$7.4\times10^{-3} = \frac{(x)^2}{1.0-x}$$

$$7.4\times10^{-3}(1.0-x) = x^2$$

$$x^2 + 7.4\times10^{-3}x - 7.4\times10^{-3} = 0$$

$$x = \frac{-7.4\times10^{-3} \pm \sqrt{(7.4\times10^{-3})^2 - 4(1)(-7.4\times10^{-3})}}{2(1)}$$

$$x = -0.0037 \pm 0.0867 = 0.0830 \text{ or } -0.0904$$

$$x = [H^+] = [H_2PO_4^-] = 0.083\,M$$

$$pH = -\log[H^+] = -\log(0.083) = 1.08$$

We expect no real additional hydrogen ion concentration to be added by the second ionization step, so we can use our current values for $[H^+]$ and $[H_2PO_4^-]$ in K_{a2} to solve for $[HPO_4^{2-}]$:

$$K_{a2} = \frac{[HPO_4^{2-}][H^+]}{[H_2PO_4^-]}$$

$$[HPO_4^{2-}] = K_{a2}\frac{[H_2PO_4^-]}{[H^+]} = 6.2\times10^{-8}\frac{0.083\,M}{0.083\,M} = 6.2\times10^{-8}\,M$$

In summary: $[HPO_4^{2-}] = 6.2\times10^{-8}\,M \qquad pH = 1.08$

61. The two reactions follow. (The conjugate bases are shown in the boxes.)

$$H_2SO_3 + H_2O \rightleftarrows H_3O^+ + \boxed{HSO_3^-}$$
$$HSO_3^- + H_2O \rightleftarrows H_3O^+ + \boxed{SO_3^{2-}}$$

63. Using N to represent nicotine, there are two stepwise hydrolysis steps:

$$N + H_2O \rightleftarrows NH^+ + OH^-$$
$$NH^+ + H_2O \rightleftarrows NH_2^+ + OH^-$$

We treat this multiple set of basic steps just as we would handle a polyprotic acid: Nearly all of the OH^- will come from the first step. Therefore,

	$N + H_2O \rightleftarrows$	NH^+ +	OH^-
Initial	0.045 *M*	0 *M*	0 *M*
Change	$-x$	$+x$	$+x$
Equilibrium	$0.045 - x$	$+x$	$+x$
With assumption	0.045	$+x$	$+x$

$$K_b = \frac{[NH^+][OH^-]}{[N]}$$

$$7.0\times10^{-7} = \frac{(x)^2}{0.045}$$

$$x = [OH^-] = \sqrt{(0.045)(7.0\times10^{-7})} = 1.8\times10^{-4}\ M$$

$$pOH = -\log(1.8\times10^{-4}) = 3.75$$

$$pH = 14 - pOH = 10.25$$

The second step starts with a much weaker acid at lower concentration and will not add to the hydroxide concentration appreciably.

65. NH_4Cl has an acidic cation (originating from the base NH_3); NaCl has no acid–base properties and should form a neutral solution; and $NaC_2H_3O_2$ has a basic anion (originating from the acid $HC_2H_3O_2$). These solutions are acidic, neutral, and basic, respectively, so in order of decreasing pH: $NaC_2H_3O_2$, NaCl, NH_4Cl.

67. Lower pK_a values correspond to stronger acids. The conjugate base of a stronger acid is itself a weaker base, so Y^- will be the stronger base and will result in the more basic solution.

69. The isoelectric pH is the point where a chemical (usually an amino acid) is in its zwitterionic form (dual positive and negative charges) and is electrically neutral. Each amino acid will have its own isoelectric pH, which enables chemists to separate the amino acids electrically in a medium that has a pH gradient. The amino acid will move in the electric field until it reaches its isoelectric pH and stops moving.

71. We can treat $Zn(H_2O)_6$ just as we would any monoprotic acid and find the pH:

	$Zn(H_2O)_6 \rightleftarrows$	H^+ +	$Zn(H_2O)_5OH^-$
Initial	0.10 *M*	0 *M*	0 *M*
Change	$-x$	$+x$	$+x$
Equilibrium	$0.10 - x$	$+x$	$+x$
With assumption	0.10	$+x$	$+x$

$$K_a = \frac{[H^+][Zn(H_2O)_6OH^-]}{[Zn(H_2O)_6]}$$

$$2.4\times10^{-10} = \frac{(x)^2}{0.10}$$

$$x = [H^+] = \sqrt{(0.10)(2.4\times10^{-10})} = 4.9\times10^{-6}\ M$$

$$pH = -\log(4.9\times10^{-6}) = 5.31$$

73. a. $K_a = \frac{K_w}{K_b} = \frac{1.00\times10^{-14}}{4.26\times10^{-5}} = 2.34\times10^{-10}$

b. $K_a = \frac{K_w}{K_b} = \frac{1.00\times10^{-14}}{8.36\times10^{-9}} = 1.20\times10^{-6}$

c. $K_a = 10^{-pK_a} = 10^{-2.85} = 1.4\times10^{-3}$

75. $HCOO^-$ is the conjugate base of HCOOH ($K_a = 1.8\times10^{-4}$), so $K_b = \frac{K_w}{K_a} = \frac{1.0\times10^{-14}}{1.8\times10^{-4}} = 5.6\times10^{-11}$.

We finish the problem treating it as simple base problem:

	$HCOO^- + H_2O \rightleftarrows$	$HCOOH +$	OH^-
Initial	0.050 M	0 M	0 M
Change	$-x$	$+x$	$+x$
Equilibrium	$0.050 - x$	$+x$	$+x$
With assumption	0.050	$+x$	$+x$

$$K_b = \frac{[HCOOH][OH^-]}{[HCOO^-]}$$

$$5.6\times10^{-11} = \frac{(x)^2}{0.050}$$

$$x = [OH^-] = \sqrt{(0.050)(5.6\times10^{-11})} = 1.7\times10^{-6}\ M$$

$$pOH = -\log(1.7\times10^{-6}) = 5.78$$

$$pH = 14 - pOH = 8.22$$

77. a. $NH_3 + HCl \rightarrow NH_4Cl$

b. The solution is acidic. NH_4^+ is the conjugate acid to the weak base NH_3 and is a weak acid. Cl^- is the conjugate base to the strong acid HCl and has no base properties.

c. We first need the K_a for NH_4^+: $K_a(NH_4^+) = \frac{K_w}{K_b(NH_3)} = \frac{1.0\times10^{-14}}{1.8\times10^{-5}} = 5.6\times10^{-10}$. We can then solve the problem as a simple acid ionization:

	$NH_4^+ \rightleftarrows$	$NH_3 +$	H^+
Initial	0.136 M	0 M	0 M
Change	$-x$	$+x$	$+x$
Equilibrium	$0.136 - x$	$+x$	$+x$
With assumption	0.136	$+x$	$+x$

$$K_a = \frac{[H^+][NH_3]}{[NH_4^+]}$$

$$5.6\times10^{-10} = \frac{(x)^2}{0.136}$$

$$x = [H^+] = \sqrt{(0.136)(5.6\times10^{-10})} = 8.7\times10^{-6}\ M$$

$$pH = -\log(8.7\times10^{-6}) = 5.06$$

79. Carbonate is the conjugate base of the weak acid HCO_3^- ($K_a = 5.6\times10^{-11}$). We first need the K_b for CO_3^{2-}: $K_b = \dfrac{K_w}{K_a} = \dfrac{1.0\times10^{-14}}{5.6\times10^{-11}} = 1.8\times10^{-4}$. We finish the problem treating it as simple base problem:

	CO_3^{2-} + H_2O ⇄	HCO_3^-	+ OH^-
Initial	0.150 M	0 M	0 M
Change	$-x$	$+x$	$+x$
Equilibrium	$0.150-x$	$+x$	$+x$
With assumption	0.150	$+x$	$+x$

$$K_b = \frac{[HCO_3^-][OH^-]}{[CO_3^{2-}]}$$

$$1.8\times10^{-4} = \frac{(x)^2}{0.150}$$

$$x = [OH^-] = \sqrt{(0.150)(1.8\times10^{-4})} = 5.2\times10^{-3}\ M$$

$$pOH = -\log(5.2\times10^{-3}) = 2.28$$

$$pH = 14 - pOH = 11.72$$

81. a. $C_6H_5COO^- + H_2O \rightarrow C_6H_5COOH + OH^-$, $K_b = \dfrac{K_w}{K_a} = \dfrac{1.0\times10^{-14}}{6.5\times10^{-5}} = 1.5\times10^{-10}$

b. Solving for the pH:

	$C_6H_5COO^-$ + H_2O ⇄	C_6H_5COOH	+ OH^-
Initial	0.010 M	0 M	0 M
Change	$-x$	$+x$	$+x$
Equilibrium	$0.010-x$	$+x$	$+x$
With assumption	0.010	$+x$	$+x$

$$K_b = \frac{[C_6H_5COO^-][OH^-]}{[C_6H_5COOH]}$$

$$1.5\times10^{-10} = \frac{(x)^2}{0.010}$$

$$x = [OH^-] = \sqrt{(0.010)(1.5\times10^{-10})} = 1.2_2\times10^{-6}\ M$$

$$pOH = -\log(1.2_2\times10^{-6}) = 5.91$$

$$pH = 14 - pOH = 8.09$$

83. $H_3\overset{\oplus}{N}-CH_2-CH_2-CH_2-CH_2-\underset{COO^{\ominus}}{\overset{H}{C}}-NH_2$

85. To determine the anhydride that creates a certain acid, we need only remove H_2O from the formula: H_2SO_4 becomes SO_3, for which the structure is O=S(=O)=O.

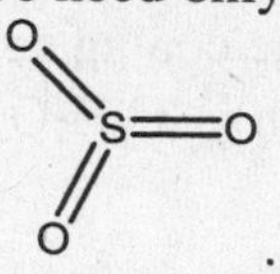

87. a. Because water is a neutral molecule, the addition of a positive proton to the structure results in a +1 charge.

b. The hydrogen ion can associate with more than one unit of water to form $H_5O_2^+$, $H_7O_3^+$, etc.

c. The Lewis structure of H_3O^+ has three bonds and one lone pair on the central oxygen. This molecule would have a tetrahedral electron geometry. Because the lone pair takes one corner of the tetrahedron, the molecular geometry is a trigonal pyramid.

89. Yes. Even though a strong acid is 100% dissociated, a low enough concentration of the acid could be made to match the acidity of a concentrated weak acid solution.

91. Acetic acid is a stronger acid than water and therefore makes a weaker base than water. The strong acids will have a more difficult time dissociating in an acetic acid solution, so the differences in the strong acids can be detected by their varying dissociation in acetic acid.

93. After the first dissociation of a diprotic acid is complete, the anion left behind has a negative charge. This extra charge has two effects: It "satisifies" the need of the electronegative atoms for the extra electrons, reducing the pull on the electrons in the remaining bond to hydrogen, thereby strengthening the bond relative to the neutral molecule. And when the second proton is removed, a −2 ion results, which has a greater attraction to the proton, making it more difficult to remove.

95. a. Ephedrine alkaloids act as a stimulant, as does caffeine. The combination of two stimulants greatly increases the chance of undesired side effects, including death.

b. The ephedrine structure:

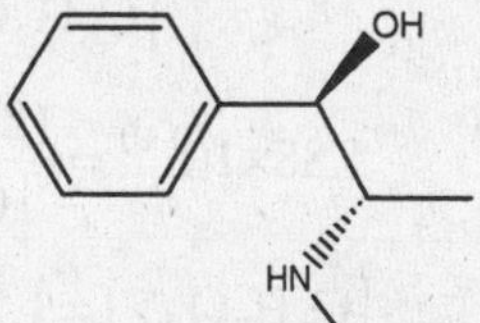

OH

HN

c. Caffeine (shown below) is an alkaloid. Alkaloids are natural nitrogen-containing molecules. The nitrogen in the structure is an amine that has basic (alkaline) properties; hence the name *alkaloid*.

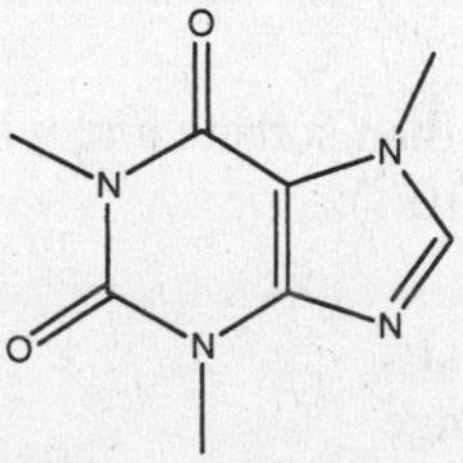

O N N O N N

Some of the common alkaloids include nicotine (shown below) and the amino acids.

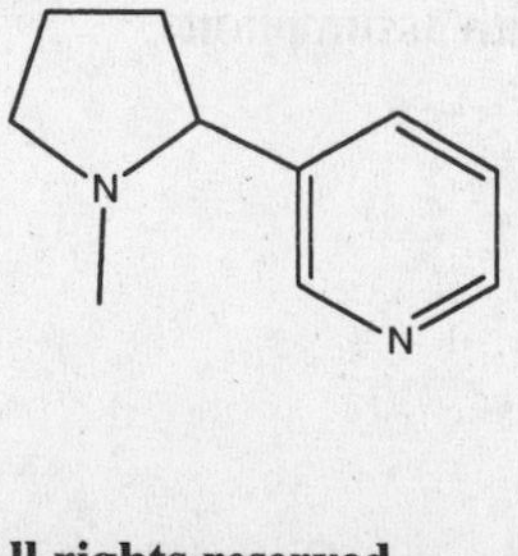

N N

97. a. Because sorbate is the conjugate base of a weak acid, it will form a basic solution.

b. $KC_6H_7O_2(s) \rightarrow K^+(aq) + C_6H_7O_2^-(aq)$ and $C_6H_7O_2^- + H_2O \rightleftarrows HC_6H_7O_2 + OH^-$.

c. Solving for the pH:

	$C_6H_7O_2^- + H_2O \rightleftarrows$	$HC_6H_7O_2 +$	OH^-
Initial	0.00100 M	0 M	0 M
Change	$-x$	$+x$	$+x$
Equilibrium	$0.00100 - x$	$+x$	$+x$
With assumption	0.00100	$+x$	$+x$

$$K_b = \frac{[HC_6H_7O_2][OH^-]}{[C_6H_7O_2^-]}$$

$$5.88\times10^{-10} = \frac{(x)^2}{0.00100}$$

$$x = [OH^-] = \sqrt{(0.0100)(5.88\times10^{-10})} = 2.42\times10^{-6}\,M$$

$$pOH = -\log(2.42\times10^{-6}) = 5.61$$

$$pH = 14 - pOH = 8.38$$

d. With the given pH, we can find the concentration of both products species (see reaction in part c) as they form in a 1:1 ratio:

$$pOH = 14 - pH = 4.56$$

$$[HC_6H_7O_2] = [OH^-] = 10^{-pOH} = 10^{-4.56} = 2.75\times10^{-5}\,M$$

Using the equilibrium constant expression, we can solve for the sorbate concentration that is needed:

$$K_b = \frac{[HC_6H_7O_2][OH^-]}{[C_6H_7O_2^-]}$$

$$5.88\times10^{-10} = \frac{(2.75\times10^{-5})^2}{[C_6H_7O_2^-]}$$

$$[C_6H_7O_2^-] = 1.29\,M$$

The mass needed is

$$0.500\,L\,C_6H_7O_2^- \times \frac{1.29\,mol\,C_6H_7O_2^-}{1L} \times \frac{1\,mol\,KC_6H_7O_2}{1\,mol\,C_6H_7O_2^-} \times \frac{150.22\,g\,KC_6H_7O_2}{1\,mol\,KC_6H_7O_2} = 96.9\,g\,KC_6H_7O_2$$

e. NH_3 is a much stronger base than sorbate and will control the pH of the solution. For 0.010 M NH_3 ($K_b = 1.8\times10^{-5}$):

	$NH_3 + H_2O \rightleftarrows$	$NH_4^+ +$	OH^-
Initial	0.010 M	0 M	0 M
Change	$-x$	$+x$	$+x$
Equilibrium	$0.010 - x$	$+x$	$+x$
With assumption	0.010	$+x$	$+x$

$$K_b = \frac{[NH_4^+][OH^-]}{[NH_3]}$$

$$1.8\times10^{-5} = \frac{(x)^2}{0.010}$$

$$x = [OH^-] = \sqrt{(0.010)(1.8\times10^{-5})} = 4.2_4\times10^{-4}\ M$$

$$pOH = -\log(4.2_4\times10^{-4}) = 3.37$$

$$pH = 14 - pOH = 10.63$$

The pH of the solution is much higher (10.63 versus 8.38) as a consequence of the presence of the ammonia.

Chapter 18: Applications of Aqueous Equilibria

The Bottom Line

- A titration is a technique used to find out how much of a substance is in a solution. (Section 18.1)
- There are several types of titrations, including reduction–oxidation, precipitation, complex-formation, and acid–base titration. (Section 18.1)
- Many reactions are pH-sensitive and require buffers to control pH. (Section 18.1)
- A buffer resists change in pH upon addition of a strong acid or strong base or upon dilution. (Section 18.1)
- Buffers are typically composed of weak conjugate acid–base pairs. (Section 18.1)
- We can solve for the pH of buffers in a straightforward way by recognizing the importance of Le Châtelier's principle and the common-ion effect. (Section 18.1)
- We can calculate the approximate ratio of conjugate acid to base in order to prepare a buffer of a known pH. Activity effects are important, and our acid-to-base ratio will probably need to be slightly adjusted to be at the desired pH. (Section 18.1)
- Solving for the pH of a buffer upon addition of strong acid or base is really solving a limiting-reactant problem. (Section 18.1)
- It is possible to exceed the buffer capacity, in which case the pH will go sharply higher (with excess base) or lower (with excess acid). (Section 18.1)
- Strong-acid–strong-base titrations show a relatively level pH until near the equivalence point, where the pH dramatically changes. (Section 18.2)
- Titration curves in which one component is weak and the other is strong contain four regions: the initial pH, the buffer region, the equivalence-point region and the post–equivalence point region. (Section 18.2)
- The buffer region contains a point at which one-half of the analyte has been converted to its conjugate. This is called the titration midpoint, and the pH is equal to the pK_a of the analyte. (Section 18.2)
- The larger the pK of the analyte, the sharper will be the change in pH at the equivalence point. (Section 18.2)
- The higher the concentration of the weak acid (or base) and the strong base (or acid), the sharper the endpoint. (Section 18.2)
- An indicator is used to visually detect the equivalence point of a titration. (Section 18.2)
- Only a few drops of an indicator are added to the titration solution so that the equivalence point and endpoint can be as close together as possible. (Section 18.2)
- Solubility equilibria can often be complex, involving several side reactions and molecular-level processes that make calculations challenging. (Section 18.2)
- The effects of ion-pairing, activity, and other thermodynamic considerations add to the challenge of properly calculating the concentration of dissolved salts in aqueous solution. (Section 18.3)
- Gravimetric analysis is based on weighing the precipitate that includes the substance of interest. (Section 18.3)
- The pH of an aqueous solution can significantly affect the solubility of the substances in that solution. (Section 18.3)
- A chemical complex typically consists of one or more metal cations bonded to one or more Lewis bases. (Section 18.4)

- The formation constant is a measure of the extent of reaction between a Lewis base and metal ion in aqueous solution. (Section 18.4)
- EDTA is the primary example of a highly effective chelating agent. (Section 18.4)
- The reaction of chelating agents and metal ions has a very high formation constant. (Section 18.4)
- The analysis of calcium in hard water by EDTA titration is an important application of complex-ion equilibrium. (Section 18.4)

Solutions to Practice Problems

18.1. Just as in the exercise, we will assume that the concentrations of ammonia and ammonium ion are changed little from the initial values: $[NH_3] = [NH_3]_0$ and $[NH_4^+] = [NH_4^+]_0$. We can solve directly for the equilibrium concentration of H^+ and for the pH:

$$K_a = \frac{[NH_3][H_3O^+]}{[NH_4^+]}$$

$$[H_3O^+] = K_a \frac{[NH_4^+]}{[NH_3]}$$

$$[H_3O^+] = 5.6\times10^{-10} \times \frac{(3.50)}{(1.50)} = 1.3\times10^{-9}\,M$$

$$pH = -\log[H_3O^+] = -\log(1.3\times10^{-9}) = 8.88$$

18.2. The K_a for formic acid is 1.8×10^{-4} while the K_b for formate ion is $K_b = \frac{K_w}{K_a} = 5.6\times10^{-11}$, indicating that the acid is much stronger than the conjugate base. The resulting mixture should be acidic. We will assume that the initial concentrations do not change much and calculate the pH:

$$K_a = \frac{[HCOO^-][H_3O^+]}{[HCOOH]}$$

$$[H_3O^+] = K_a \frac{[HCOOH]}{[HCOO^-]}$$

$$[H_3O^+] = 1.8\times10^{-4} \times \frac{(0.300)}{(0.400)} = 1.3_5\times10^{-4}\,M$$

$$pH = -\log[H_3O^+] = -\log(1.3_5\times10^{-4}) = 3.87$$

18.3. From the given information, we can use the acid dissociation expression to solve for the necessary concentration of formate ion and use that concentration to find the mass of sodium formate needed to create the buffer. We begin by calculating the concentration of the hydrogen ion from the pH:

$$[H^+] = 10^{-pH} = 10^{-3.95} = 1.1_2\times10^{-4}\,M$$

Solving for the formate ion concentration:

$$K_a = \frac{[HCOO^-][H_3O^+]}{[HCOOH]}$$

$$[HCOO^-] = K_a \frac{[HCOOH]}{[H_3O^+]}$$

$$[HCOO^-] = 1.8\times10^{-4} \times \frac{(0.150)}{(1.1_2 \times 10^{-4})} = 0.24\,M$$

The mass of formate needed can be found by a string of conversions from the concentration of formate ion:

$$500\,\text{mL soln} \times \frac{1\,\text{L}}{1000\,\text{mL}} \times \frac{0.24\,\text{mol HCOO}^-}{1\,\text{L soln}} \times \frac{1\,\text{mol HCOONa}}{1\,\text{mol HCOO}^-} \times \frac{68.01\,\text{g HCOONa}}{1\,\text{mol HCOONa}}$$
$$= 8.2\,\text{g HCOONa}$$

18.4. In this case, we create the buffer by converting some of the ammonia to ammonium ion by adding the strong acid HCl. (The HCl reacts fully with the ammonia to form ammonium ion until the HCl is consumed.) To solve for the amount of HCl needed, we first need to find the equilibrium amounts of the ammonia and ammonium ion. We are forced to use the molar amounts instead of concentrations, because the addition of the HCl solution will change the overall volume of the solution. However, the number of moles will not be affected by the dilution. The total amount of ammonia and ammonium ion in the solution is the same as the initial amount of ammonia:

$$\text{mol NH}_3 + \text{mol NH}_4^+ = 0.2000\,M \times 0.0250\text{L} = 0.00500\,\text{mol}$$

We can solve for the ratio of the two species by using the acid dissociation expression. Note that because both species are in the same volume, their concentration ratio will be the same as the mole ratio.

$$K_a = \frac{[NH_3][H_3O^+]}{[NH_4^+]}$$

$$\frac{\text{mol NH}_3}{\text{mol NH}_4^+} = \frac{[NH_3]}{[NH_4^+]} = \frac{K_a}{[H_3O^+]} = \frac{K_a}{10^{-pH}} = \frac{5.6\times10^{-10}}{10^{-8.80}} = 0.35_3$$

We can use both results to find the concentrations at equilibrium:

$$\text{mol NH}_3 + \text{mol NH}_4^+ = 0.00500$$

$$\text{mol NH}_3 = 0.35_3(\text{mol NH}_4^+)$$

$$0.35_3(\text{mol NH}_4^+) + (\text{mol NH}_4^+) = 0.00500$$

$$1.35_3(\text{mol NH}_4^+) = 0.00500$$

$$\text{mol NH}_4^+ = \frac{0.00500}{1.35_3} = 0.00370\,\text{mol}$$

$$\text{mol NH}_3 = 0.00130$$

The amount of NH_4^+ created is equivalent to the amount of HCl added to the solution:

$$0.00370\,\text{mol}\,NH_4^+ = 0.00370\,\text{mol}\,HCl$$

$$0.00370\,\text{mol}\,HCl \times \frac{1\,\text{L}}{0.10\,\text{mol}\,HCl} \times \frac{1000\,\text{mL}}{1\,\text{L}} = 37\,\text{mL}\,HCl$$

18.5. This problem is quickly solved using the Henderson–Hasselbalch equation. The pK_a of acetic acid is $pK_a = -\log K_a = -\log(1.8\times10^{-5}) = 4.74$. The pH of the solution is

$$\text{pH} = pK_a + \log\left(\frac{[CH_3COO^-]}{[CH_3COOH]}\right)$$

$$\text{pH} = 4.74 + \log\left(\frac{0.250M}{0.500M}\right) = 4.44$$

18.6. Starting with the system given in Exercise 18.6 that contains 0.100 mol NH_3 and 0.400 mol NH_4^+, we find the new pH after the addition of 30.0 mL of 0.100 *M* HCl (= 0.003 mol added). The reaction between HCl and NH_3 goes to completion and sets the initial amounts for the new buffer:

	H^+ +	NH_3 →	NH_4^+
moles initial	0.00 *M*	0.100 *M*	0.040 *M*
moles added	0.00300	0	0
change	−0.00300	−0.00300	+0.00300
moles at equil.	≈0	0.097	0.043

If we use the Henderson–Hasselbalch equation to solve for the pH, we can take advantage of the fact that the ratio of the moles of ammonia to the moles of ammonium is the same as the ratio of the concentrations. (Recall that they share the same volume.)

$$\text{pH} = pK_a + \log\frac{[NH_3]}{[NH_4^+]} = 9.26 + \log\left(\frac{0.097}{0.043}\right) = 9.61$$

We solve for the pH after the addition of 50.0 mL of 1.50 *M* HCl (= 0.075 mol) in the same manner as above:

	H^+ +	NH_3 →	NH_4^+
moles initial	0.00	0.100	0.0400
moles added	0.075	0	0
change	−0.075	−0.075	+0.075
moles at equil.	≈0	0.025	0.115

$$\text{pH} = pK_a + \log\frac{[NH_3]}{[NH_4^+]} = 9.26 + \log\left(\frac{0.025}{0.115}\right) = 8.59$$

18.7. We start by using the equation to solve for the ratio of the base concentration to the acid concentration:

$$\mathrm{pH} = \mathrm{p}K_\mathrm{a} + \log\frac{[\mathrm{HCOO^-}]}{[\mathrm{HCOOH}]}$$

$$3.54 = 3.74 + \log\frac{[\mathrm{HCOO^-}]}{[\mathrm{HCOOH}]}$$

$$\log\frac{[\mathrm{HCOO^-}]}{[\mathrm{HCOOH}]} = -0.20$$

$$\frac{[\mathrm{HCOO^-}]}{[\mathrm{HCOOH}]} = 10^{-0.20} = 0.631$$

$$\text{or}\ \frac{[\mathrm{HCOOH}]}{[\mathrm{HCOO^-}]} = \frac{1}{0.631} = 1.58$$

For the rest of the problem, we proceed in exactly the same way as in the exercise by using the two equations to solve for the two concentrations:

$$[\mathrm{HCOOH}] = 1.58[\mathrm{HCOO^-}]$$

$$\text{so mol HCOOH} = 1.58 \times \text{mol HCOO}^-$$

$$\text{and mol HCOOH} + \text{mol HCOO}^- = 0.0300$$

then

$$1.58 \times \text{mol HCOO}^- + \text{mol HCOO}^- = 0.0300$$

$$2.58 \times \text{mol HCOO}^- = 0.0300$$

$$\text{mol HCOO}^- = \frac{0.0300}{2.58} = 0.0116\ \text{mol}$$

$$\text{mol HCOOH} = 0.0300 - 0.0116 = 0.0184\ \text{mol}$$

The change in the concentration of the acid is 0.0184 mol – 0.0150 mol = 0.0034 mol. The number of moles of HCl added is equal to this change. The volume of HCl required is

$$0.0034\ \text{mol HCl} \times \frac{1.0\ \text{L HCl}}{0.1000\ \text{mol HCl}} = 0.034\ \text{L} = 34\ \text{mL}$$

This answer is essentially the same as the answer obtained in the exercise, differing only as a consequence of rounding errors in the pK_a and the concentration ratio. (When we use values without rounding throughout the calculation, the ratio of the concentration of the acid to that of the base is 1.60, which is the same as in the exercise.)

18.8. At any volume of titrant added prior to the equivalence point, some of the analyte will remain in excess and will determine the pH of the solution. Because the analyte is H^+ and the titrant is OH^-, we can summarize the reaction tables in the exercise with the following equation. The numerator represents the number of moles of H^+ that remain after subtracting the number of moles of added OH^-. The denominator is the total volume that exists after the two solutions are mixed.

$$[\mathrm{H^+}] = \frac{M_{\mathrm{H^+}} \times V_{\mathrm{H^+}} - M_{\mathrm{OH^-}} \times V_{\mathrm{OH^-}}}{V_{\mathrm{H^+}} + V_{\mathrm{OH^-}}}$$

Starting with 25.00 mL of 0.2500 M H^+ and adding 0.2500 M OH^-:
At 0.00mL OH^- added, the initial concentration of HCl is the H^+ concentration, 0.2500 M. The pH is

$$\text{pH} = -\log\left[\text{H}^+\right] = -\log(0.2500) = 0.6021$$

At 10.00 mL OH^- added:

$$\left[\text{H}^+\right] = \frac{M_{\text{H}^+} \times V_{\text{H}^+} - M_{\text{OH}^-} \times V_{\text{OH}^-}}{V_{\text{H}^+} + V_{\text{OH}^-}} = \frac{0.2500\frac{\text{mol}}{\text{L}} \times 25.00\,\text{mL} - 0.2500\frac{\text{mol}}{\text{L}} \times 10.00\,\text{mL}}{(25.00 + 10.00)\text{mL}}$$

$$\left[\text{H}^+\right] = 0.1071\,M$$

$$\text{pH} = -\log\left[\text{H}^+\right] = -\log(0.1071) = 0.9700$$

At 20.00 mL OH^- added:

$$\left[\text{H}^+\right] = \frac{M_{\text{H}^+} \times V_{\text{H}^+} - M_{\text{OH}^-} \times V_{\text{OH}^-}}{V_{\text{H}^+} + V_{\text{OH}^-}} = \frac{0.2500\frac{\text{mol}}{\text{L}} \times 25.00\,\text{mL} - 0.2500\frac{\text{mol}}{\text{L}} \times 20.00\,\text{mL}}{(25.00 + 20.00)\text{mL}}$$

$$\left[\text{H}^+\right] = 0.02778\,M$$

$$\text{pH} = -\log\left[\text{H}^+\right] = -\log(0.02778) = 1.5563$$

At 25.00 mL OH^- added, the number of moles of H^+ added and the number of moles of OH^- added are equal. The pH of the solution is determined by the self-ionization of water. The solution is neutral with a pH = 7.

At 30.00 mL OH^- added, all of the H^+ has been consumed and the OH^- is in excess. We change the equation to solve for the excess OH^- and obtain the pH via the pOH:

$$\left[\text{OH}^-\right] = \frac{M_{\text{OH}^-} \times V_{\text{OH}^-} - M_{\text{H}^+} \times V_{\text{H}^+}}{V_{\text{OH}^-} + V_{\text{H}^+}} = \frac{0.2500\frac{\text{mol}}{\text{L}} \times 30.00\,\text{mL} - 0.2500\frac{\text{mol}}{\text{L}} \times 25.00\,\text{mL}}{(25.00 + 30.00)\text{mL}}$$

$$\left[\text{OH}^-\right] = 0.02273\,M$$

$$\text{pOH} = -\log\left[\text{OH}^-\right] = -\log(0.02273) = 1.6435$$

$$\text{pH} = 14 - \text{pOH} = 12.3565$$

At 40.00 mL OH^- added:

$$\left[\text{OH}^-\right] = \frac{M_{\text{OH}^-} \times V_{\text{OH}^-} - M_{\text{H}^+} \times V_{\text{H}^+}}{V_{\text{OH}^-} + V_{\text{H}^+}} = \frac{0.2500\frac{\text{mol}}{\text{L}} \times 40.00\,\text{mL} - 0.2500\frac{\text{mol}}{\text{L}} \times 25.00\,\text{mL}}{(25.00 + 40.00)\text{mL}}$$

$$\left[\text{OH}^-\right] = 0.05769\,M$$

$$\text{pOH} = -\log\left[\text{OH}^-\right] = -\log(0.05769) = 1.2389$$

$$\text{pH} = 14 - \text{pOH} = 12.7611$$

Plotting these pH values versus the volume of OH^- added yields the titration curve:

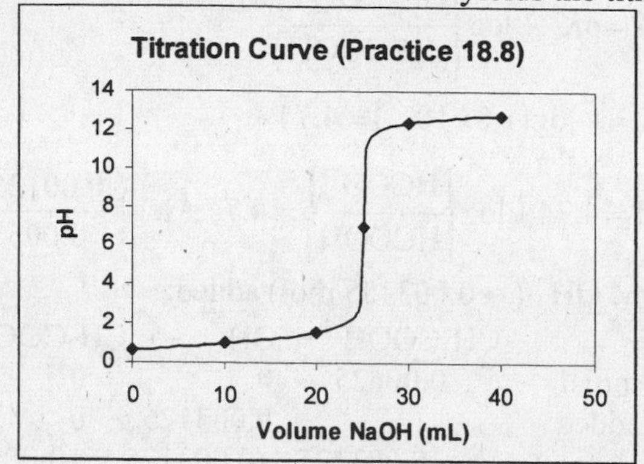

18.9. We break the calculations into four regions: the initial point, where the acidity of the acid alone determines the pH; a pre-equivalence region, where both the acid and conjugate base are present and the buffer determines the pH; the equivalence point, where the acid has been completely converted into its conjugate base, which determines the pH; and a post-equivalence region, where the excess titrant determines the pH.

At 0.00 mL OH^- added:

	$CH_3COOH \rightleftarrows$	CH_3COO^- +	H^+
initial	0.250 *M*	0 *M*	0 *M*
change	$-x$	$+x$	$+x$
equilibrium	$0.250 - x$	x	x
assumption	0.250	x	x

$$K_a = \frac{[CH_3COO^-][H^+]}{[CH_3COOH]}$$

$$1.8\times10^{-5} = \frac{x^2}{0.2500}$$

$$x = [H^+] = \sqrt{(1.8\times10^{-5})(0.2500)} = 2.1\times10^{-3}\,M$$

$$pH = -\log[H^+] = -\log(2.1\times10^{-3}) = 2.67$$

At 5.00 mL of 0.2500M OH^- (= 0.001250 mol) added, some of the acid is converted into the conjugate base. We can take advantage of the Henderson–Hasselbalch equation and use only the moles of acid and base in the equation to determine the pH. We first determine the number of moles of each from the reaction:

	CH_3COOH +	OH^- →	CH_3COO^- + H_2O
moles initial	0.00625	0	0
moles added		0.00125	0
change	−0.00125	−0.00125	+0.00125
moles at equil.	0.005	≈0	0.00125

$$pH = pK_a + \log\frac{[CH_3COO^-]}{[CH_3COOH]}$$

$$pK = -\log(1.8\times10^{-5}) = 4.74$$

$$pH = 4.74 + \log\frac{[HCOO^-]}{[HCOOH]} = 4.74 + \log\left(\frac{0.00125}{0.005}\right) = 4.14$$

At 12.50 mL of 0.2500M OH^- (= 0.003125 mol) added:

	CH_3COOH	+ OH^-	→ CH_3COO^- + H_2O
moles initial	0.00625	0	0
moles added		0.003125	0
change	–0.003125	–0.003125	+0.003125
moles at equil.	0.003125	≈0	0.003125

$$pH = 4.74 + \log\frac{[HCOO^-]}{[HCOOH]} = 4.74 + \log\left(\frac{0.003125}{0.003125}\right) = 4.74$$

At 24.00 mL of 0.2500M OH^- (= 0.006000 mol) added:

	CH_3COOH	+ OH^-	→ CH_3COO^- + H_2O
moles initial	0.00625	0	0
moles added		0.00600	0
change	–0.00600	–0.00600	+0.00600
moles at equil.	0.00025	≈0	0.00600

$$pH = 4.74 + \log\frac{[HCOO^-]}{[HCOOH]} = 4.74 + \log\left(\frac{0.00600}{0.00025}\right) = 6.12$$

At 25.00 of 0.2000M OH^- (= 0.006250 mol) added:

	CH_3COOH	+ OH^-	→ CH_3COO^- + H_2O
moles initial	0.00625	0	0
moles added		0.00625	0
change	–0.00625	–0.00625	+0.00625
moles at equil.	0	≈0	0.00625

The concentration of the acetate ion is $[CH_3COO^-] = \dfrac{0.006250\,\text{mol}}{(0.02500\text{L} + 0.02500\text{L})} = 0.1250\,M$.

The pH is determined by the hydrolysis of the acetate ion:

	CH_3COO^- + H_2O →	CH_3COOH	+ OH^-
initial	0.1250 M	0 M	0 M
change	$-x$	$+x$	$+x$
equilibrium	$0.1250 - x$	x	x
assumption	0.1250	x	x

$$K_b = \frac{[CH_3COOH][OH^-]}{[CH_3COO^-]}$$

$$K_b = \frac{K_w}{K_a} = \frac{1.0 \times 10^{-14}}{1.8 \times 10^{-5}} = 5.6 \times 10^{-10}$$

$$5.6 \times 10^{-10} = \frac{x^2}{0.1250}$$

$$x = [OH^-] = \sqrt{(5.6 \times 10^{-10})(0.1250)} = 8.4 \times 10^{-6}\,M$$

$$pOH = -\log[OH^-] = -\log(8.4 \times 10^{-6}) = 5.08$$

$$pH = 14 - pOH = 8.92$$

At 40.00 mL OH^- added, the hydroxide is in excess and determines the pH. We can use the same equation that was used in Practice 18.8:

$$[OH^-] = \frac{M_{OH^-} \times V_{OH^-} - M_{HA} \times V_{HA}}{V_{OH^-} + V_{H^+}} = \frac{0.2500\frac{mol}{L} \times 40.00\,mL - 0.2500\frac{mol}{L} \times 25.00\,mL}{(40.00 + 25.00)mL}$$

$$[OH^-] = 0.05769\,M$$

$$pOH = -\log[OH^-] = -\log(0.05769) = 1.2389$$

$$pH = 14 - pOH = 12.7611$$

The titration curve is

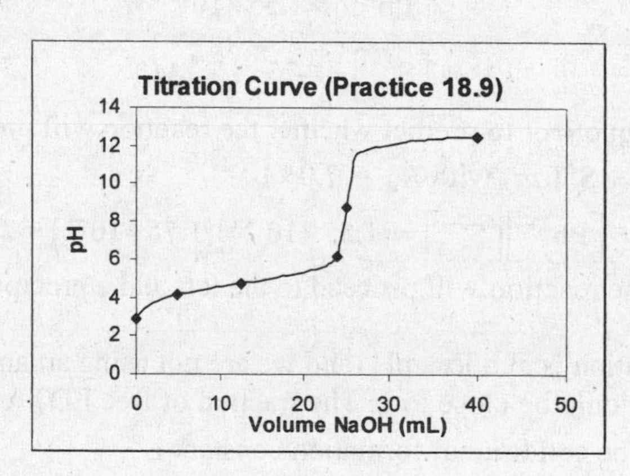

18.10. At the equivalence point of a titration of a strong base by a strong acid, both species are completely consumed. The solution has a neutral pH of 7. The best indicator for this titration would be bromothymol blue, whose color change is centered around pH = 7.

18.11. Barium fluoride dissociates according to the equation $BaF_2(s) \rightleftarrows Ba^{2+}(aq) + 2\,F^-(aq)$. Setting up the ICE table, we can solve for the solubility:

	$BaF_2(s) \rightleftarrows$	$Ba^{2+}(aq)$	$+ 2\,F^-(aq)$
initial		0 M	0 M
change		$+s$	$+2s$
equilibrium		s	$2s$

$$K_{sp} = [Ba^{2+}][F^-]^2$$
$$2.4\times10^{-5} = (s)(2s)^2 = 4s^3$$
$$s = \sqrt[3]{\frac{2.4\times10^{-5}}{4}} = 1.82\times10^{-2}\ \text{mol/L}$$

18.12. Lead bromide dissociates according to the equation $PbBr_2(s) \rightleftarrows Pb^{2+}(aq) + 2\ Br^-(aq)$. Setting up the ICE table, we can solve for K_{sp} given the bromide concentration (0.021 mol/L). Note that the solubility, *s*, will be half the value of the bromide concentration: $s = 0.010_5$ mol/L.

	$PbBr_2(s) \rightleftarrows$	$Pb^{2+}(aq)$	$+\ 2\ Br^-(aq)$
initial		0 *M*	0 *M*
change		$+s$	$+2s$
equilibrium		s	$2s$

$$K_{sp} = [Ba^{2+}][F^-]^2$$
$$K_{sp} = (s)(2s)^2 = 4s^3$$
$$K_{sp} = 4(1.0_5\times10^{-2})^3 = 4.6\times10^{-6}$$

18.13. Mixing equal volumes of the solutions will dilute the concentration of each by a factor of 2. The concentration of lead (II) and sulfide are

$$[Pb^{2+}] = 1.55\times10^{-10}\ M$$
$$[S^{2-}] = 1.75\times10^{-4}\ M$$

We use the reaction quotient to predict whether the reaction will proceed to the left in the reaction $PbS(s) \rightleftarrows Pb^{2+}(aq) + S^{2-}(aq)$ with $K_{sp} = 7.0\times10^{-29}$:

$$Q_{sp} = [Pb^{2+}][S^{2-}] = (1.55\times10^{-10})(1.75\times10^{-4}) = 2.71\times10^{-14}$$

Because $Q_{sp} > K_{sp}$, the reaction will proceed to the left and a precipitate will form.

18.14. Because the solution is at a low pH (and we are not using an ammonia buffer at pH = 3.0), the fraction of free Zn should be close to 1. The fraction of free EDTA at pH = 3.0 was given in the text as 2.0×10^{-11}. The conditional formation constant is

$$K' = K_f \alpha_{Zn^{2+}} \alpha_{EDTA^{4-}} = (3.0\times10^{16})(1)(2\times10^{-11}) = 6\times10^5.$$

18.15. Converting the concentration of Ca^{2+} into molarity units (recall that ppm = mg/L):

$$\frac{123.8\ \text{mg Ca}^{2+}}{\text{L}} \times \frac{1\ \text{g}}{1000\ \text{mg}} \times \frac{1\ \text{mol Ca}}{40.08\ \text{g Ca}} = 3.089\times10^{-3}\ M$$

Because EDTA reacts in a 1:1 ratio with all metal cations, we will need the same number of moles of each:

$$0.05000\ \text{L Ca}^{2+} \times \frac{3.089\times10^{-3}\ \text{mol Ca}^{2+}}{1\ \text{L Ca}^{2+}} \times \frac{1\ \text{mol EDTA}}{1\ \text{mol Ca}^{2+}} \times \frac{1\ \text{L EDTA}}{0.01944\ \text{mol EDTA}} \times \frac{1000\ \text{mL}}{1\ \text{L}}$$
$$= 7.945\ \text{mL EDTA}$$

Solutions to Student Problems

1. In each case, there is the acid or base equilibrium in addition to the self-ionization of water.

a. $NH_3 + H_2O \rightleftarrows NH_4^+ + OH^-$

$2\ H_2O \rightleftarrows H_3O^+ + OH^-$

b. $Fe(OH)_3 \rightleftarrows Fe^{3+} + 3\ OH^-$

$Fe^{3+} + 6\ H_2O \rightleftarrows Fe(H_2O)_6^{3+}$

$Fe(H_2O)_6^{3+} + H_2O \rightleftarrows Fe(H_2O)_5(OH)^{2+} + H_3O^+$

$2\ H_2O \rightleftarrows H_3O^+ + OH^-$

c. $HCOO^- + H_2O \rightleftarrows HCOOH + OH^-$

$2\ H_2O \rightleftarrows H_3O^+ + OH^-$

3. a. This combination has both an acid (HF) and a conjugate base (F^-) and would form a buffer.

b. These two are not conjugates and would not form a buffer.

c. These two are not conjugates and would not form a buffer.

d. Even though this combination appears to have both an acid (H_2SO_4) and a conjugate base (HSO_4^-), H_2SO_4 is a strong acid and would not form a buffer.

e. This combination has both an acid (NH_4^+) and a conjugate base (NH_3), and would form a buffer.

5. In each case we are safe in assuming that there will be little change in the concentrations of the acid and its conjugate base. We can use the mass-action expression to solve directly for the hydrogen ion concentration.

a. For 1.00 *M* NH_4^+ and 1.00 *M* NH_3:

$$K_a(NH_4^+) = \frac{K_w}{K_b(NH_3)} = \frac{1.0\times10^{-14}}{1.8\times10^{-5}} = 5.6\times10^{-10}$$

$$K_a = \frac{[NH_3][H^+]}{[NH_4^+]}$$

$$[H^+] = K_a\frac{[NH_4^+]}{[NH_3]} = 5.6\times10^{-10}\frac{(1.00)}{(1.00)} = 5.6\times10^{-10}\,M$$

$$pH = -\log[H^+] = -\log(5.6\times10^{-10}) = 9.25$$

b. For 4.50 *M* NH_4^+ and 0.50 *M* NH_3:

$$[H^+] = K_a\frac{[NH_4^+]}{[NH_3]} = 5.6\times10^{-10}\frac{(4.50)}{(0.50)} = 5.0\times10^{-9}\,M$$

$$pH = -\log[H^+] = -\log(5.0\times10^{-9}) = 8.30$$

c. For 2.50 M CH_3COOH and 0.75 M CH_3COO^-:

$$K_a(CH_3COOH) = 1.8\times10^{-5}$$

$$K_a = \frac{[CH_3COO^-][H^+]}{[CH_3COOH]}$$

$$[H^+] = K_a\frac{[CH_3COOH]}{[CH_3COO^-]} = 1.8\times10^{-5}\frac{(2.50)}{(0.75)} = 6.0\times10^{-5}\,M$$

$$pH = -\log[H^+] = -\log(6.0\times10^{-5}) = 4.22$$

7. a. Adding the conjugate base to a buffer will shift the reaction HA $\rightleftarrows$ H^+ + A^- toward the acid, which lowers the concentration of hydrogen ion and makes the solution more basic. The pH will increase.

b. Adding the acid to a buffer will shift the reaction HA $\rightleftarrows$ H^+ + A^- toward the base, which increases the concentration of hydrogen ion and makes the solution more acidic. The pH will decrease.

c. Adding Cl^- will have no effect; it does not have any acid or base properties.

9. Acetic acid has a $K_a = 1.8\times10^{-5}$. We can use the mass-action expression to solve for the ratio of acid to base in each solution.

$$K_a = \frac{[CH_3COO^-][H^+]}{[CH_3COOH]}$$

a. pH = 3.74:

$$\frac{[CH_3COOH]}{[CH_3COO^-]} = \frac{[H^+]}{K_a} = \frac{10^{-pH}}{K_a} = \frac{10^{-3.74}}{1.8\times10^{-5}} = 10$$

b. pH = 4.74:

$$\frac{[CH_3COOH]}{[CH_3COO^-]} = \frac{[H^+]}{K_a} = \frac{10^{-pH}}{K_a} = \frac{10^{-4.74}}{1.8\times10^{-5}} = 1.0$$

c. pH = 5.74:

$$\frac{[CH_3COOH]}{[CH_3COO^-]} = \frac{[H^+]}{K_a} = \frac{10^{-pH}}{K_a} = \frac{10^{-5.74}}{1.8\times10^{-5}} = 0.10$$

11. The reaction of HCl with NH_3 generates the NH_4^+ necessary to create a buffer with ammonia. This reaction (HCl + NH_3 $\rightarrow$ NH_4^+ + Cl^-) has two consequences: First, the amount of NH_4^+ at equilibrium will be the same as the amount of HCl added. Second, the total amount of NH_3 and NH_4^+ at equilibrium will equal the amount of NH_3 initially (0.1000 L × 0.2500 M = 0.02500 mol). We can use the mass-action expression for the acid ionization (NH_4^+ $\rightleftarrows$ H^+ + NH_3) to solve for the ratio of the amount of ammonium ion to the amount of ammonia and then set up a pair of equations to solve for the amount of NH_4^+.

a. For pH = 9.26:

$$K_a = \frac{K_w}{K_b(NH_3)} = \frac{1.0\times10^{-14}}{1.8\times10^{-5}} = 5.6\times10^{-10}$$

$$K_a = \frac{[NH_3][H^+]}{[NH_4^+]}$$

$$\frac{[NH_3]}{[NH_4^+]} = \frac{K_a}{[H^+]} = \frac{K_a}{10^{-pH}} = \frac{5.6\times10^{-10}}{10^{-9.26}} = 1.0$$

so $[NH_3] = [NH_4^+]$ or $n_{NH_3} = 1.0 n_{NH_4^+}$ because the volume is the same for both

$$n_{NH_4^+} + n_{NH_3} = 0.02500\,\text{mol}$$

substituting: $n_{NH_4^+} + 1.0 n_{NH_4^+} = 0.02500\,\text{mol}$

$$2.0 n_{NH_4^+} = 0.02500\,\text{mol}$$

$$n_{NH_4^+} = 0.012_5\,\text{mol} = n_{HCl}$$

Solving for the volume of HCl:

$$0.012_5\,\text{mol HCl} \times \frac{1\,\text{L soln}}{0.200\,\text{mol HCl}} \times \frac{1000\,\text{mL}}{1\,\text{L}} = 62\,\text{mL}$$

b. For pH = 10.5:

$$K_a = \frac{K_w}{K_b(NH_3)} = \frac{1.0\times10^{-14}}{1.8\times10^{-5}} = 5.6\times10^{-10}$$

$$K_a = \frac{[NH_3][H^+]}{[NH_4^+]}$$

$$\frac{[NH_3]}{[NH_4^+]} = \frac{K_a}{[H^+]} = \frac{K_a}{10^{-pH}} = \frac{5.6\times10^{-10}}{10^{-10.5}} = 17._7$$

so $[NH_3] = 17._7[NH_4^+]$ or $n_{NH_3} = 17._7 n_{NH_4^+}$ because the volume is the same for both

$$n_{NH_4^+} + n_{NH_3} = 0.02500\,\text{mol}$$

substituting: $n_{NH_4^+} + 17._7 n_{NH_4^+} = 0.02500\,\text{mol}$

$$18._7 n_{NH_4^+} = 0.02500\,\text{mol}$$

$$n_{NH_4^+} = 0.0013\,\text{mol} = n_{HCl}$$

Solving for the volume of HCl:

$$0.0013\,\text{mol HCl} \times \frac{1\,\text{L soln}}{0.200\,\text{mol HCl}} \times \frac{1000\,\text{mL}}{1\,\text{L}} = 6.7\,\text{mL}$$

c. For pH = 8.5:

$$K_a = \frac{K_w}{K_b(NH_3)} = \frac{1.0\times10^{-14}}{1.8\times10^{-5}} = 5.6\times10^{-10}$$

$$K_a = \frac{[NH_3][H^+]}{[NH_4^+]}$$

$$\frac{[NH_3]}{[NH_4^+]} = \frac{K_a}{[H^+]} = \frac{K_a}{10^{-pH}} = \frac{5.6\times10^{-10}}{10^{-8.5}} = 0.17_7$$

so $[NH_3] = 0.17_7[NH_4^+]$ or $n_{NH_3} = 0.17_7 n_{NH_4^+}$ because the volume is the same for both

$n_{NH_4^+} + n_{NH_3} = 0.02500\,\text{mol}$

substituting: $n_{NH_4^+} + 0.17_7 n_{NH_4^+} = 0.02500\,\text{mol}$

$1.17_7 n_{NH_4^+} = 0.02500\,\text{mol}$

$n_{NH_4^+} = 0.0212\,\text{mol} = n_{HCl}$

Solving for the volume of HCl:

$$0.0212\,\text{mol HCl} \times \frac{1\,\text{L soln}}{0.200\,\text{mol HCl}} \times \frac{1000\,\text{mL}}{1\,\text{L}} = 106\,\text{mL}$$

13. In general, for any weak acid HA and base A^- related by HA $\rightleftarrows$ $H^+ + A^-$, we can write the mass-action expression and solve for the hydrogen ion concentration:

$$K_a = \frac{[H^+][A^-]}{[HA]}$$

$$[H^+] = K_a\frac{[HA]}{[A^-]}$$

When the concentration of an acid and that of its conjugate base are equal, $[H^+] = K_a$, and $pH = -\log K_a = pK_a$.

a. $K_a = 5.6\times10^{-10}$ and $pH = -\log 5.6\times10^{-10} = 9.25$

b. $K_a = 1.8\times10^{-5}$ and $pH = -\log 1.8\times10^{-5} = 4.74$

c. $K_a = 1.8\times10^{-4}$ and $pH = -\log 1.8\times10^{-4} = 3.74$

15. The final pH of a buffer is determined by the pK_a of the acid that sets the center of the possible range of pH for the buffer, while the exact ratio of the concentration of a conjugate base to its acid determines how far away the final pH is from the pK_a of the acid. When the amount of base is higher, the pH is more basic (higher) than the pK_a; when the amount of acid is greater, the pH is more acidic (lower) than the pK_a.

17. a. The most important equilibria are the ionization of the acid ($ClCH_2COOH \rightleftarrows ClCH_2COO^- + H^+$, $K_a = \frac{[H^+][ClCH_2COO^-]}{[ClCH_2COOH]}$), and the base hydrolysis ($ClCH_2COO^- + H_2O \rightleftarrows ClCH_2COOH + OH^-$, $K_b = \frac{[OH^-][ClCH_2COOH]}{[ClCH_2COO^-]}$).

b. We can use the Henderson–Hasselbalch equation to solve for the pH. Because the base concentration and the acid concentration are in a ratio, we can simply use the molar amounts and do not need to calculate the concentrations.

$$pH = pK_a + \log\frac{[ClCH_2COO^-]}{[ClCH_2COOH]} = pK_a + \log\frac{n_{ClCH_2COO^-}}{n_{ClCH_2COOH}}$$

$$pH = -\log\left(1.4\times10^{-3}\right) + \log\left(\frac{1.5\,g\,ClCH_2COOK \times \frac{1\,mol\,ClCH_2COOK}{132.59\,g\,ClCH_2COOK}}{0.1000\,L\,ClCH_2COOH \times \frac{0.10\,mol\,ClCH_2COOH}{1\,L\,ClCH_2COOH}}\right)$$

$$pH = 2.91$$

19. The acid ionization reaction is $CH_3NH_3^+ \rightleftarrows CH_3NH_2 + H^+$. Using the K_a approach:

$$K_a = \frac{[H^+][CH_3NH_2]}{[CH_3NH_3^+]}$$

$$[H^+] = K_a\frac{[CH_3NH_3^+]}{[CH_3NH_2]} = \left(2.3\times10^{-11}\right)\frac{(0.20\,mol/V)}{(0.10\,mol/V)} = 4.6\times10^{-11}\,M$$

$$pH = -\log[H^+] = -\log\left(4.6\times10^{-11}\right) = 10.34$$

Note that the volume, V, was not provided, but cancels from each concentration expression. Using the K_b approach, the reaction is $CH_3NH_2 + H_2O \rightleftarrows CH_3NH_3^+ + OH^-$.

$$K_b = \frac{[OH^-][CH_3NH_3^+]}{[CH_3NH_2]}$$

$$[OH^-] = K_b\frac{[CH_3NH_2]}{[CH_3NH_3^+]} = \left(4.3\times10^{-4}\right)\frac{(0.10\,mol/V)}{(0.20\,mol/V)} = 2.1_5\times10^{-4}\,M$$

$$pOH = -\log[OH^-] = -\log\left(2.1_5\times10^{-4}\right) = 3.67$$

$$pH = 14 - pOH = 10.33$$

These two approaches give the same value within rounding error, but generally the K_a approach is more direct when we are solving for the pH.

21. When we are given the concentrations of the acid and conjugate base in a buffer, it is often quicker to use the Henderson–Hasselbalch equation to find the pH. The pK_a of ammonium ion is

$$\mathrm{p}K_a = -\log K_a = -\log\left(5.56\times10^{-10}\right) = 9.26$$

a. $$\mathrm{pH} = \mathrm{p}K_a + \log\left(\frac{[NH_3]}{[NH_4^+]}\right) = 9.26 + \log\left(\frac{0.10}{0.10}\right) = 9.26$$

b. $$\mathrm{pH} = \mathrm{p}K_a + \log\left(\frac{[NH_3]}{[NH_4^+]}\right) = 9.26 + \log\left(\frac{0.20}{0.050}\right) = 9.86$$

c. $$\mathrm{pH} = \mathrm{p}K_a + \log\left(\frac{[NH_3]}{[NH_4^+]}\right) = 9.26 + \log\left(\frac{1.50}{0.10}\right) = 10.43$$

d. $$\mathrm{pH} = \mathrm{p}K_a + \log\left(\frac{[NH_3]}{[NH_4^+]}\right) = 9.26 + \log\left(\frac{0.050}{0.750}\right) = 8.08$$

23. The ionization reaction involving these two species is $H_2PO_4^- \rightleftarrows H^+ + HPO_4^{2-}$. We can use the mass-action expression to solve for the ratio:

$$K_a = \frac{[H^+][HPO_4^{2-}]}{[H_2PO_4^-]}$$

$$\frac{[H_2PO_4^-]}{[HPO_4^{2-}]} = \frac{[H^+]}{K_a} = \frac{10^{-\mathrm{pH}}}{K_a} = \frac{10^{-7.40}}{6.2\times10^{-8}} = 0.64$$

25. a. The ionization reaction involving these two species is $HCOOH \rightleftarrows H^+ + HCOO^-$. We can use the Henderson–Hasselbalch equation to solve for the pH:

$$\mathrm{pH} = \mathrm{p}K_a + \log\frac{[HCOO^-]}{[HCOOH]} = \mathrm{p}K_a + \log\frac{n_{HCOO^-}}{n_{HCOOH}} = -\log\left(1.8\times10^{-4}\right) + \log\frac{n_{HCOO^-}}{n_{HCOOH}}$$

$$\mathrm{pH} = 3.74 + \log\left(\frac{0.45\,\mathrm{g\,HCOONa} \times \dfrac{1\,\mathrm{mol\,HOONa}}{68.01\,\mathrm{g\,HOONa}}}{0.5000\,\mathrm{L\,HCOOH} \times \dfrac{0.20\,\mathrm{mol\,HCOOH}}{1\,\mathrm{L\,HCOOH}}}\right)$$

$$\mathrm{pH} = 2.56$$

b. Because the buffer is more acidic than the pK_a, there is more acid in the buffer solution than there is base. This buffer would then be more able to react with added base than with added acid.

27. The acid ionization reaction is $CH_3CH(OH)COOH \rightleftarrows CH_3CH(OH)COO^- + H^+$. We can use the Henderson–Hasselbalch equation to solve for the pH:

$$\text{pH} = \text{p}K_a + \log\frac{[CH_3CH(OH)COO^-]}{[CH_3CH(OH)COOH]} = \text{p}K_a + \log\frac{n_{CH_3CH(OH)COO^-}}{n_{CH_3CH(OH)COOH}}$$

$$\text{pH} = -\log(1.4\times10^{-4}) + \log\left(\frac{0.015}{0.020}\right) = 3.73$$

29. If the buffer were equally able to resist acid changes and base changes, the acid concentration and the conjugate base concentration would be the same. Any buffer with equal concentrations of each species will have pH = pK_a.

$$\text{pH} = \text{p}K_a = -\log K_a = -\log(1.3\times10^{-5}) = 4.88$$

31. In each case we find the total moles of the analyte, use a stoichiometric ratio to convert to moles of analyte, and then find the volume using the concentration of the titrant.

a. $$0.045\,\text{L HCl}\times\frac{0.23\,\text{mol HCl}}{1\,\text{L HCl}}\times\frac{1\,\text{mol NaOH}}{1\,\text{mol HCl}}\times\frac{1\,\text{L NaOH}}{0.15\,\text{mol NaOH}} = 0.069\,\text{L}$$

b. $$0.0500\,\text{L NaOH}\times\frac{0.50\,\text{mol NaOH}}{1\,\text{L NaOH}}\times\frac{1\,\text{mol HCl}}{1\,\text{mol NaOH}}\times\frac{1\,\text{L HCl}}{0.23\,\text{mol HCl}} = 0.11\,\text{L}$$

c. $$0.0200\,\text{L H}_2\text{SO}_4\times\frac{0.20\,\text{mol H}_2\text{SO}_4}{1\,\text{L H}_2\text{SO}_4}\times\frac{2\,\text{mol KOH}}{1\,\text{mol H}_2\text{SO}_4}\times\frac{1\,\text{L KOH}}{0.15\,\text{mol KOH}} = 0.053\,\text{L}$$

d. $$0.050\,\text{L NaOH}\times\frac{0.10\,\text{mol NaOH}}{1\,\text{L NaOH}}\times\frac{1\,\text{mol H}_2\text{SO}_4}{2\,\text{mol NaOH}}\times\frac{1\,\text{L H}_2\text{SO}_4}{0.23\,\text{mol H}_2\text{SO}_4} = 0.011\,\text{L}$$

33. The endpoint of this titration is expected at

$$0.075\,\text{L NaOH}\times\frac{0.137\,\text{mol NaOH}}{1\,\text{L NaOH}}\times\frac{1\,\text{mol HCl}}{1\,\text{mol NaOH}}\times\frac{1\,\text{L HCl}}{0.2055\,\text{mol HCl}} = 0.050\,\text{L} = 50.0\,\text{mL}$$

a. At the initial pH, no reaction has taken place, so $[OH^-] = 0.137\ M$. Then

$$\text{pOH} = -\log[OH^-] = -\log(0.137) = 0.863$$
$$\text{pH} = 14 - \text{pOH} = 13.137$$

b. For a reaction of a strong base with strong acid, the amount of excess hydroxide will determine the pH. We can calculate the excess base by subtracting the number of moles of acid from the starting moles of base. We divide the result by the total volume to get the concentration. At 10.0 mL added:

$$[OH^-] = \frac{\text{mol OH}^- - \text{mol H}^+}{\text{total volume}} = \frac{M_{OH^-}\times V_{OH^-} - M_{H^+}\times V_{H^+}}{V_{OH^-} + V_{H^+}}$$

$$[OH^-] = \frac{(0.137M)(75\text{mL}) - (0.2055M)(10.0\,\text{mL})}{75\,\text{mL} + 10.0\,\text{mL}} = 0.096_7\ M$$

$$\text{pOH} = -\log[OH^-] = -\log(0.096_7) = 1.01$$

$$\text{pH} = 14 - \text{pOH} = 12.99$$

c. At 25.0 mL:

$$[OH^-] = \frac{(0.137M)(75\text{mL}) - (0.2055M)(25.0\text{mL})}{75\text{mL} + 25.0\text{mL}} = 0.051_4\ M$$

$$pOH = -\log[OH^-] = -\log(0.051_4) = 1.29$$

$$pH = 14 - pOH = 12.71$$

d. At 50.0 mL: This is the equivalence point of the reaction. All of the H^+ and OH^- have fully reacted. Only the self-ionization of water controls the pH. The solution is neutral: pH = 7.

e. Beyond the equivalence point, the pH is controlled by the excess titrant, HCl. We can modify the above equation to solve for the hydrogen ion concentration. At 100.0 mL:

$$[H^+] = \frac{\text{mol}\,H^+ - \text{mol}\,OH^-}{\text{total volume}} = \frac{M_{H^+} \times V_{H^+} - M_{OH^-} \times V_{OH^-}}{V_{H^+} + V_{OH^-}}$$

$$[H^+] = \frac{(0.2055M)(100.0\text{mL}) - (0.137M)(75\text{mL})}{100.0\text{mL} + 75\text{mL}} = 0.0587\ M$$

$$pH = -\log[H^+] = -\log(0.0587) = 1.231$$

35. This titration has a weak base as an analyte. The initial point is a solution of a pure weak acid. All other points prior to the equivalence point will involve finding the pH of a buffer solution, for which the Henderson–Hasselbalch equation is appropriate. At the equivalence point, all of the weak base will have been converted into the conjugate acid. Finding the pH there will involve solving for the pH of a solution of a pure weak acid. Beyond the equivalence point, the titrant will be in excess and will control the pH. We expect the equivalence point at

$$0.050\text{L}\,NH_3 \times \frac{0.100\text{mol}\,NH_3}{1\text{L}\,NH_3} \times \frac{1\text{mol}\,HCl}{1\text{mol}\,NH_3} \times \frac{1\text{L}\,HCl}{0.125\text{mol}\,HCl} = 0.0400\text{L} = 40.0\text{mL}$$

a. For 0.100 M NH_3 ($K_b = 1.8 \times 10^{-5}$):

	NH_3 +	H_2O ⇄	NH_4^+ +	OH^-
Initial	0.100M		0 M	0 M
Change	$-x$		$+x$	$+x$
Equilibrium	$0.10 - x$		$+x$	$+x$
With assumption	0.10		$+x$	$+x$

$$K_b = \frac{[NH_4^+][OH^-]}{[NH_3]}$$

$$1.8\times10^{-5} = \frac{(x)^2}{0.100}$$

$$x = [OH^-] = \sqrt{(0.100)(1.8\times10^{-5})} = 1.3_4\times10^{-3}\ M$$

$$pOH = -\log(1.3_4\times10^{-3}) = 2.87$$

$$pH = 14 - pOH = 11.13$$

b. For the buffer calculations, we can take advantage of the fact that the acid and base concentrations are in a ratio; therefore, we can use molar amounts rather than concentrations. The moles of ammonia are reduced by the amount of acid added, while the amount of ammonium ion is equivalent to the amount of acid added. (*Note*: $M \times$ mL = mmol = millimole = 10^{-3} mol). At 10.0 mL added:

$$\text{mol NH}_3 = M_{\text{NH}_3} \times V_{\text{NH}_3} - M_{\text{HCl}} \times V_{\text{HCl}}$$
$$= 50.0\text{mL}(0.100\text{mol/L}) - 10.0\text{mL}(0.125\text{mol/L}) = 3.75\text{mmol}$$
$$\text{mol NH}_4{}^+ = M_{\text{HCl}} \times V_{\text{HCl}} = 10.0\text{mL}(0.125\text{mol/L}) = 1.25\,\text{mmol}$$

$$K_\text{a}(\text{NH}_4{}^+) = \frac{K_\text{w}}{K_\text{b}(\text{NH}_3)} = \frac{1.0\times10^{-14}}{1.8\times10^{-5}} = 5.6\times10^{-10}$$

$$\text{p}K_\text{a} = -\log K_\text{a} = -\log\left(5.6\times10^{-10}\right) = 9.26$$

$$\text{pH} = \text{p}K_\text{a} + \log\left(\frac{[\text{NH}_3]}{[\text{NH}_4{}^+]}\right) = 9.26 + \log\left(\frac{3.75}{1.25}\right) = 9.74$$

c. At 20.0 mL added:

$$\text{mol NH}_3 = M_{\text{NH}_3} \times V_{\text{NH}_3} - M_{\text{HCl}} \times V_{\text{HCl}}$$
$$= 50.0\text{mL}(0.100\text{mol/L}) - 20.0\text{mL}(0.125\text{mol/L}) = 2.5\text{mmol}$$
$$\text{mol NH}_4{}^+ = M_{\text{HCl}} \times V_{\text{HCl}} = 20.0\text{mL}(0.125\text{mol/L}) = 2.5\,\text{mmol}$$
$$\text{pH} = \text{p}K_\text{a} + \log\left(\frac{[\text{NH}_3]}{[\text{NH}_4{}^+]}\right) = 9.26 + \log\left(\frac{2.5}{2.5}\right) = 9.26$$

d. At 40.0 mL added, all of the ammonia has been converted into ammonium. The concentration of NH_4^+ is

$$[\text{NH}_4{}^+] = \frac{M_{\text{NH}_3} \times V_{\text{NH}_3}}{V_\text{total}} = \frac{(0.100\text{mol/L})(50.0\text{mL})}{50.0\,\text{mL} + 40.0\,\text{mL}} = 0.0556\,\text{M}$$

Solving for the pH of this solution:

	NH_4^+ $\rightleftarrows$	NH_3 +	H^+
Initial	0.0556 M	0 M	0 M
Change	$-x$	$+x$	$+x$
Equilibrium	$00556 - x$	$+x$	$+x$
With assumption	0.0556	$+x$	$+x$

$$K_\text{a} = \frac{[\text{H}^+][\text{NH}_3]}{[\text{NH}_4{}^+]}$$

$$5.6\times10^{-10} = \frac{(x)^2}{0.0556}$$

$$x = [\text{H}^+] = \sqrt{(0.0556)(5.6\times10^{-10})} = 5.5_6\times10^{-6}\,M$$

$$\text{pH} = -\log(5.5_6\times10^{-6}) = 5.25$$

e. At 50.0 mL HCl, the excess H^+ controls the pH:

$$\left[H^+\right]=\frac{\text{mol}\,H^+-\text{mol}\,NH_3}{\text{total volume}}=\frac{M_{H^+}\times V_{H^+}-M_{NH_3}\times V_{NH_3}}{V_{H^+}+V_{NH_3}}$$

$$\left[H^+\right]=\frac{(0.125\,M)(50.0\,\text{mL})-(0.100\,M)(50.0\text{mL})}{50.0\,\text{mL}+50.0\,\text{mL}}=0.0125\,\text{M}$$

$$\text{pH}=-\log\left[H^+\right]=-\log(0.0125)=1.903$$

37. This is a titration of a strong base by a strong acid, and each point can be solved in the manner of Problem 33. The endpoint of this titration is expected at

$$0.0250\,\text{L KOH}\times\frac{0.250\,\text{mol KOH}}{1\,\text{L KOH}}\times\frac{1\,\text{mol HNO}_3}{1\,\text{mol KOH}}\times\frac{1\,\text{L HNO}_3}{0.150\,\text{mol HNO}_3}=0.0417\,\text{L}=41.7\text{mL}$$

a. At the initial pH, no reaction has taken place, so $[OH^-] = 0.25\ M$. Then

$$\text{pOH}=-\log[OH^-]=-\log(0.250)=0.602$$

$$\text{pH}=14-\text{pOH}=13.398$$

b. At 2.00 mL added:

$$[OH^-]=\frac{\text{mol}\,OH^- - \text{mol}\,H^+}{\text{total volume}}=\frac{M_{OH^-}\times V_{OH^-}-M_{H^+}\times V_{H^+}}{V_{OH^-}+V_{H^+}}$$

$$[OH^-]=\frac{(0.250M)(25.0\text{mL})-(0.150\,M)(2.00\,\text{mL})}{25.0\,\text{mL}+2.00\,\text{mL}}=0.220\,M$$

$$\text{pOH}=-\log[OH^-]=-\log(0.220)=0.658$$

$$\text{pH}=14-\text{pOH}=13.342$$

c. At 20.0 mL:

$$[OH^-]=\frac{(0.250M)(25.0\text{mL})-(0.150\,M)(20.0\,\text{mL})}{25.0\,\text{mL}+20.00\,\text{mL}}=0.0722\,M$$

$$\text{pOH}=-\log[OH^-]=-\log(0.0722)=1.141$$

$$\text{pH}=14-\text{pOH}=12.859$$

d. At 40.0 mL:

$$[OH^-]=\frac{(0.250M)(25.0\text{mL})-(0.150\,M)(40.0\,\text{mL})}{25.0\,\text{mL}+40.00\,\text{mL}}=0.0038_5\,M$$

$$\text{pOH}=-\log[OH^-]=-\log(0.0038_5)=2.41$$

$$\text{pH}=14-\text{pOH}=11.59$$

e. At 41.7 mL: This is the equivalence point of the reaction. All of the H^+ and all of the OH^- have fully reacted. Only the self-ionization of water controls the pH. The solution is neutral: pH = 7.

f. Beyond the equivalence point, the pH is controlled by the excess titrant, HNO_3. We can modify the above equation to solve for the hydrogen ion concentration. At 43.0 mL:

$$\left[H^+\right] = \frac{\text{mol}\,H^+ - \text{mol}\,OH^-}{\text{total volume}} = \frac{M_{H^+} \times V_{H^+} - M_{OH^-} \times V_{OH^-}}{V_{H^+} + V_{OH^-}}$$

$$\left[H^+\right] = \frac{(0.150\,M)(43.0\,\text{mL}) - (0.250M)(25.0\text{mL})}{43.0\,\text{mL} + 25.0\,\text{mL}} = 0.0029_4\,M$$

$$\text{pH} = -\log\left[H^+\right] = -\log\left(0.0029_4\right) = 2.53$$

g. At 50.0 mL:

$$\left[H^+\right] = \frac{\text{mol}\,H^+ - \text{mol}\,OH^-}{\text{total volume}} = \frac{M_{H^+} \times V_{H^+} - M_{OH^-} \times V_{OH^-}}{V_{H^+} + V_{OH^-}}$$

$$\left[H^+\right] = \frac{(0.150\,M)(50.0\,\text{mL}) - (0.250M)(25.0\text{mL})}{50.0\,\text{mL} + 25.0\,\text{mL}} = 0.0167\,M$$

$$\text{pH} = -\log\left[H^+\right] = -\log(0.0167) = 1.777$$

The titration curve:

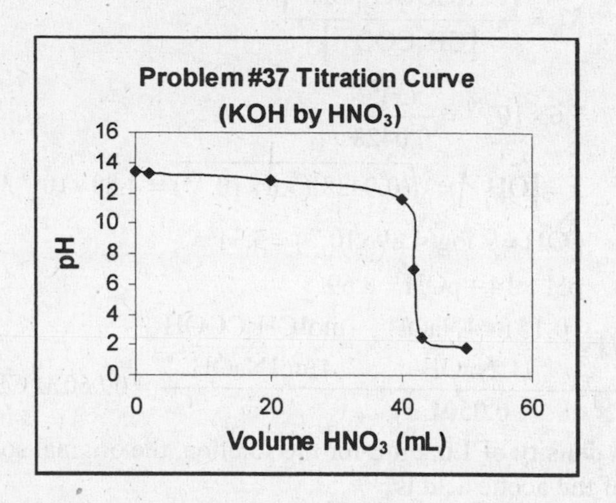

39. a. This titration is similar to that in Problem 35, with an equivalence point in the acidic region. Depending on the actual final pH, either methyl orange (pH ≈ 4.5) or methyl red (pH ≈ 5).

b. This titration is similar to that in Problem 38, with an equivalence point in the basic region. Phenolphthalein is the most common indicator for this type of titration.

c. This titration will have a neutral pH at the equivalence point. Bromthymol blue is the most appropriate indicator. (Because of the sharpness and large pH range in most strong-acid–strong-base titrations, phenolphthalein is often used; however, the color-change range is in the basic region.)

41. a. This pH is to the acid side of the phenolphthalein color change. It will be colorless.

b. At neutral pH, bromothymol blue is at the center of its color range and will be a mix of its two colors (yellow and blue): green.

c. This solution has a pH = –log(0.0056) = 2.25. At this pH, methyl orange will be red.

d. An ammonia solution will have a basic pH. Methyl violet changes color at acidic pH. It will be blue.

43. Before finding the pH of the solution at the end of the reaction, we need to find the concentration of the acetate ion. Acetic acid reacts in a 1:1 ratio with NaOH, so the number of moles of acetate will equal the number of moles of NaOH used. We divide that value by the total volume of the solution after the reaction.

$$\frac{0.0200\,\text{L NaOH} \times \frac{0.15\,\text{mol NaOH}}{1\,\text{L NaOH}} \times \frac{1\,\text{mol CH}_3\text{COO}^-}{1\,\text{mol NaOH}}}{0.050\,\text{L} + 0.020\,\text{L}} = 0.0428\,M\,\text{CH}_3\text{COO}^-$$

Solving for the pH of this solution:

	$CH_3COO^- + H_2O \rightleftarrows$	$CH_3COOH +$	OH^-
Initial	$0.0428\ M$	$0\ M$	$0\ M$
Change	$-x$	$+x$	$+x$
Equilibrium	$0.0428 - x$	$+x$	$+x$
With assumption	0.0428	$+x$	$+x$

$$K_b = \frac{K_w}{K_a} = \frac{1.0\times10^{-14}}{1.8\times10^{-5}} = 5.6\times10^{-10}$$

$$K_b = \frac{[\text{CH}_3\text{COOH}][\text{OH}^-]}{[\text{CH}_3\text{COO}^-]}$$

$$5.6\times10^{-10} = \frac{(x)^2}{0.0428}$$

$$x = [\text{OH}^-] = \sqrt{(0.0428)(5.6\times10^{-10})} = 4.89\times10^{-6}\,M$$

$$\text{pOH} = -\log(4.89\times10^{-6}) = 5.31$$

$$\text{pH} = 14 - \text{pOH} = 8.69$$

b. $$\frac{0.0200\,\text{L NaOH} \times \frac{0.15\,\text{mol NaOH}}{1\,\text{L NaOH}} \times \frac{1\,\text{mol CH}_3\text{COOH}}{1\,\text{mol NaOH}}}{0.050\,\text{L}} = 0.060\,M\,\text{CH}_3\text{COOH}$$

c. If we assume a density of 1.0 g/mL for the solution, the original solution has a mass of 50.0 g and the mass of the acetic acid is

$$0.050\,\text{L CH}_3\text{COOH} \times \frac{0.060\,\text{mol CH}_3\text{COOH}}{1\,\text{L CH}_3\text{COOH}} \times \frac{60.05\,\text{g CH}_3\text{COOH}}{1\,\text{mol CH}_3\text{COOH}} = 0.180\,\text{g CH}_3\text{COOH}$$

The mass percent is $\frac{\text{mass CH}_3\text{COOH}}{\text{mass total}} \times 100\% = \frac{0.180\,\text{g}}{50.0\,\text{g}} \times 100\% = 0.36\%$.

45. a. The typical color change in an indicator occurs one pH unit to either side of the pK_a of the indicator. The center of the given range is 8.85 and should be the pK_a.

b. The acid form of the indicator is red, and the base form is green. We can write the acid ionization reaction as Red ⇄ green + H^+. We can use the mass-action expression for this reaction to solve for the ratio of the red form to green form:

$$K_a = \frac{[H^+][\text{green}]}{[\text{red}]}$$

$$\frac{[\text{red}]}{[\text{green}]} = \frac{[H^+]}{K_a} = \frac{10^{-pH}}{10^{-pK_a}} = \frac{10^{-9.50}}{10^{-8.85}} = 0.22$$

c. The endpoint of an HCl with NaOH titration is neutral, pH = 7. Because the color-change range for this indicator is from 7.85 to 9.85, the indicator will remain red well past the endpoint of the titration. It is not well suited for this titration.

47. a. $AgI(s) \rightleftarrows Ag^+(aq) + I^-(aq)$; $K_{sp} = [Ag^+][I^-]$

b. $Ag_2CrO_4(s) \rightleftarrows 2\,Ag^+(aq) + CrO_4^{2-}(aq)$; $K_{sp} = [Ag^+]^2[CrO_4^{2-}]$

c. $Al_2S_3(s) \rightleftarrows 2\,Al^{3+}(aq) + 3\,S^{2-}(aq)$; $K_{sp} = [Al^{3+}]^2[S^{2-}]^3$

d. $Ca_3(PO_4)_2(s) \rightleftarrows 3\,Ca^{2+}(aq) + 2\,PO_4^{3-}(aq)$; $K_{sp} = [Ca^{2+}]^3[PO_4^{3-}]^2$

49. We could set up a complete ICE table for each reaction (as is done for part a), but when a solid sample is dissolved into a solution containing no other ions, the equilibrium concentration of each ion will always be given by its stoichiometric coefficient times the solubility *s*. This enables us to quickly write K_{sp} in terms of the molar solubility (as will be done for the remaining problems).

a. Solving for *s*:

	$CuS(s) \rightleftarrows$	$Cu^{2+}(aq)$	$+\ S^{2-}(aq)$
Initial		0 *M*	0 *M*
Change		+ *s*	+ *s*
Equilibrium		+ *s*	+ *s*

$$K_{sp} = [Cu^{2+}][S^{2-}] = s^2 = 8.5\times10^{-45}$$

$$s = \sqrt{8.5\times10^{-45}} = 9.2\times10^{-23}\ \text{mol/L}$$

b. $Ag_3(PO_4)(s) \rightleftarrows 3\,Ag^+(aq) + PO_4^{3-}(aq)$

$$K_{sp} = [Ag^+]^3[PO_4^{3-}] = (3s)^3(s) = 27s^4 = 1.8\times10^{-18}$$

$$s = \sqrt[4]{\frac{1.8\times10^{-18}}{27}} = 1.6\times10^{-5}\ \text{mol/L}$$

c. $FeCO_3(s) \rightleftarrows Fe^{2+}(aq) + CO_3^{2-}(aq)$;

$$K_{sp} = [Fe^{2+}][CO_3^{2-}] = (s)(s) = s^2 = 2.1\times10^{-11}$$

$$s = \sqrt{2.1\times10^{-11}} = 4.6\times10^{-6}\ \text{mol/L}$$

51. a. $NiS(s) \rightleftarrows Ni^{2+}(aq) + S^{2-}(aq)$

$$K_{sp} = [Ni^{2+}][S^{2-}] = (s)(s) = s^2 = (5.5\times10^{-11})^2 = 3.0\times10^{-21}$$

b. $PbCrO_4(s) \rightleftarrows Pb^{2+}(aq) + CrO_4^{2-}(aq)$

$$K_{sp} = [Pb^{2+}][CrO_4^{2-}] = (s)(s) = s^2 = (1.41\times10^{-8})^2 = 2.0\times10^{-16}$$

c. $Ag_2CO_3(s) \rightleftarrows 2\ Ag^+(aq) + CO_3^{2-}(aq)$

$$K_{sp} = [Ag^+]^2[CO_3^{2-}] = (2s)^2(s) = 4s^3 = 4(2.0\times10^{-4})^3 = 3.2\times10^{-11}$$

53. Because all three have the same basic form (one cation, two anions), all will have the same relationship between K_{sp} and s. We can directly compare the values for K_{sp} with the largest value having the largest solubility. BaF_2 has the largest K_{sp} and molar solubility.

55. After recalculating the concentrations, we can find the value for the reaction quotient, Q_{sp}. Comparing this value to K_{sp} will enable us to determine whether a precipitate will form. For each set of solutions, we can use the solubility rules to predict the precipitation reaction. (Note that we will write the reaction as a dissolution to match the format for the K_{sp}, even though the precipitation reaction is actually the reverse.)

a. The reaction will be $BaSO_4(s) \rightleftarrows Ba^{2+}(aq) + SO_4^{2-}(aq)$. The concentrations and Q_{sp} are

$$[Ba^{2+}] = \frac{125\text{mL}(0.100\,M)}{135\text{mL}} = 0.0926\,M$$

$$[SO_4^{2-}] = \frac{10\text{mL}(0.050M)}{135\text{mL}} = 0.0037\,M$$

$$Q_{sp} = [Ba^{2+}][SO_4^{2-}] = (0.0926)(0.0037) = 3.4\times10^{-4}$$

Because $K_{sp}(BaSO_4) = 1.5 \times 10^{-9}$, we have $Q_{sp} > K_{sp}$ and the reaction will shift left, forming a precipitate.

b. The reaction is $AgCl(s) \rightleftarrows Ag^+(aq) + Cl^-(aq)$. The concentrations and Q_{sp} are

$$[Ag^+] = \frac{35\text{mL}(0.0045\,M)}{135\text{mL}} = 0.0012\,M$$

$$[Cl^-] = \frac{100\text{mL}(0.0038M)}{135\text{mL}} = 0.0028\,M$$

$$Q_{sp} = [Ag^+][Cl^-] = (0.0012)(0.0028) = 3.3\times10^{-6}$$

Because $K_{sp}(AgCl) = 1.6 \times 10^{-10}$, we have $Q_{sp} > K_{sp}$ and the reaction will shift left, forming a precipitate.

57. Lowering the pH results in more H^+ available to react with the OH^- from the iron(III) hydroxide. Removing some of the OH^- from the solution makes it possible for more salt to dissolve. Raising the pH will increase the OH^- concentration in the solution, which will force more salt to remain as a solid, lowering its solubility.

59. $CaF_2(s) \rightleftarrows Ca^{2+}(aq) + 2\,F^-(aq)$

$$K_{sp} = \left[Ca^{2+}\right]\left[F^-\right]^2 = (s)(2s)^2 = 4s^3 = 4.0\times10^{-11}$$

$$s = \sqrt[3]{\frac{4.0\times10^{-11}}{4}} = 2.1_5\times10^{-4}\,\text{mol/L}$$

$$\left[F^-\right] = 2s = 4.3\times10^{-4}\,\text{mol/L}$$

61. Solving for the solubility of each:

$CaCO_3(s) \rightleftarrows Ca^{2+}(aq) + CO_3^{2-}(aq)$;

$$K_{sp} = \left[Ca^{2+}\right]\left[CO_3^{2-}\right] = (s)(s) = s^2 = 8.7\times10^{-9}$$

$$s = \sqrt{8.7\times10^{-9}} = 9.3\times10^{-5}\,\text{mol/L} = \left[Ca^{2+}\right]$$

$Ca(IO_3)_2(s) \rightleftarrows Ca^{2+}(aq) + 2\,IO_3^-(aq)$;

$$K_{sp} = \left[Ca^{2+}\right]\left[IO_3^-\right]^2 = (s)(2s)^2 = 4s^3 = 7.1\times10^{-7}$$

$$s = \sqrt[3]{\frac{7.1\times10^{-7}}{4}} = 5.6\times10^{-3}\,\text{mol/L} = \left[Ca^{2+}\right]$$

$Ca_3(PO_4)_2(s) \rightleftarrows 3\,Ca^{2+}(aq) + 2\,PO_4^{3-}(aq)$

$$K_{sp} = \left[Ca^{2+}\right]^3\left[PO_4^{3-}\right]^2 = (3s)^3(2s)^2 = 108s^5 = 1.3\times10^{-32}$$

$$s = \sqrt[5]{\frac{1.3\times10^{-32}}{108}} = 1.6_4\times10^{-7}\,\text{mol/L}$$

$$\left[Ca^{2+}\right] = 3s = 4.9\times10^{-7}\,\text{mol/L}$$

$Ca(IO_3)_2$ provides the highest calcium ion concentration. Each of the anions in the above salts is a weak base. Acidic solutions will convert some of the base into acid, allowing more salt to dissolve.

63. a. $Ag^+(aq) + NH_3(aq) \rightleftarrows Ag(NH_3)^+$

$Ag(NH_3)^+(aq) + NH_3(aq) \rightleftarrows Ag(NH_3)_2^+$

b. $Ni^{2+}(aq) + NH_3(aq) \rightleftarrows Ni(NH_3)^{2+}(aq)$

$Ni(NH_3)^{2+}(aq) + NH_3(aq) \rightleftarrows Ni(NH_3)_2^{2+}(aq)$

$Ni(NH_3)_2^{2+}(aq) + NH_3(aq) \rightleftarrows Ni(NH_3)_3^{2+}(aq)$

$Ni(NH_3)_3^{2+}(aq) + NH_3(aq) \rightleftarrows Ni(NH_3)_4^{2+}(aq)$

$Ni(NH_3)_4^{2+}(aq) + NH_3(aq) \rightleftarrows Ni(NH_3)_5^{2+}(aq)$

$Ni(NH_3)_5^{2+}(aq) + NH_3(aq) \rightleftarrows Ni(NH_3)_6^{2+}(aq)$

65. These calculations are very similar to those we use to find the equivalence-point volume in a titration:

a. $$0.100\,\text{L}\,Zn^{2+} \times \frac{0.150\,\text{mol}\,Zn^{2+}}{1\,\text{L}\,Zn^{2+}} \times \frac{1\,\text{mol}\,EDTA^{4-}}{1\,\text{mol}\,Zn^{2+}} \times \frac{1\,\text{L}\,EDTA^{4-}}{0.0156\,\text{mol}\,EDTA^{4-}} \times \frac{1000\,\text{mL}}{1\,\text{L}}$$

$$= 962\,\text{mL}\,EDTA^{4-}$$

b. $$0.0500\,\text{L}\,Ca^{2+} \times \frac{0.740\,\text{mol}\,Ca^{2+}}{1\,\text{L}\,Ca^{2+}} \times \frac{1\,\text{mol}\,EDTA^{4-}}{1\,\text{mol}\,Ca^{2+}} \times \frac{1\,\text{L}\,EDTA^{4-}}{0.0156\,\text{mol}\,EDTA^{4-}} \times \frac{1000\,\text{mL}}{1\,\text{L}}$$

$$= 2370\,\text{mL}\,EDTA^{4-} = 2.37\times10^{3}\,\text{mL}\,EDTA^{4-}$$

c. $$0.0200\,\text{L}\,Mg^{2+} \times \frac{0.050\,\text{mol}\,Mg^{2+}}{1\,\text{L}\,Mg^{2+}} \times \frac{1\,\text{mol}\,EDTA^{4-}}{1\,\text{mol}\,Mg^{2+}} \times \frac{1\,\text{L}\,EDTA^{4-}}{0.0156\,\text{mol}\,EDTA^{4-}} \times \frac{1000\,\text{mL}}{1\,\text{L}}$$

$$= 64\,\text{mL}\,EDTA^{4-}$$

67. We write the conditional formation constant as

$$K_f' = \alpha_{Zn^{2+}}\alpha_{Y^{4-}}K_f = (0.05)(1.0\times10^{-4})(3.0\times10^{16}) = 1.5\times10^{11}$$

69. a. Because the K_{sp} for $Cu(OH)_2$ is low (1.6×10^{-19}), the formation constant must be quite large to overcome the low solubility of the salt.

b. $Cu^{2+}(aq) + NH_3(aq) \rightleftarrows Cu(NH_3)^{2+}(aq)$
$Cu(NH_3)^{2+}(aq) + NH_3(aq) \rightleftarrows Cu(NH_3)_2^{2+}(aq)$
$Cu(NH_3)_2^{2+}(aq) + NH_3(aq) \rightleftarrows Cu(NH_3)_3^{2+}(aq)$
$Cu(NH_3)_3^{2+}(aq) + NH_3(aq) \rightleftarrows Cu(NH_3)_4^{2+}(aq)$

71. In order for the lead to be complexed by the EDTA and removed from the patient, the lead must be more strongly complexed to EDTA than calcium is. Otherwise, the lead would remain in solution, and the calcium would remain complexed to the EDTA.

73. Your answers may vary. Some possible answers include blood chemistry, EDTA titrations of certain metals, and growth of certain bacteria in a lab culture.

75. a. The best buffer will be created with an acid with a pK_a close to the desired pH. The K_a and pK_a values are H_3PO_4 ($K_a = 7.5 \times 10^{-3}$, pK_a = 2.12); $H_2PO_4^-$ ($K_a = 6.2 \times 10^{-8}$, pK_a = 7.21); and HPO_4^{2-} ($K_a = 4.8 \times 10^{-13}$, p$K_a$ = 12.32). The best acid to use would be H_2PO_4.

b. The important reaction is the ionization of $H_2PO_4^-$ ($H_2PO_4^- \rightleftarrows HPO_4^{2-} + H^+$). We can use the mass-action expression for this reaction to solve for the ratio:

$$K_a = \frac{[H^+][HPO_4^{2-}]}{[H_2PO_4^-]}$$

$$\frac{[HPO_4^{2-}]}{[H_2PO_4^-]} = \frac{K_a}{[H^+]} = \frac{6.2\times10^{-8}}{10^{-8.20}} = 9.8$$

77. a. Each of these metals is found in Period 4 of the periodic table. Zinc and cobalt ions have more protons than calcium ion and form a smaller ion that may be not be able to coordinate with EDTA in all six sites. Calcium, with its larger size, can fully coordinate with the EDTA in all six sites, resulting in a much stronger complex and a higher formation constant.

b. EDTA has a –4 charge when it is in the form that complexes with metals. That –4 charge will interact more strongly with a +3 charge than with a +2 charge; therefore, Fe^{3+} will form a stronger complex with EDTA than will Fe^{2+}.

79. Formic acid reacts in a 1:1 ratio with NaOH, so we can solve easily for the concentration of the diluted formic acid solution:

$$\frac{\text{moles formic acid}}{\text{volume}} = \frac{0.025\,\text{L NaOH} \times \frac{0.0010\,\text{mol NaOH}}{1\,\text{L NaOH}} \times \frac{1\,\text{mol HCOOH}}{1\,\text{mol NaOH}}}{0.050\,\text{L HCOOH}}$$

$$= 0.00050\,M\ \text{HCOOH}$$

The number of moles in the original is the same as the number of moles in the entire diluted solution: $\frac{0.00050\,\text{mol HCOOH}}{1\,\text{L HCOOH}} \times 0.050\,\text{L} = 2.5\times10^{-5}\,\text{mol HCOOH}$

81. a. The formation reaction, $Fe^{2+}(aq) + EDTA^{4-}(aq) \rightleftarrows FeEDTA^{2-}(aq)$, has the mass-action expression $K_f = \frac{[FeEDTA^{2-}]}{[Fe^{2+}][EDTA^{4-}]}$. If we assume that nearly all of the original iron is in the complexed form and that the volume change is negligible, we can solve for the concentration of the uncomplexed iron:

$$[Fe^{2+}] = \frac{[FeEDTA^{2-}]}{K_f[EDTA^{4-}]} = \frac{(0.20)}{(2.1\times10^{14})(0.10)} = 9.5\times10^{-15}\,M$$

b. The K_{sp} for $Fe(OH)_2$ is 1.8×10^{-15} and the dissolution reaction is $Fe(OH)_2(aq) \rightleftarrows Fe^{2+}(aq) + 2\,OH^-(aq)$. We can solve for the ion product for this reaction to determine whether precipitation will occur:

$$Q_{sp} = [Fe^{2+}][OH^-]^2 = (9.5\times10^{-15})(0.20)^2 = 3.8\times10^{-16}$$

Because $Q_{sp} < K_{sp}$, no precipitation will occur.

82. a. At the midpoint of a titration of a weak acid, $pH = pK_a$, so we can solve for the K_a:

$$K_a = 10^{-pK_a} = 10^{-pH} = 10^{-3.58} = 2.6\times10^{-4}$$

b. To solve for the initial pH, we need the concentration of the original acid (HA). The number of moles in the original acid solution will be twice that of the hydroxide used at the midpoint of the titration:

$$2\times\frac{0.0256\,\text{L NaOH}\times\frac{0.100\,\text{mol NaOH}}{1\,\text{L NaOH}}\times\frac{1\,\text{mol HA}}{1\,\text{mol NaOH}}}{0.100\,\text{L HA}} = 0.0512\,M$$

At 0.00 mL, we have a solution of pure weak acid. With a large K_a and low concentration, we should expect to need the quadratic equation to solve the problem. For 0.0512 M HA ($K_a = 2.6 \times 10^{-4}$):

	HA	⇄	H^+ +	A^-
Initial	0.0512 M		0 M	0 M
Change	$-x$		$+x$	$+x$
Equilibrium	$0.0512 - x$		$+x$	$+x$

$$K_a = \frac{[A^-][H^+]}{[HA]}$$

$$2.6\times10^{-4} = \frac{(x)^2}{0.0512 - x}$$

$$2.6\times10^{-4}(0.0512 - x) = x^2$$

$$1.3_3\times10^{-5} - 2.6\times10^{-4}x = x^2$$

$$x^2 + 2.6\times10^{-4}x - 1.3_3\times10^{-5} = 0$$

$$x = \frac{-2.6\times10^{-4} \pm \sqrt{(2.6\times10^{-4})^2 - 4(1)(-1.3_3\times10^{-5})}}{2(1)}$$

$$x = -1.3\times10^{-4} \pm 0.0036_5 = 0.0035 \text{ or } -0.0038$$

$$x = [H^+] = 0.0035M$$

$$pH = -\log(0.0035) = 2.46$$

At 10.00 mL:

$$\text{mol HA} = M_{HA}\times V_{HA} - M_{NaOH}\times V_{NaOH}$$
$$= 100.0\text{mL}(0.0512\,\text{mol/L}) - 10.0\text{mL}(0.100\,\text{mol/L}) = 4.12\,\text{mmol}$$

$$\text{mol A}^- = M_{NaOH}\times V_{NaOH} = 10.0\text{mL}(0.100\,\text{mol/L}) = 1.00\,\text{mmol}$$

$$pH = pK_a + \log\left(\frac{[HCOO^-]}{[HCOOH]}\right) = 3.58 + \log\left(\frac{1.00}{4.12}\right) = 2.97$$

At 25.6 mL, we are at the midpoint of the titration, so pH = pK_a = 3.58.

At 50.00 mL:

$$\text{mol HA} = M_{HA}\times V_{HA} - M_{NaOH}\times V_{NaOH}$$
$$= 100.0\text{mL}(0.0512\,\text{mol/L}) - 50.0\text{mL}(0.100\,\text{mol/L}) = 0.12\,\text{mmol}$$

$$\text{mol A}^- = M_{NaOH}\times V_{NaOH} = 50.0\text{mL}(0.100\,\text{mol/L}) = 5.00\,\text{mmol}$$

$$pH = pK_a + \log\left(\frac{[HCOO^-]}{[HCOOH]}\right) = 3.58 + \log\left(\frac{5.00}{0.12}\right) = 5.20$$

At 75.0 mL, we are past the equivalence point of the titration, and the pH is controlled by the excess OH^-:

$$[OH^-] = \frac{\text{mol OH}^- - \text{mol HA}}{\text{total volume}} = \frac{M_{OH^-}\times V_{OH^-} - M_{HA}\times V_{HA}}{V_{OH^-} + V_{HA}}$$

$$[OH^-] = \frac{(0.100M)(75.0\text{mL}) - (0.0512M)(100.0\text{mL})}{75.0\,\text{mL} + 100.0\,\text{mL}} = 0.0136M$$

$$pOH = -\log[OH^-] = -\log(0.0136) = 1.866$$

$$pH = 14 - pOH = 12.134$$

At 100.0 mL:

$$\left[OH^-\right] = \frac{\text{mol}\,OH^- - \text{mol}\,HA}{\text{total volume}} = \frac{M_{OH^-} \times V_{OH^-} - M_{HA} \times V_{HA}}{V_{OH^-} + V_{HA}}$$

$$\left[OH^-\right] = \frac{(0.100M)(100.0\,\text{mL}) - (0.0512\,M)(100.0\,\text{mL})}{100.0\,\text{mL} + 100.0\,\text{mL}} = 0.0244\,M$$

$$\text{pOH} = -\log\left[OH^-\right] = -\log(0.0244) = 1.613$$

$$\text{pH} = 14 - \text{pOH} = 12.387$$

The titration curve:

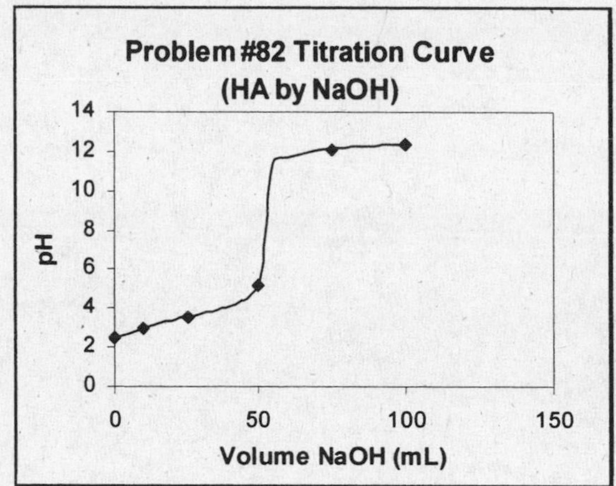

c. The endpoint of the titration is very near pH = 8.0. Thymol blue's upper color-change range would be very appropriate for this titration. Phenolphthalein would be acceptable, though not ideal.

d. The number of moles of acid in this volume is given by $0.0512\,M \times 0.100\,\text{L} = 0.00512\,\text{mol}$.

The molecular weight would be $\dfrac{0.123\,\text{g}}{0.00512\,\text{mol}} = 24.0\,\text{g/mol}$.

Chapter 19: Electrochemistry

The Bottom Line

- Redox reactions involve both a reduction and an oxidation half-reaction.
- Redox reactions can be identified by determining the oxidation state of the atoms involved in a reaction.
- Redox reactions can be balanced by summation of balanced half-reactions.
- Positive cell potentials indicate a spontaneous reaction and are related to the free energy change by $\Delta G = -nFE°$.
- Electrochemical cells require both a path for the electrons and a path for other ions.
- The oxidation reaction takes place at the anode. The reduction reaction takes place at the cathode.
- The Nernst equation relates the actual potential of a redox reaction to conditions other than the standard conditions.
- Half-reaction potentials can be used to determine the relative reactivity of metals. Organization of the metals in this fashion is known as a chemical activity series.
- Cell potentials enable us to calculate equilibrium constants.
- Electrowinning and electroplating are examples of electrolysis reactions. In electrolysis, a positive potential that includes the overpotential is applied to force the reaction to run in reverse.

Solutions to Practice Problems

19.1. From the rules given in this chapter, we know that the total of all the oxidation states for all the atoms will be zero for both compounds because they are neutral species. For $C_6H_{12}O_6$, we assign +1 to the hydrogen and −2 to the oxygen. We then have only one species left to assign (the carbon). Adding the oxidation states yields 12 H = 12 × (+1) = +12 and 6 O = 6 × (−2) = −12. These two total zero. Because the compound is neutral, carbon has an oxidation number of 0. For CO, we can assign −2 to oxygen. Because the compound is neutral, the carbon must have an oxidation state of +2.

19.2. Using the rules for assigning oxidation states, we find the order of increasing oxidation states of sulfur to be Li_2S (S = −2) < S_8 (S = 0) < SO_2 (S = +4) < H_2SO_4 (S = +6). These oxidation states are quickly obtained by realizing that all compounds are neutral and that the other elements have a simply assigned oxidation state: Li = +1; O = −2; H = +1.

19.3. Because we are given the cell potential for the reaction, we need only determine the number of electrons that are exchanged in the reaction in order to use the expression $\Delta G° = -nFE°$. If we look at the oxidation state of zinc on each side of the reaction, we see that as an element on the left it has an oxidation state of 0, while on the right it has an oxidation state of +2 (balanced by the −2 on the oxygen.) The change from 0 to +2 required a two-electron exchange, so n = 2 and the free energy change is

$$\Delta G° = -nFE° = -(2\,\text{mole}^-)(96485\,\text{C/mole}^-)\left(1.86\,\text{V} \times \frac{1\,\text{J/C}}{1\,\text{V}}\right) \times \frac{1\,\text{kJ}}{1000\,\text{J}} = -359\,\text{kJ}$$

The reaction is spontaneous.

19.4. Following the method in the text, we can balance the reaction:

Step 1 Determine the oxidation state on each element in each compound in the chemical equation. Do we in fact have a redox reaction?

$$ClO^{-}(aq) + H^{+}(aq) + Cu(s) \rightarrow Cl^{-}(aq) + H_2O(l) + Cu^{2+}(aq)$$

oxidation states: +1 −2 (ClO⁻), +1 (H⁺), 0 (Cu), −1 (Cl⁻), +1 −2 (H₂O), +2 (Cu²⁺)

Copper is being oxidized from 0 to +2
Chlorine is being reduced from +1 to −1
This is a redox reaction.

Step 2 Separate the redox reaction into two parts: an oxidation reaction and a reduction reaction. Include just the compounds that have a change in the oxidation state.

$ClO^{-}(aq) \rightarrow Cl^{-}(aq)$ (reduction)
$Cu(s) \rightarrow Cu^{2+}(aq)$ (oxidation)

Step 3 Balance all atoms except oxygen and hydrogen.
For these reactions, they are already balanced

Step 4 Balance the oxygen atoms by adding H_2O to one side in each half-reaction.

$ClO^{-}(aq) \rightarrow Cl^{-}(aq) + H_2O(l)$
$Cu(s) \rightarrow Cu^{2+}(aq)$

Step 5 Balance the hydrogen atoms by adding H^{+} to one side in each half-reaction.

$2\,H^{+}(aq) + ClO^{-}(aq) \rightarrow Cl^{-}(aq) + H_2O(l)$
$Cu(s) \rightarrow Cu^{2+}(aq)$

Step 6 Balance the charges in each half-reaction by adding electrons to one side.

$2\,e^{-} + 2\,H^{+}(aq) + ClO^{-}(aq) \rightarrow Cl^{-}(aq) + H_2O(l)$
$Cu(s) \rightarrow Cu^{2+}(aq) + 2e^{-}$

Step 7 As needed, multiply one or both of the half-reactions by some coefficient so that both half-reactions will have the same number of electrons.
Both equations have two electrons, so no changes are needed.

Step 8 Add the reactions together and simplify. Note that the electrons on each side mathematically cancel, indicating the same number of electrons gained in the reduction as lost in the oxidation.

$2\,e^{-} + 2\,H^{+}(aq) + ClO^{-}(aq) \rightarrow Cl^{-}(aq) + H_2O(l)$
$Cu(s) \rightarrow Cu^{2+}(aq) + 2e^{-}$

$\cancel{2e^{-}} + 2\,H^{+}(aq) + ClO^{-}(aq) + Cu(s) \rightarrow Cl^{-}(aq) + H_2O(l) + Cu^{2+}(aq) + \cancel{2e^{-}}$

Balanced: $2\,H^{+}(aq) + ClO^{-}(aq) + Cu(s) \rightarrow Cl^{-}(aq) + H_2O(l) + Cu^{2+}(aq)$

19.5. Following the method in the text, we can balance the reaction. The process will proceed as in Practice 19.4, but we adjust the final balanced equation to convert from acidic to basic solution.

Step 1 Determine the oxidation state on each element in each compound in the chemical equation. Do we in fact have a redox reaction?

$$MnO_4^{-}(aq) + Mn^{2+}(aq) \rightarrow MnO_2(s)$$

oxidation states: +7 −2 (MnO₄⁻), +2 (Mn²⁺), +4 −2 (MnO₂)

Manganese (as the ion) is being oxidized from +2 to +4.
Manganese (in MnO_4^{-}) is being reduced from +7 to +4.
This is a redox reaction.

Step 2 Separate the redox reaction into two parts: an oxidation reaction and a reduction reaction. Include just the compounds that have a change in the oxidation state.

$Mn^{2+}(aq) \rightarrow MnO_2(s)$ (oxidation)

$MnO_4^-(aq) \rightarrow MnO_2(s)$ (reduction)

Step 3 Balance all atoms except oxygen and hydrogen.

For these reactions, they are already balanced.

Step 4 Balance the oxygen atoms by adding H_2O to one side in each half-reaction.

$2\ H_2O(l) + Mn^{2+}(aq) \rightarrow MnO_2(s)$

$MnO_4^-(aq) \rightarrow MnO_2(s) + 2\ H_2O(l)$

Step 5 Balance the hydrogen atoms by adding H^+ to one side in each half-reaction.

$2\ H_2O(l) + Mn^{2+}(aq) \rightarrow MnO_2(s) + 4\ H^+(aq)$

$4\ H^+(aq) + MnO_4^-(aq) \rightarrow MnO_2(s) + 2\ H_2O(l)$

Step 6 Balance the charges in each half-reaction by adding electrons to one side.

$2\ H_2O(l) + Mn^{2+}(aq) \rightarrow MnO_2(s) + 4\ H^+(aq) + 2\ e^-$

$3\ e^- + 4\ H^+(aq) + MnO_4^-(aq) \rightarrow MnO_2(s) + 2\ H_2O(l)$

Step 7 As needed, multiply one or both of the half-reactions by some coefficient so that both half-reactions will have the same number of electrons.

$[2\ H_2O(l) + Mn^{2+}(aq) \rightarrow MnO_2(s) + 4\ H^+(aq) + 2\ e^-] \times 3$

$[3\ e^- + 4\ H^+(aq) + MnO_4^-(aq) \rightarrow MnO_2(s) + 2\ H_2O(l)] \times 2$

Step 8 Add the reactions together and simplify. Note that the electrons on each side mathematically cancel, indicating the same number of electrons gained in the reduction as lost in the oxidation.

$6\ H_2O(l) + 3\ Mn^{2+}(aq) \rightarrow 3\ MnO_2(s) + 12\ H^+(aq) + 6\ e^-$

$6\ e^- + 8\ H^+(aq) + 2\ MnO_4^-(aq) \rightarrow 2\ MnO_2(s) + 4\ H_2O(l)$

$$\cancel{6e^-} + \cancel{8H^+} + 2\ MnO_4^- + \cancel{6}\,(2)H_2O(l) + 3\ Mn^{2+} \rightarrow 3\ MnO_2 + \cancel{12}\,(4)\ H^+ + \cancel{6e^-} + 2\ MnO_2 + \cancel{4H_2O(l)}$$

Balanced in acid: $2\ MnO_4^-(aq) + 2\ H_2O(l) + 3\ Mn^{2+}(aq) \rightarrow 5\ MnO_2(s) + 4\ H^+(aq)$

Converting to base: $+ 4\ OH^-$ $\qquad + 4\ OH^-$

$2\ MnO_4^-(aq) + \cancel{2\ H_2O(l)} + 3\ Mn^{2+}(aq) + 4\ OH^-(aq) \rightarrow 5\ MnO_2(s) + \cancel{4}\,(2)\ H_2O(l)$

Balanced in base: $2\ MnO_4^-(aq) + 3\ Mn^{2+}(aq) + 4\ OH^-(aq) \rightarrow 5\ MnO_2(s) + 2\ H_2O(l)$

19.6. We split the reactions into half-reactions and balance the equation, while determining the cell potential:

$CH_4(g) + H_2O(l) \rightarrow CO(g) + 6\ H^+(aq) + 6\ e^-$ $\quad E^\circ_{ox} = -E^\circ_{red} = -(0.25\ V) = -0.25\ V$

$[\,2\ H^+(aq) + 2\ e^- \rightarrow H_2(g)\,] \times 3$ $\quad E^\circ_{red} = 0$

$CH_4(g) + H_2O(l) \rightarrow CO(g) + 3\ H_2(g)$ $\quad E^\circ_{cell} = -0.25\ V$

$2 \times [Ag(s) \rightarrow Ag^+(aq) + e^-\,]$ $\quad E^\circ_{ox} = -E^\circ_{red} = -(0.80\ V) = -0.80\ V$

$2\ H^+(aq) + H_2O_2(aq) + 2\ e^- \rightarrow 2\ H_2O(l)$ $\quad E^\circ_{red} = 1.78\ V$

$2\ Ag(s) + 2\ H^+(aq) + H_2O_2(aq) \rightarrow 2\ Ag^+(aq) + 2\ H_2O(l)$ $\quad E^\circ_{cell} = 0.98\ V$

19.7. For each reaction, the oxidation will take place at the anode, usually drawn on the left, and the reduction occurs at the cathode, usually drawn on the right.

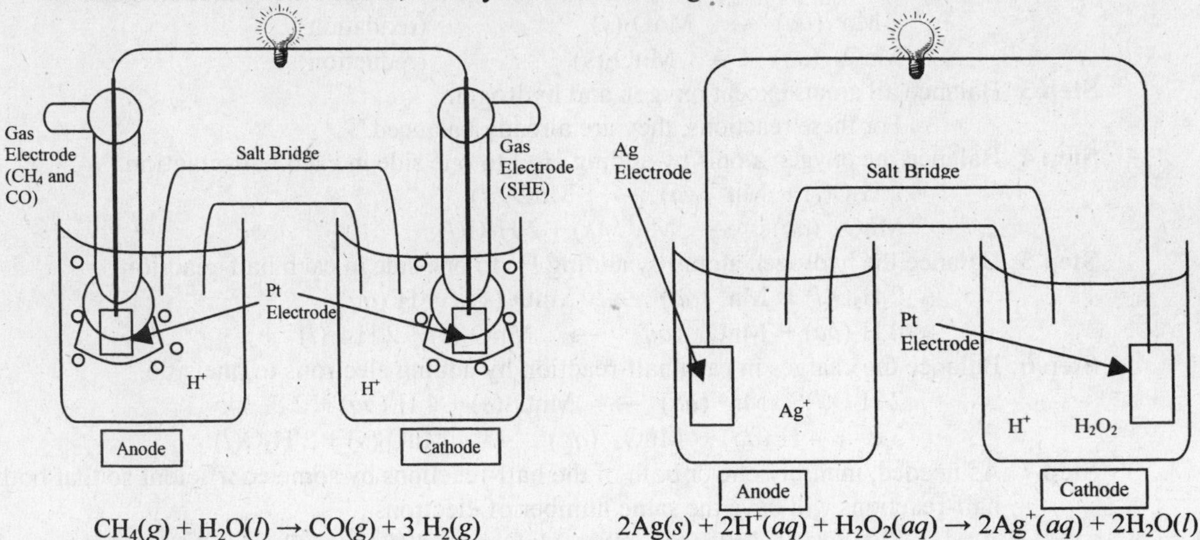

$CH_4(g) + H_2O(l) \rightarrow CO(g) + 3\ H_2(g)$ $2Ag(s) + 2H^+(aq) + H_2O_2(aq) \rightarrow 2Ag^+(aq) + 2H_2O(l)$

The potentials remain the same as those calculated in Practice 19.6. The methane cell is electrolytic with a cell potential of −0.25 V. The silver cell is galvanic with a cell potential of +0.98 V.

19.8. The iron reduction has a standard reduction potential of −0.44 V, and the aluminum reduction has a standard potential of −1.66 V. In order to get a galvanic cell (one with a positive overall potential), we need to reverse the aluminum reaction, making it an oxidation that hence occurs at the anode:

$$Fe^{2+}(aq) + 2e^- \rightarrow Fe(s) \qquad E° = -0.44\ V$$
$$Al(s) \rightarrow Al^{3+}(aq) + 3e^- \qquad E° = +1.66\ V$$
$$3Fe^{2+}(aq) + 2Al(s) \rightarrow 3Fe(s) + 2Al^{3+}(aq) \qquad E°_{cell} = 1.22\ V$$

Writing the cell notation, we write the anode first then the cathode:

$$Al(s)\ |\ Al^{3+}(aq)\ \|\ Fe^{2+}(aq)\ |\ Fe(s)$$

19.9. We can use the reactivity series of the metals to answer the question. As shown in Table 19.6, the metals at the top of the table are the most reactive. Therefore, in order of decreasing reactivity, we have Ca > Na > Al > Cu.

19.10. The reactions in this cell are shown below. The left side is the anode and the right side is the cathode:

$$\text{Anode: } H_2(g,\ 1\ atm) \rightarrow 2\ H^+(aq,\ 1.0\ M) + 2e^-$$
$$\text{Cathode: } 2\ H^+(aq,\ 0.10\ M) + 2e^- \rightarrow H_2(g,\ 1\ atm)$$
$$\text{Net: } 2\ H^+\ (0.10\ M) + H_2(g,\ 1\ atm) \rightarrow 2\ H^+\ (1.0\ M) + H_2(g,\ 1\ atm)$$

Because both reactions are the same, the standard cell potential is zero. Writing the Nernst equation and solving for the voltage:

$$E_{cell} = E°_{cell} - \frac{0.0592}{2}\log\frac{\left[H^+\right]^2_{Anode} P_{H_2,Cathode}}{\left[H^+\right]^2_{Cathode} P_{H_2,Anode}}$$

$$E_{cell} = 0 - \frac{0.0592}{2}\log\frac{[1]^2(1)}{[0.10]^2(1)} = -0.0592\ V$$ This cell is electrolytic.

19.11. The cell potential for the reaction is +1.10 V, and there are two electrons exchanged in the reaction. The equilibrium constant is

$$K = 10^{\frac{nE^\circ}{0.0592\,\text{V}}} = 10^{(2)(1.10\,\text{V})/0.0592\,\text{V}} = 1.45\times10^{37}$$

19.12. This type of problem is easily handled by treating it as a unit conversion problem starting with the current and time. The reduction of tin uses two electrons for each tin atom: $Sn^{2+} + 2e^- \rightarrow Sn$.

$$0.45\,\text{A}\times\frac{1\text{C/s}}{1\text{A}}\times1.5\,\text{h}\times\frac{3600\,\text{s}}{1\,\text{h}}\times\frac{1\,\text{mol}\,e^-}{96485\,\text{C}}\times\frac{1\,\text{mol Sn}}{2\,\text{mol}\,e^-}\times\frac{118.71\,\text{g Sn}}{1\,\text{mol Sn}} = 1.5\,\text{g Sn}$$

Solutions to Student Problems

1. A *battery* is a series of cells, while most "batteries" purchased at stores are a single cell.

3. The two reactions are a reduction reaction, which involves the gain of electrons, and an oxidation reaction, which involves the loss of electrons.

5. DMSO is $H_3C-S(=O)-CH_3$. Using the rules, we assign a +1 oxidation state to each H and −2 to O. Sulfur is a Group VI element, so 6 − 8 = −2 oxidation state for S. Totaling the oxidation states, we have 6(+1) + 1(−2) +1(−2) = +2. The two carbons must total −2 to make the molecule neutral, giving them an oxidation number of −1.

7. The oxidation states of chlorine in these molecules are 0 in Cl_2, +4 in ClO_2, +7 in $NaClO_4$, and −1 in HCl. The oxidation state is most positive in $NaClO_4$ and most negative in HCl.

9. The order from lowest to highest oxidation number for nitrogen is NH_3 (−3), N_2H_4 (−2), N_2O (+1), NO (+2), and NO_2 (+4).

11. a. $KMnO_4$ has K = +1, O = −2, and Mn = +7
    b. $LiMnO_2$ has Li = +1, O = −2, and Mn = +3
    c. NH_4ClO_4 has H = +1, O = −2, N = −3, and Cl = +7

13. a. An atom that has gained an electron is reduced and is an oxidizing agent.
    b. When an atom's oxidation number increases, it is oxidized and is a reducing agent.
    c. When the oxidation number of an atom changes from −2 to −3, the atom is reduced and is an oxidizing agent.

15.

	$3\,Mg(s)$	+	$2\,H_3PO_4(aq)$	→	$Mg_3(PO_4)_2(aq)$	+	$3\,H_2(g)$
oxidation states	0		+1 +5 −2		+2 +5 −2		0

17. a. Oxygen has an oxidation state of −2; therefore, Cl must have an oxidation state of +1 to allow the total charge to be −1 on the ion.
    b. If ClO^- is a good oxidizing agent, it must be itself reduced which is a gain of electrons.

19. a. $SrCl_2(l) \rightarrow Sr(s) + Cl_2(g)$

oxidation states: $+2\ -1$ | 0 | 0

b. Because the oxidation state on Sr changes from +2 to 0, it is gaining electrons and is reduced.

21. a. Reduction is the gain of electrons, so the electrons appear on the reactant side of the reaction. Only the first reaction is a reduction. Oxidation is the loss of electrons, so the electrons appear on the product side of the reaction. The second and fourth reactions are oxidations.

b. The third reaction is a precipitation reaction.

c. Combining the first two reactions:

$$O_2(aq) + 2H_2O(l) + 4e^- \rightarrow 4OH^-(aq)$$
$$[Fe(s) \rightarrow Fe^{2+}(aq) + 2e^-] \times 2$$
$$\overline{2\ Fe(s) + O_2(aq) + 2\ H_2O(l) \rightarrow 4\ OH^-(aq) + 2\ Fe^{2+}(aq)}$$

Combining this reaction with the third:

$$2\ Fe(s) + O_2(aq) + 2\ H_2O(l) \rightarrow 4\ OH^-(aq) + 2\ Fe^{2+}(aq)$$
$$[Fe^{2+}(aq) + 2OH^-(aq) \rightarrow Fe(OH)_2(s)] \times 2$$
$$\overline{2\ Fe(s) + O_2(aq) + 2\ H_2O(l) \rightarrow 2\ Fe(OH)_2(s)}$$

Combining the first and fourth reactions:

$$O_2(aq) + 2H_2O(l) + 4e^- \rightarrow 4OH^-(aq)$$
$$[Fe(OH)_2(s) + OH^-(aq) \rightarrow FeO(OH)(s) + H_2O(l) + e^-] \times 4$$
$$\overline{O_2(aq) + 4\ Fe(OH)_2(s) \rightarrow 4\ FeO(OH)(s) + 2\ H_2O(l)}$$

We then combine the final reactions shown in the previous two additions:

$$[2\ Fe(s) + O_2(aq) + 2\ H_2O(l) \rightarrow 2\ Fe(OH)_2(s)] \times 2$$
$$O_2(aq) + 4\ Fe(OH)_2(s) \rightarrow 4\ FeO(OH)(s) + 2\ H_2O(l)$$
$$\overline{4\ Fe(s) + 3\ O_2(aq) + 2\ H_2O(l) \rightarrow 4\ FeO(OH)\ (s)}$$

23. Standard conditions specify that all gases are at 1 bar pressure and all solutions have a concentration of 1 *M*.

25. a. When E° is negative, the electrochemical reaction is nonspontaneous.

b. When E° is positive, the value for ΔG° is negative.

c. When a reaction is spontaneous, the values for ΔG° will be negative and the values for E° will be positive.

d. Nonspontaneous redox reactions will have a negative value for E°.

27. In order to create a galvanic cell, we need to reverse the most negative (or least positive) reduction, giving one oxidation and one reduction:

$$[Fe^{3+} + e \rightarrow Fe^{2+}] \times 2 \qquad E^\circ = +0.77\ V$$
$$Fe \rightarrow Fe^{2+} + 2e^- \qquad E^\circ = +0.44\ V$$
$$\overline{2\ Fe^{3+} + Fe \rightarrow 3\ Fe^{2+} \qquad E^\circ_{cell} = +1.21\ V}$$

29. The free energy change is related to the cell potential by $\Delta G^\circ = -nFE^\circ$. There are two moles of electrons exchanged in the reaction; therefore,

$$\Delta G^\circ = -nFE^\circ = -\left(2\,\text{mol}\,e^-\right)\left(96485\,\text{C/mol}\,e^- \times \frac{1\,\text{J/V}}{1\,\text{C}}\right)(1.21\,\text{V}) = -233\,\text{kJ}$$

31. The oxidation number on lead changes from +4 to +2, indicating a transfer of two electrons. The free energy is $\Delta G° = -nFE° = -\left(2\,\text{mol}\,e^-\right)\left(96485\,\text{C/mol}\,e^- \times \frac{1\,\text{J/V}}{1\,\text{C}}\right)(1.12\,\text{V}) = -216\,\text{kJ}$.

33. The redox atoms must be balanced, the nonredox atoms must be balanced, and the electrons (used to balance the charges) involved in the oxidation and reduction half-reactions must cancel.

35. a. The balanced reaction (done using the technique shown in Practice 19.4) is

$$2\,CO_2 + 2\,H^+ + 2e^- \rightarrow H_2C_2O_4$$

This reaction is a reduction. Because the oxidation number of carbon changes from +4 to +3, it is the species that is reduced.

b. The balanced reaction is $Np^{4+} + 2\,H_2O \rightarrow NpO_2^+ + 4\,H^+ + e^-$. This reaction is an oxidation. Because the oxidation number of Np changes from +4 to +5, it is the species that is oxidized.

37. a. The balanced half-reactions and overall reaction are

$$Sn^{2+} \rightarrow Sn^{4+} + 2e^-$$
$$[Cu^{2+} + e^- \rightarrow Cu^+] \times 2$$
$$\overline{Sn^{2+} + 2\,Cu^{2+} \rightarrow Sn^{4+} + 2\,Cu^+}$$

b. The balanced half-reactions and overall reaction are

$$2\,S_2O_3^{2-} \rightarrow S_4O_6^{2-} + 2\,e^-$$
$$I_3^- + 2\,e^- \rightarrow 3\,I^-$$
$$\overline{I_3^- + 2\,S_2O_3^{2-} \rightarrow S_4O_6^{2-} + 3\,I^-}$$

c. The balanced half-reactions and overall reaction are

$$SO_3^- + H_2O \rightarrow SO_4^{2-} + 2\,H^+ + e^-$$
$$Fe^{3+} + e^- \rightarrow Fe^{2+}$$
$$\overline{Fe^{3+} + SO_3^- + H_2O \rightarrow SO_4^{2-} + Fe^{2+} + 2\,H^+}$$

39. The balanced half-reactions and overall reaction, first in acidic solution, are

$$[Cu \rightarrow Cu^{2+} + 2\,e^-] \times 3$$
$$Cr_2O_7^{2-} + 14\,H^+ + 6\,e^- \rightarrow 2\,Cr^{3+} + 7\,H_2O$$
$$\overline{Cr_2O_7^{2-} + 14\,H^+ + 3\,Cu \rightarrow 3\,Cu^{2+} + 2\,Cr^{3+} + 7\,H_2O}$$

For each H^+ we add one OH^- to each side. The H^+ and OH^- combine to create H_2O, some of which cancel water from the opposite side:

$$+14\,OH^- \qquad\qquad +14\,OH^-$$
$$\overline{Cr_2O_7^{2-} + 7\,H_2O + 3\,Cu \rightarrow 3\,Cu^{2+} + 2\,Cr^{3+} + 14\,OH^-}$$

41. The balanced half-reactions and overall reaction, first in acidic solution, are

$$ClO_4^- + 6\,H^+ + 6\,e^- \rightarrow ClO^- + 3\,H_2O$$
$$I^- + 3\,H_2O \rightarrow IO_3^- + 6\,H^+ + 6\,e^-$$
$$\overline{ClO_4^- + I^- \rightarrow IO_3^- + ClO^-}$$

Because there are no H^+ remaining in the balanced reaction, this redox reaction is the same for acidic and basic solutions.

43. a. The reaction and its potential are $Zn^{2+} + 2\,e^- \rightarrow Zn$, $E° = -0.76$ V,

b. $\Delta G° = -nFE° = -\left(2\,\text{mol}\,e^-\right)\left(96485\,\text{C/mol}\,e^- \times \dfrac{1\,\text{J/V}}{1\,\text{C}}\right)\left(-0.76\,\text{V}\right) = +1.5\times10^2\ \text{kJ}$

c. There is no change in the cell potential. When the reaction is doubled, the free energy doubles, but the number of electrons also doubles. The two changes cancel each other, causing the potential to remain unchanged.

45. a. Li^+ has the most negative reduction potential, indicating the least spontaneous reaction of the three.

b. The strongest oxidizing agent will be the species that is most likely to be reduced. F_2 has the most positive reduction potential; it is the strongest oxidizing agent.

c. To create the most positive cell potential, we combine the most positive reduction with the reverse of the most negative reduction.

oxidation:	$Zn \rightarrow Zn^{2+} + 2\,e^-$	$E° = +0.76$ V
reduction:	$Cu^{2+} + 2\,e^- \rightarrow Cu$	$E° = +0.34$ V
net reaction:	$Zn + Cu^{2+} \rightarrow Cu + Zn^{2+}$	$E°_{cell} = 1.10$ V

47. The balanced half-reactions and overall reaction are

$$Cu + H_2O \rightarrow CuO + 2\,H^+ + 2\,e^-$$
$$2\,CO_2 + 2\,e^- \rightarrow C_2O_4^{2-}$$
$$\overline{2\,CO_2 + Cu + H_2O \rightarrow CuO + 2\,H^+ + C_2O_4^{2-}}$$

49. The balanced half-reactions and overall reaction are

$$[C_2O_4^{2-} \rightarrow 2\,CO_2 + 2\,e^-] \times 5$$
$$[MnO_4^- + 8\,H^+ + 5\,e^- \rightarrow Mn^{2+} + 4\,H_2O] \times 2$$
$$\overline{2\,MnO_4^- + 16\,H^+ + 5\,C_2O_4^{2-} \rightarrow 10\,CO_2 + 2\,Mn^{2+} + 8\,H_2O}$$

51. The balanced half-reactions and overall reaction are

$$C_6H_8O_6 \rightarrow C_6H_6O_6 + 2\,H^+ + 2\,e^-$$
$$I_2 + 2\,e^- \rightarrow 2\,I^-$$
$$\overline{I_2 + C_6H_8O_6 \rightarrow C_6H_6O_6 + 2\,H^+ + 2\,I^-}$$

53. A *cell* is an electrochemical device consisting of an anode and cathode allowing for the exchange between chemical and electrical energy. A *half-reaction* is the equation that describes the reduction part or the oxidation part of a redox reaction. A *galvanic cell* (also known as a *voltaic cell*) is a cell that produces electricity from a chemical reaction. The *electromotive force* is a measure of how strongly a species pulls electrons toward itself in a redox process.

55. The anode is given by the left side of the cell notation and is an oxidation reaction:

$$Zn \rightarrow Zn^{2+} + 2\,e^-$$

The species that serves as the oxidizing agent is itself reduced. This reaction happens at the cathode: $Cu^{2+} + 2\,e^- \rightarrow Cu$. Cu^{2+} is the oxidizing agent.

57. The reduction potentials for lead ion and silver ion are −0.13 V and 0.80 V, respectively. For a spontaneous reaction, the silver ion reaction will remain as a reduction and lead will be oxidized. If the nitrate ions are moving left to right through the salt bridge, the electrons are moving from the right cell to the left cell, which is the opposite of the standard convention. This requires that the left beaker be the cathode (gaining electrons) and the right beaker be the anode (losing electrons.) Therefore, the oxidation of lead will occur in the right beaker, and reduction of silver will occur in the left beaker.

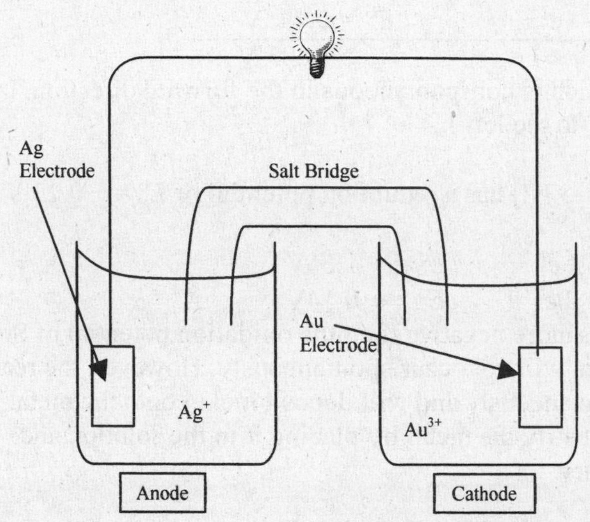

59. a.

The two reduction reactions are

$$Ag^+ + e^- \rightarrow Ag \qquad E° = 0.80 \text{ V}$$
$$Au^{3+} + 3\,e^- \rightarrow Au \qquad E° = 1.50 \text{ V}$$

To create a spontaneous reaction, we reverse the reaction with the smaller potential—the silver reaction. Therefore, the balanced reaction and cell potential are

anode:	$[Ag \rightarrow Ag^+ + e^-] \times 3$	$E° = -0.80$ V
cathode:	$Au^{3+} + 3\,e^- \rightarrow Au$	$E° = 1.50$ V
net:	$3\,Ag + Au^{3+} \rightarrow 3\,Ag^+ + Au$	$E°_{cell} = 0.70$ V

b. Three electrons are exchanged in the reaction.

c. Shown above.

d. An oxidizing agent is reduced; Au^{3+} is the oxidizing agent.

e. The cell notation is $Ag \mid Ag^+ \parallel Au^{3+} \mid Au$. Note that the anode is always written on the left, followed by the cathode.

61. The balanced reactions are (in acidic solution first):

$$Zn + H_2O \rightarrow ZnO + 2\,H^+ + 2\,e^-$$
$$HgO + 2\,H^+ + 2\,e^- \rightarrow Hg + H_2O$$

Converting each to basic solution separately:

$$Zn + 2\,OH^- \rightarrow ZnO + H_2O + 2\,e^-$$
$$HgO + H_2O + 2\,e^- \rightarrow Hg + 2\,OH^-$$

The cell notation is $Zn(s) \mid ZnO(s) \mid OH^- \parallel OH^- \mid HgO(s) \mid Hg(l)$

Because there are equal numbers of OH^- on each side, the net cell reaction does not include OH^-. We could also write the cell notation as $Zn(s) \mid ZnO(s) \parallel HgO(s) \mid Hg(l)$.

63. a. Written as oxidations (because reducing agents are oxidized), Ba has the more positive half-cell potential.

b. Written as reductions, the lead reaction has a more positive (less negative) half-cell potential; therefore, Pb^{2+} is the better oxidizing agent.

c. As written, the half-cell reactions are

$$\begin{array}{ll} Ag \rightarrow Ag^{+} + e^{-} & E° = -0.80\ V \\ Fe^{3+} + e^{-} \rightarrow Fe^{2+} & E° = 0.77\ V \\ \hline Ag + Fe^{3+} \rightarrow Ag^{+} + Fe^{2+} & E° = -0.03\ V \end{array}$$

Under standard conditions, this reaction is nonspontaneous in the forward direction. It will be spontaneous in the reverse direction (to the left.)

65. The reduction of nickel(II) [$Ni^{2+} + 2\,e^{-} \rightarrow Ni$] has a reduction potential of $E° = -\,0.23$ V. The oxidations of aluminum and tin are

$$\begin{array}{ll} Al \rightarrow Al^{3+} + 3\,e^{-} & E° = +\,1.66V \\ Sn \rightarrow Sn^{2+} + 2\,e^{-} & E° = +\,0.14V \end{array}$$

Because the reduction potential of Ni^{2+} is more negative than the oxidation potential of Sn is positive, the reaction of nickel(II) with tin will not occur spontaneously. However, the reaction of nickel(II) with aluminum will occur spontaneously and will deposit nickel onto the metal if it was placed in the nickel solution. One can identify the metal by placing it in the solution and observing whether or not a reaction occurs.

67. The constants in the natural logarithm form of the Nernst equation are

$$\frac{RT}{F} = \frac{(8.3145\,\mathrm{J/mol\cdot K})(298.15\,\mathrm{K})}{\left(96485\,\mathrm{C/mol} \times \frac{1\,\mathrm{J/V}}{1\,\mathrm{C}}\right)} = 0.0257\mathrm{V}$$. Natural logarithms and base 10 logarithms are related by $\ln x = 2.303 \log x$, so we must multiply the value of 0.0257 V by 2.303, obtaining 0.0592 V.

69. We first need to find the standard cell potential and the balanced reaction. The iron reaction is at the anode (oxidation), and the copper reaction is at the cathode (reduction):

$$\begin{array}{ll} Fe \rightarrow Fe^{2+} + 2e^{-} & E° = +0.44V \\ Cu^{2+} + 2\,e^{-} \rightarrow Cu & E° = +0.34\ V \\ \hline Fe + Cu^{2+} \rightarrow Fe^{2+} + Cu & E°_{cell} = 0.78\ V \end{array}$$

The Nernst equation for this cell will be

$$E_{cell} = E°_{cell} - \frac{0.0257}{n} \ln \frac{\left[Fe^{2+}\right]}{\left[Cu^{2+}\right]} = 0.78 - \frac{0.0257}{2} \ln \frac{\left[Fe^{2+}\right]}{\left[Cu^{2+}\right]}$$

a. $E_{cell} = 0.78 - \dfrac{0.0257}{2} \ln \dfrac{(0.10)}{(0.10)} = 0.78\mathrm{V}$

b. $E_{cell} = 0.78 - \dfrac{0.0257}{2} \ln \dfrac{(1.5)}{(0.10)} = 0.75\mathrm{V}$

c. $E_{cell} = 0.78 - \dfrac{0.0257}{2} \ln \dfrac{(0.10)}{(1.5)} = 0.81\mathrm{V}$

71. You could state that the free energy change for the reaction is zero, $\Delta G = 0$, or that the battery has reached equilibrium, or that the cell potential is zero, $E_{cell} = 0$ V.

73. We can treat each electrolysis problem as a unit conversion exercise. The reduction of gold [$Au^{3+} + 3\,e^{-} \rightarrow Au$] uses 3 moles of electrons for each mole of gold.

a. $1.25\,\text{C/s} \times 60\,\text{s} \times \frac{1\,\text{mol}\,e^-}{96485\,\text{C}} \times \frac{1\,\text{mol Au}}{3\,\text{mol}\,e^-} \times \frac{196.97\,\text{g Au}}{1\,\text{mol Au}} = 0.051\,\text{g}$

b. $2.11\,\text{C/s} \times 2.33\,\text{h} \times \frac{3600\,\text{s}}{1\,\text{h}} \times \frac{1\,\text{mol}\,e^-}{96485\,\text{C}} \times \frac{1\,\text{mol Au}}{3\,\text{mol}\,e^-} \times \frac{196.97\,\text{g Au}}{1\,\text{mol Au}} = 12.0\,\text{g}$

c. $0.75\,\text{C/s} \times 1\,\text{day} \times \frac{24\,\text{h}}{1\,\text{day}} \times \frac{3600\,\text{s}}{1\,\text{h}} \times \frac{1\,\text{mol}\,e^-}{96485\,\text{C}} \times \frac{1\,\text{mol Au}}{3\,\text{mol}\,e^-} \times \frac{196.97\,\text{g Au}}{1\,\text{mol Au}} = 44\,\text{g}$

75. In order to plate the spoon with silver, silver ion must be reduced [$Ag^+ + e^- \rightarrow Ag$]. The teaspoon would be the cathode. The current required is

$$\frac{0.33\,\text{g Ag} \times \frac{1\,\text{mol Ag}}{107.87\,\text{g Ag}} \times \frac{1\,\text{mol}\,e^-}{1\,\text{mol Ag}} \times \frac{96485\,\text{C}}{1\,\text{mol}\,e^-}}{15\,\text{min} \times \frac{60\,\text{s}}{1\,\text{min}}} = 0.32_8\,\text{C/s} = 0.33\,\text{A}$$

77. The reaction is $PbSO_4 + 2\,e^- \rightarrow Pb + SO_4^{2-}$. The amount of lead produced is

$$8.00\,\text{C/s} \times 30.0\,\text{min} \times \frac{60\,\text{s}}{1\,\text{min}} \times \frac{1\,\text{mol}\,e^-}{96485\,\text{C}} \times \frac{1\,\text{mol Pb}}{2\,\text{mol}\,e^-} \times \frac{207.2\,\text{g Pb}}{1\,\text{mol Pb}} = 15.5\,\text{g Pb}$$

79. If we are to plate out copper, we need the steel to serve as the cathode, where the copper will be reduced onto the pan. We can use a copper anode to provide the copper for the reduction, Because the oxidation of the anode will create Cu^{2+} ions.

81. The balanced reactions and standard cell potential are:

$$\begin{array}{ll} [FeO_4^{2-} + 8\,H^+ + 3\,e^- \rightarrow Fe^{3+} + 4\,H_2O] \times 4 & E° = 2.20\text{ V} \\ [2\,H_2O \rightarrow O_2 + 4\,H^+ + 4\,e^-] \times 3 & E° = -1.23\text{ V} \\ \hline 4\,FeO_4^{2-} + 20\,H^+ \rightarrow 10\,H_2O + 4\,Fe^{3+} + 3\,O_2 & E° = 0.97\text{ V} \end{array}$$

a. The standard cell potential is 0.97 V. The equilibrium constant is

$$K = 10^{\frac{nE°}{0.0592\,\text{V}}} = 10^{(12)(0.97\,\text{V})/0.0592\,\text{V}} = 10^{196.6} = 10^{0.6} \times 10^{196} = 3.98 \times 10^{196}$$

b. In basic solution we add one OH^- for each H^+ in the reaction, converting them into water and canceling any that appear on both sides. We add 20 OH^- to each side and cancel 10 H_2O to get

$$4\,FeO_4^{2-} + 10\,H_2O \rightarrow 20\,OH^- + 4\,Fe^{3+} + 3\,O_2$$

c. Using the Nernst equation, we find the cell potential:

$$E_{\text{cell}} = E°_{\text{cell}} - \frac{0.0257}{n} \ln \frac{(P_{O_2})^3 [Fe^{3+}]^4}{[H^+]^{20} [FeO_4^{2-}]^4}$$

$$= 0.97 - \frac{0.0257}{12} \ln \frac{(8.3 \times 10^{-5})^3 (1.1 \times 10^{-3})^4}{(10^{-2.8})^{20} (1.5 \times 10^{-3})^4} = 0.76\text{V}$$

d. At high pH (low H^+), $Fe(OH)_3$ forms. This solid removes Fe^{3+} from the water and forces the reaction to completion.

83. a. Before drawing the diagram, we must determine the direction of the spontaneous reaction, which tells us which metal serves as the anode and which serves as the cathode.
The standard reduction potentials are

$$Fe^{2+} + 2\,e^- \rightarrow Fe \qquad E° = -0.44\text{ V}$$
$$Zn^{2+} + 2\,e^- \rightarrow Zn \qquad E° = -0.76\text{ V}$$

To make a galvanic cell, we reverse the zinc reaction:

reduction / cathode:	$Fe^{2+} + 2\,e^- \rightarrow Fe$	$E° = -0.44$ V
oxidation / anode:	$Zn \rightarrow Zn^{2+} + 2\,e^-$	$E° = +0.76$ V
	$Zn + Fe^{2+} \rightarrow Zn^{2+} + Fe$	$E°_{cell} = 0.32$ V

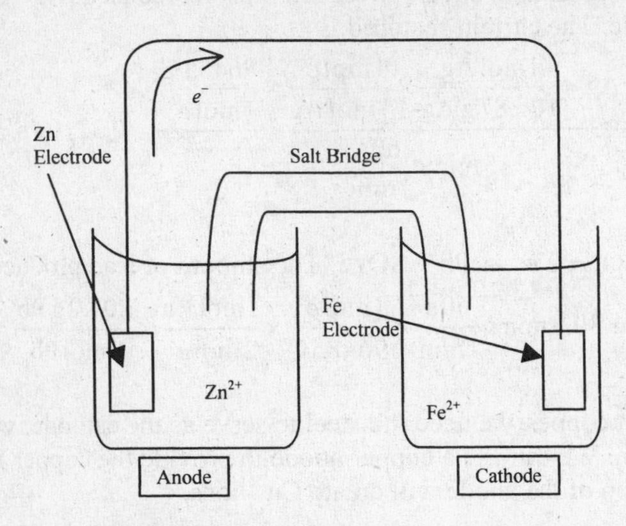

b. As shown above, the zinc reaction is the oxidation.

c. The cell potential is $E_{cell} = E°_{cell} - \frac{0.0257}{n}\ln\frac{[Zn^{2+}]}{[Fe^{2+}]} = 0.32 - \frac{0.0257}{2}\ln\frac{(0.25)}{(0.25)} = 0.32\text{V}$

d. As long as the concentrations of Zn^{2+} and Fe^{2+} remain equal, the cell potential will always be 0.32 V.

e. Assuming that the standard potential does not change with temperature, there will be no change in the potential with a change in the temperature. The temperature enters into the equation in the constant preceding the natural logarithm (ln). Because the concentrations are equal, we have $\ln(1) = 0$, and that term disappears.

f. One advantage would be that these metals are less toxic than the lead or mercury used in many batteries.

g. The largest disadvantage would be the spontaneous corrosion of the metals. Iron will rust in solution, and zinc reacts directly with the H^+ in water. After some time, there would be no metal left to serve as the electrodes.

Chapter 20: Coordination Complexes

The Bottom Line

- Coordination complexes are present in simple metal ions in solution and as the reaction center in many biological molecules. (Section 20.1)
- A Lewis base that donates a lone pair of electrons to a metal to form a coordinate covalent bond acts as a ligand in producing a coordination complex. (Section 20.2)
- The coordination numbers of various metal centers are commonly observed to be 2, 4 or 6. (Section 20.3)
- Common coordination geometries are linear, tetrahedral, square planar, and octahedral. (Section 20.4)
- A variety of isomer classes are observed for metal complexes. (Section 20.5)
- The proper names and formulas for coordination complexes are specified by IUPAC rules. (Section 20.6)
- In transition metal complexes, the *d* orbitals are no longer degenerate but split into two or more energy levels, depending on coordination geometry. (Section 20.7)
- The electron configuration for octahedral complexes gives rise to high-spin and low-spin complexes for d^4 to d^7 metal centers. (Section 20.7)
- The color of many transition metal compounds arises when a photon of visible light is absorbed and an electron is excited to a higher-energy *d* orbital. (Section 20.7)
- The order of ligands in terms of their influence on the magnitude of the *d* orbital splitting and the energy of the photon of light absorbed is defined as the spectrochemical series. (Section 20.7)
- The number of unpaired electrons determines the magnetic moment of the complex. (Section 20.7)
- Metal centers that exchange ligands rapidly are called labile; those that exchange ligands slowly are called inert. (Section 20.8)
- Chelate complexes have high formation constants because of an entropy effect. (Section 20.8)
- Because transition metal complexes often exhibit several oxidation states, transition metal complexes are good electron transfer (redox) agents. (Section 20.8)

Solutions to Practice Problems

20.1. a. $[FeCl_6]^{4-}$
b. $[Ni(NH_3)_4]^{2+}$
c. $[Zn(H_2O)_6]^{2+}$

20.2. a. Chromium is the central atom of the complex. There are six carbonyl ligands bound to the metal, so the coordination number is 6. With six ligands, the complex has octahedral geometry.
b. Iron is the central atom of the complex. There are six aqua ligands bound to the metal, so the coordination number is 6. With six ligands, the complex has octahedral geometry.
c. Titanium is the central atom of the complex. There are four chloro ligands, so the coordination number is 4. The geometry of the complex will be either tetrahedral or square planar. Because the valence electron configuration for Ti^{2+} is not d^8, the tetrahedral geometry is preferred.
d. Copper is the central atom of the complex. There are four ammine ligands, so the coordination number is 4. The geometry of the complex will be either tetrahedral or square

planar. Because the valence electron configuration for Cu^+ is d^{10} and not d^8, the tetrahedral geometry is preferred.

20.3. a. The complex cation has two chloro and four ammine ligands. Sulfate is the anion and carries a −2 charge. The two chloro ligands contribute an additional −2 charge. The ammine ligands are neutral, so the metal must have a +4 oxidation state to make the compound neutral. We name the ligands in alphabetical order, so the name is tetraamminedichlorochromium(IV) sulfate.

b. The complex anion has four cyano ligands. The two cation sodium ions have a +2 total charge, while the four cyano ligands contribute a charge of −4. To make the compound neutral, the nickel atom has a +2 oxidation state. The name is sodium tetracyanonickelate(II).

c. The complex ion is $FeCl_6^{-4}$. The −4 charge results from six −1 charges on the fluoro ligands and the +2 charge on the metal. To balance the charge, the compound will require two calcium ions. The formula is $Ca_2[FeF_6]$.

d. There are four ammine ligands and two carbonyl ligands, all of which are neutral. With the +2 charge of the metal, there only needs to be one sulfate to make the compound neutral. The formula is $[Mn(NH_3)_4(CO)_2]SO_4$.

20.4. a. The$[Fe(CN)_6]^{4-}$ complex has a central Fe^{2+} atom. In the +2 oxidation state, iron will have a d^6 configuration. Because the coordination number of the iron is 6, the complex will have an octahedral geometry. This geometry has three lower energy levels and two higher energy levels for the *d* electrons. In the low-spin case, all six electrons fill the lower three levels. No electrons are left unpaired.

Solutions to Student Problems

1. a. A ligand is a Lewis base that donates a lone pair of electrons to a metal center to form a coordinate covalent bond.

b. A coordinate covalent bond is a covalent bond that results when one atom donates both electrons needed to form the bond.

c. A Lewis acid accepts a pair of electrons from another atom in the formation of a coordinate covalent bond.

3. a. In $[Fe(H_2O)_6]^{2+}$, there are six ligands and six coordinate covalent bonds, one for each aqua ligand. Because aqua ligands are neutral, the +2 charge of the complex is entirely from the metal. Its oxidation number is +2.

b. In $[Co(NH_3)_6]^{2+}$, there are six ligands and six coordinate covalent bonds, one for each ammine ligand. Because ammine ligands are neutral, the +2 charge of the complex is entirely from the metal. Its oxidation number is +2.

c. In $[Zn(NH_3)_4]^{2+}$, there are four ligands and four coordinate covalent bonds, one for each ammine ligand. Because ammine ligands are neutral, the +2 charge of the complex is entirely from the metal. Its oxidation number is +2.

d. In $[Pt(CO)_4]$, there are four ligands and four coordinate covalent bonds, one for each carbonyl ligand. Because carbonyl ligands are neutral and the complex is neutral overall, there is no charge on the metal. Its oxidation number is 0.

5. For a complex to be symmetrically arranged, the complex would be either tetrahedral or square planar. Both are depicted below:

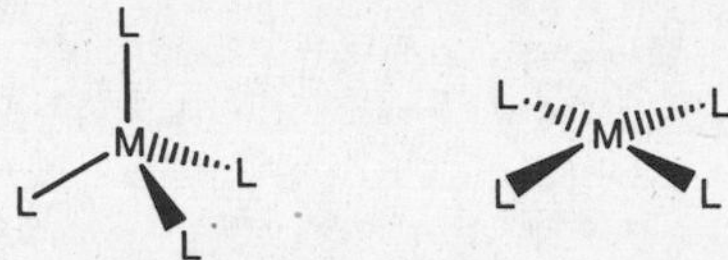

7. H–N–H (with a lone pair on N and a third H below N); Because NH_3 has an electron lone pair, it is capable of donating both to form a coordinate covalent bond. It is therefore a Lewis base.

9. This molecule is $NH_2CH_2CH_2CH_2CH_2CH_2NH_2$. Both NH_2 groups will have a lone pair and be able to act as a Lewis base. The carbon backbone is long enough to allow both nitrogen atoms to form a bond with the same metal atom. It will be bidentate.

11. Ca^{2+} and C_2H_6 have no lone pairs and would be unable to donate an electron pair to form a metal–ligand bond. They would be unlikely to act as ligands.

13. The structure of ethylenediammine is $NH_2CH_2CH_2NH_2$. Both NH_2 groups will have a lone pair and be able to act as a Lewis base. The carbon backbone is long enough to allow both nitrogen atoms to form a bond with the same metal atom. Although N_2 does have two lone pairs, they are found at opposite ends of a linear molecule. It is not possible for both nitrogen atoms to form separate bonds with the same metal atom.

15. If there is a total of six bonds, and the two A ligands take only one apiece, there remain four bonds for the two B ligands. Each B ligand must form two bonds—the B ligands are bidentate.

17. a. There are six monodentate ligands, so the coordination number is 6. The ammine ligands are neutral, and the bromo ligand has a −1 charge. For the complex to have an overall charge of +3, the chromium must have an oxidation state of +4.

b. The three ligands are bidentate ethylenediammine, so the coordination number is 6. The ligands are neutral, so the charge on the complex is due to the metal. Its oxidation number is +2.

c. There are four monodentate methylamine ligands, so the coordination number is 4. The ligands are neutral, so the charge on the complex is due to the metal. Its oxidation number is +2.

19. In the first case, there are three bidentate ligands for a coordination number of 6. In the second case, there is only one bidentate ligand but there are four additional monodentate ligands, for a coordination number of 6. Because different ligands can have differing numbers of bonds, it is possible to have equivalent coordination numbers with differing numbers of ligands.

21. Oxinate has two potential binding sites, one on the nitrogen and one on the oxygen. If two of these ligands complexed with lead, there would be a total of four bonds. The coordination number of lead would be 4.

23. a. The structure is octahedral and all bond angles are 90°.

b. The structure is octahedral and all bond angles are 90°.

c. The structure is tetrahedral and all bond angles are 109.5°.

d. The structure is tetrahedral and all bond angles are 109.5°.

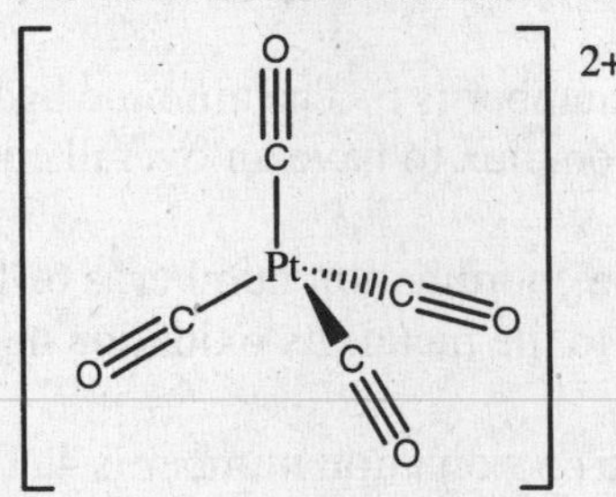

25. The complex has two bidentate ligands (en) and two monodentate ligands (Cl^-, NO_2^-), for a total coordination number of 6. This complex will be octahedral. Because the ethylenediammine is neutral and the other two ligands each have a –1 charge, the cobalt ion must be +3 to give the complex cation a +1 overall charge to balance the chloride anion charge of –1.

27. With four carbonyl ligands, the coordination number of nickel is 4. With this coordination number, the complex would be either square planar or tetrahedral.

29. a. Because the ligands remain complexed to the same metal center, they can differ only in their geometry. These complexes are geometric isomers.

b. These two complexes have exchanged some ligands for the anions. These complexes are ionization isomers.

c. The compounds are composed of two complexes that have interchanged their ligands. These complexes are coordination sphere isomers.

31. The tetrahedral structure and the two square planar isomers are shown below. If the platinum were placed at the center of a hypothetical tetrahedron, each ligand would reside at one of the vertices. Because each vertex is connected to the other three, it is not possible to create an arrangement that is different from the one shown. There can be no isomers. In a square planar complex, the metal lies at the center of the square with the ligands at the corners. Because the square is flat, each corner is not directly adjacent to all three other corners. It is possible to create different arrangements such that there are different ligands directly opposite each other. There are two different isomers for the square planar geometry.

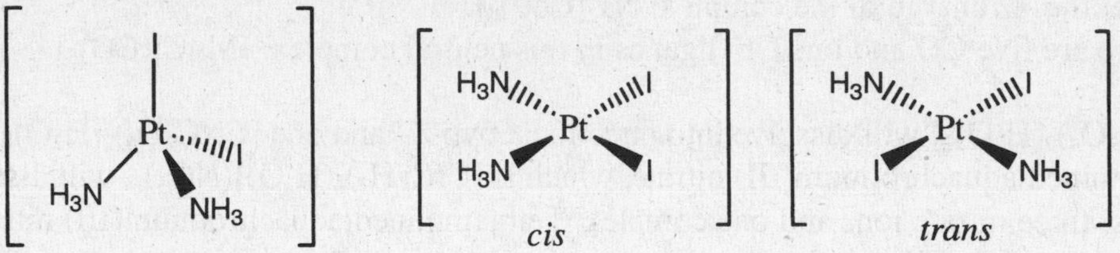

cis *trans*

33. This complex is octahedral, and there are only two possible isomers: one with all three chloro ligands in the same plane and one where they are not in the same plane.

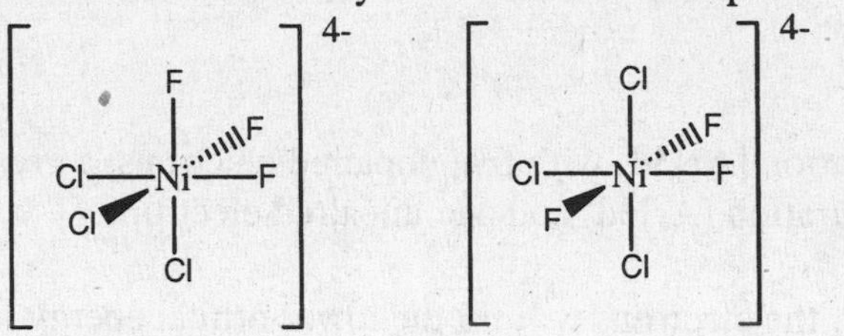

35. Because ethylenediammine is bidentate, it will always bind on two adjacent sites. The only way to get isomers is when the two binding sites can be separated. The only structure is

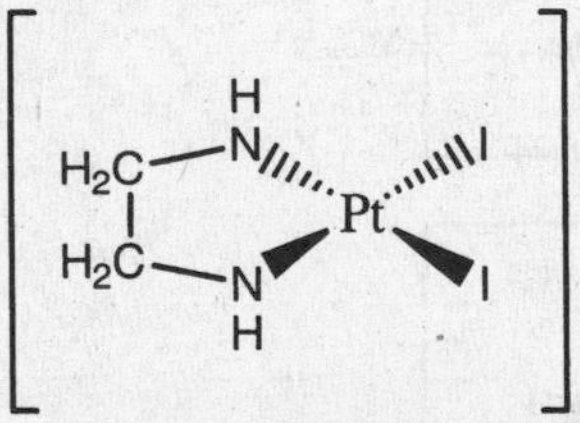

Fe^{+3} will tend for form octahedral complexes. With (en) the complex would be $[Fe(en)_3]^{3+}$, which has two isomers. The isomers are mirror images and are enantiomers (optically active).

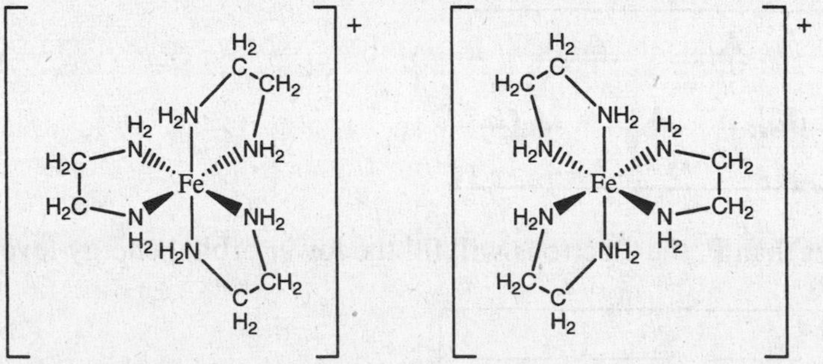

37. a. This complex is a cation, so the metal name is unaltered: amminecyanobis(ethylenediamine)cobalt(III).

b. This complex is an anion, so the metal name takes the -ate ending: diamminedioxalatochromate(III).

c. This complex is an anion, so the metal name takes the -ate ending: hexanitroferrate(III).
d. This complex is neutral, so the metal name is unaltered: triaquatrichlorocobalt(III).

39. a. There are four NH_3, one H_2O, and one Cl^- ligands in this complex cation: $[CoCl(NH_3)_4(H_2O)]^{2+}$.
b. There are two H_2O and two (en) ligands in this complex cation, which is paired with two chloride anions to offset the +2 charge of the complex: *trans*-$[Cu(H_2O)_2(en)_2]Cl_2$.
c. There are four Cl^- ligands in this complex anion, which is paired with two sodium cations to offset the −2 charge of the complex: $Na_2[CoCl_4]$.
d. There are five CO and one Cl^- ligands in this neutral complex: $[MnCl(CO)_5]$.

41. $K_2[Cr(C_2O_4)_2(H_2O)_2]$ will dissolve into three ions: two K^+ and one $[Cr(C_2O_4)_2(H_2O)_2]^{2-}$. Tetraamminediaquachromium(III) nitrate, which is $[Cr(NH_3)_4(H_2O)](NO_3)_3$, will dissolve into four ions: three nitrate ions and one complex. Tetraamminediaquachromium(III) nitrate produces the greater number of ions.

43. a. $[Ar]3d^6$
b. $[Ar]3d^4$
c. $[Ar]3d^{10}$

45. Fe^{3+} has the configuration $[Ar]3d^5$ with five unpaired electrons, a greater number than Cu^{2+}, which has the configuration $[Ar]3d^9$ and one unpaired electron.

47. With Δ_o larger than P, the electrons will fill the lower orbital energy levels first.
a. Fe^{3+} is d^5:

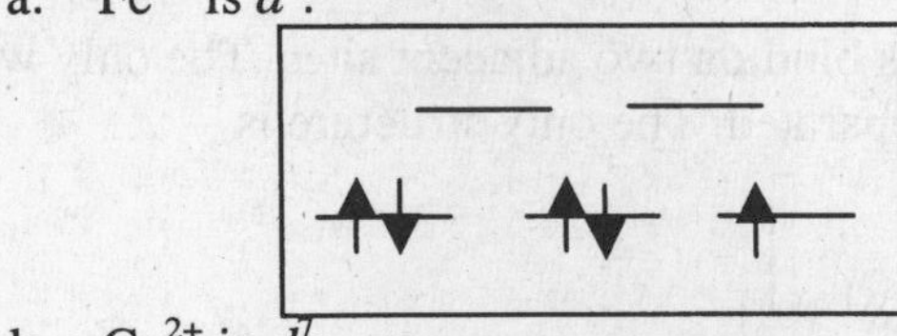

b. Co^{2+} is d^7:

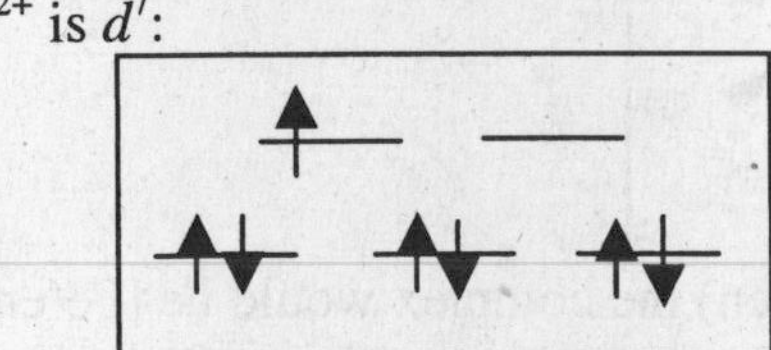

c. Ni^{2+} is d^8:

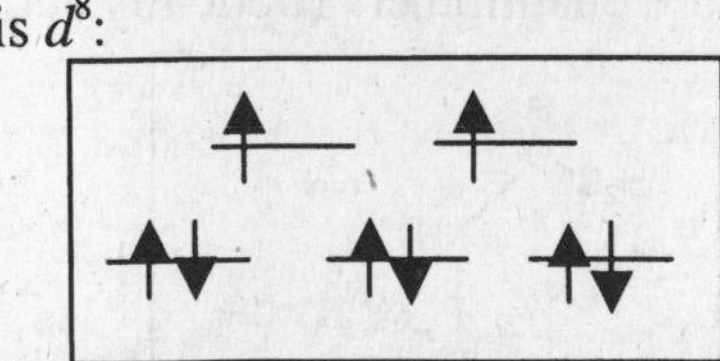

49. With Δ_o larger than P, the electrons will fill the lower orbital energy levels first.
a. Fe^{3+} is d^5:

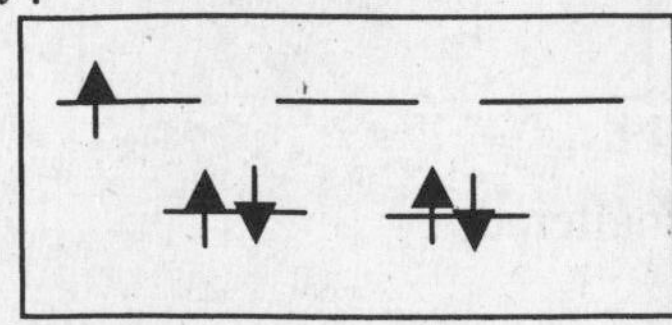

b. Co^{2+} is d^7:

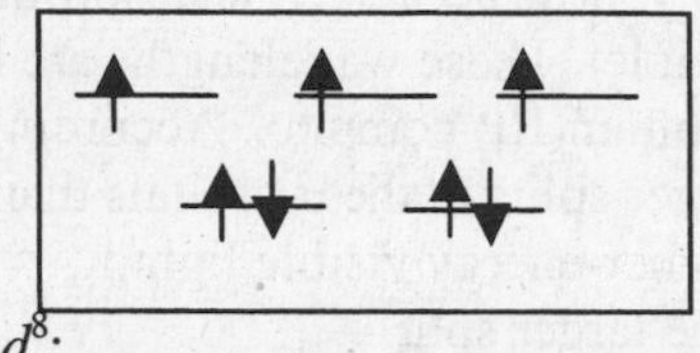

c. Ni^{2+} is d^8:

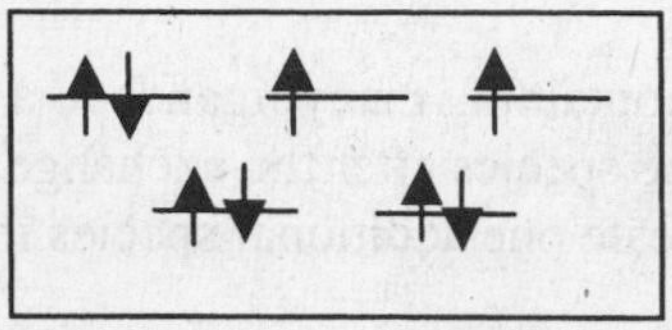

51. a. The metal here is tetrahedral Fe^{3+} (d^5). The chlorine will have a small Δ_o, forcing the electrons to fill the upper levels before pairing (high spin):

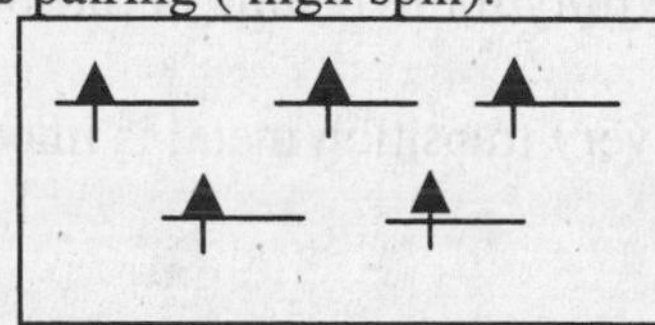

b. The metal here is octahedral Co^{3+} (d^6). The CN^- ligands will cause a large Δ_o, forcing the electrons to pair before filling the upper levels (low spin):

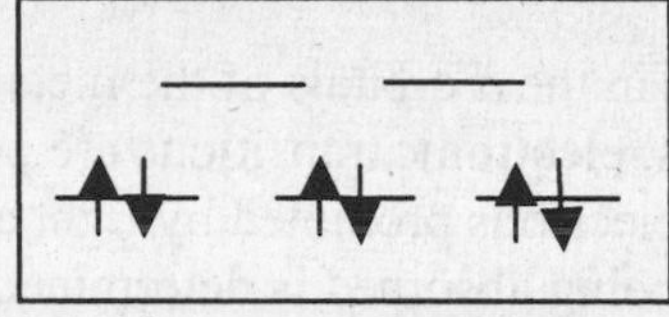

c. The metal here is octahedral Mn^+ (d^6). The CO ligands will cause a large Δ_o, forcing the electrons to pair before filling the upper levels (low spin):

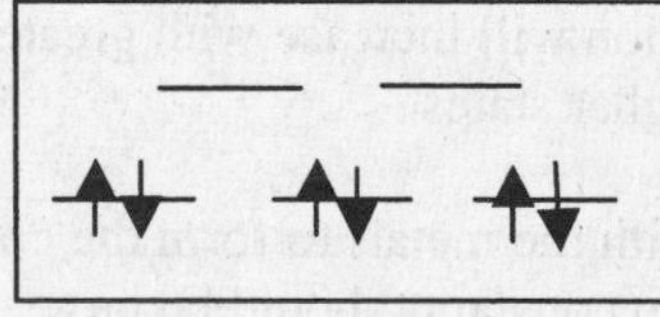

53. All of the compounds have unpaired electrons and are paramagnetic.

55. To absorb blue light (high energy, small wavelength), the complex will need a large splitting between the *d* orbital energy levels. According to the spectrochemical series, the ligand with the largest splitting in the group is CN^-; therefore, $M(CN)_6^{2-}$ will be the most likely to absorb blue light.

57. The chemical species that form the majority of the common gemstones is the same for most stones. They should all appear somewhat similar in infrared light. If two gemstones have a different chemical composition, they will appear different under infrared light. For example, diamond (consisting of network-covalent carbon) and quartz (SiO_2) behave differently because their atoms interact with different wavelengths of infrared radiation.

59. a. If the solution appears to be violet, the colors on the opposite end of the spectrum are absorbing light—mostly yellow.

b. The chromium ion is in the +3 oxidation state, which gives it a d^3 configuration. There will be three unpaired electrons in an octahedral geometry.

c. The hexaamminechromium(III) complex appears yellow because it is absorbing light at the opposite end of the spectrum—mostly blue and violet. These wavelengths are higher in energy than those absorbed by the hexaaquachromium(III) complex. According to the electrochemical series, ammine ligands cause larger splits in the *d* orbitals than does water; therefore, the ammine complex should absorb higher-energy visible light.

d. As discussed in part c, the ammonia is causing the larger split.

61. When a complex goes from a nonchelated complex containing many ligands to a chelated complex containing fewer ligands, there are more free species after the exchange. Exchanging a bidentate ligand for two monodentate ligands will create one additional species in solution, which increases the entropy of the system.

63. Because the reaction happens quickly, the aqua ligands would be classified as labile. The exchange of oxygen is also (thankfully for us) labile because the oxygen is transferred quickly.

65. Your answers will vary. Nearly every transition metal is important as a catalyst in at least one chemical reaction.

67. The total charge would be $(+3) + 2 \times (-1) + 4 \times (0) = +1$. The complex would be the cation $[CoCl_2(NH_3)_4]^+$.

69. a. If there are some electrons in the *d* orbitals of the metal center in a complex, but not enough electrons to fill the orbitals, electronic transitions are possible. Any configuration from d^1 to d^9 will be able to have its electrons promoted by light absorption.

b. The energy of the photon being absorbed is determined by the splitting between the *d* orbital levels. This splitting is controlled by the ligands of the complex. The splitting increases in order with the spectrochemical series: $Cl^- < F^- < OH^- < H_2O < NH_3 < NO_2^- < CN^- < CO$.

c. The intensity of the transition will increase with greater population in the lower states and lower population in the higher states.

71. If the binding of the ligands with the metals to form the complex is strong, showing a preference for the product, the equilibrium constants should be large.

72. a. This ligand will behave much as ethylenediammine does:

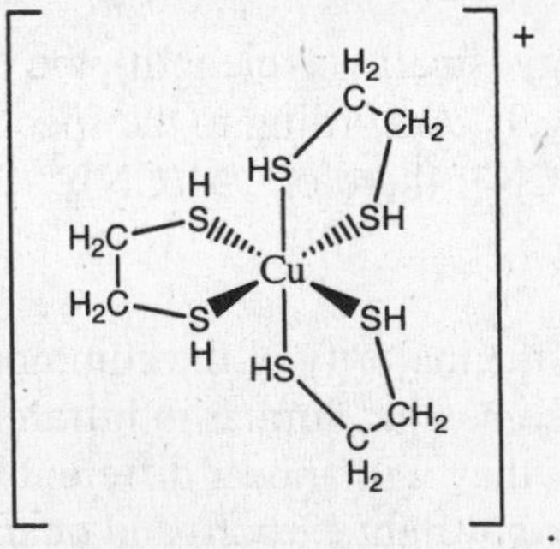

b. Cu^{2+} has a d^9 configuration:

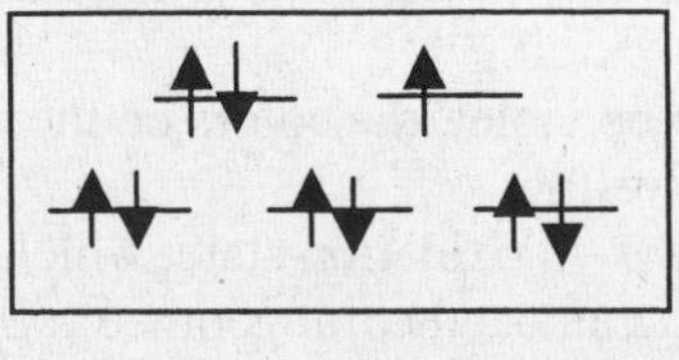

c. Because the ammine complex has a violet color and the ethanedithiol complex is blue, the absorbed colors are green and yellow, respectively. Green light lies at a higher energy than yellow; therefore, the splitting caused by the ethanedithiol is less than that caused by ammine. The new ligand will fall before NH_3 in the spectrochemical series.

d. If 1.00 g of copper produced 3.96 g of complex, then 2.69 g of the ligand was used. The numbers of moles of each are

$$1.00\,g\,Cu \times \frac{1\,mol\,Cu}{63.55\,g\,Cu} = 0.0157\,mol\,Cu$$

$$2.96\,g\,C_2H_8S_2 \times \frac{1\,mol\,C_2H_6S_2}{94.20\,g\,C_2H_6S_2} = 0.0314\,mol\,C_2H_6S_2$$

The mole ratio of ligand to copper is

$$\frac{0.0314\,mol\,C_2H_8S_2}{0.0157\,mol\,Cu} = \frac{2.0\,mol\,C_2H_8S_2}{1.0\,mol\,Cu}$$

Because the ligand is bidentate and there are two for each copper atom, the coordination number is 4.

e. Because the exchange constant is so small, the new ligand binds very strongly to copper. In other words, it could be considered an inert ligand.

Chapter 21: Nuclear Chemistry

The Bottom Line

- Each element is composed of atoms containing the same number of protons. These may contain isotopes with differing numbers of neutrons. (Section 21.1)
- Some nuclear configurations are unstable. They decay in a stepwise progression toward stable nuclei. (Section 21.2)
- There are three main types of radioactive decay: alpha-particle emission, beta-particle emission, and gamma-ray emission. (Section 21.2)
- Ionizing radiation can interact with living tissue and cause damage to the DNA of a cell. This damage may be repaired and cause no harm or, in some cases, may lead to cancer. (Section 21.3)
- Radioactive decay occurs via first-order kinetics. (Section 21.4)
- Energy is released in radioactive decay processes as a consequence of the mass defect in nuclei. (Section 21.5)
- Nuclei with a "magic" number of protons and/or neutrons (2, 8, 20, 28, 50, or 82) are stable. Nuclei with even numbers of protons and/or neutrons are also more likely to be stable. (Section 21.6)
- Nuclear fission is the splitting of heavier nuclei into lighter ones. Nuclear fusion results when smaller nuclei combine into heavier nuclei. (Section 21.7)
- Radioisotopes can be used in medicine for imaging the body and for treating and eliminating cancerous tissues. (Section 21.8)

Solutions to Practice Problems

21.1. a. Sulfur has an atomic number of 16, so there are 16 protons. The mass number of this nuclide is 32, so there are 32 – 16 = 16 neutrons.

b. ${}^{23}_{11}\text{Na}$ has 11 protons, because the atomic number is 11. The number of neutrons is given by subtracting the atomic number (11) from the mass number (23) such that 23 – 11 = 12 neutrons.

c. Radon-222 has a mass number of 222 and an atomic number of 86. There are 86 protons and 222 – 86 = 136 neutrons.

d. Technetium has a mass number of 98 and an atomic number of 43. There are 43 protons and 98 – 43 = 55 neutrons.

21.2. Recall that a beta emission results in the net conversion of a neutron into a proton; therefore, the atomic number of the nuclide will increase by 1 (changing the identity of the nuclide) while the mass number remains the same:

$$ {}^{60}_{27}\text{Co} \rightarrow {}^{0}_{-1}\beta + {}^{60}_{28}\text{Ni} $$

21.3. Release of an alpha particle, ${}^{4}_{2}\text{He}$, will result in a new nuclide with a mass number that is 4 lower than before and an atomic number that is 2 lower than before.

$$ {}^{218}_{84}\text{Po} \rightarrow {}^{4}_{2}\text{He} + {}^{214}_{82}\text{Pb} + {}^{0}_{0}\gamma \qquad {}^{230}_{90}\text{Th} \rightarrow {}^{4}_{2}\text{He} + {}^{226}_{88}\text{Ra} + {}^{0}_{0}\gamma $$

21.4. We first need to solve for the rate constant using the half-life formula:

$$k = \frac{0.693}{t_{1/2}} = \frac{0.693}{2.69\,\text{days}} = 0.258\,\text{day}^{-1}$$

The fraction of the original that remains is 0.01. We can rearrange the first-order integrated rate law to solve for the decay time:

$$\ln\frac{[A]_t}{[A]_0} = -kt$$

$$t = -\frac{\ln\frac{[A]_t}{[A]_0}}{k} = \frac{\ln\frac{0.01}{1.00}}{0.258\,\text{day}^{-1}} = 17.8\,\text{days}$$

21.5. There are equal numbers of protons on each side of the equation. This fact is important because the masses given in most tables of isotope masses are for neutral atoms, not bare nuclei. Because the number of protons is the same, the number of electrons around those nuclei is the same, and we can use the atomic (rather than nuclear) masses directly. The total mass of the products ${}^{220}_{86}\text{Rn} + {}^{4}_{2}\text{He}$ is 220.011368 u + 4.00260 u = 224.013968 u, and the total mass of the reactant ${}^{224}_{88}\text{Ra}$ is 224.020186 u. The mass defect is 224.013968 – 224.020186 = – 0.006218 u = – 0.006218 g/mol. The energy equivalent of this mass is

$$\Delta E = |\Delta m|\,c^2 = \left|-0.006218\,\text{g/mol} \times \frac{1\,\text{kg}}{1000\,\text{g}}\right| \left(2.9979 \times 10^8\,\text{m/s}\right)^2$$

$$= 5.588 \times 10^{11}\,\text{J/mol} = 5.588 \times 10^8\,\text{kJ/mol}$$

21.6. There are several factors that we need to consider to determine whether the nuclide is stable. First, nuclides tend to be stable when the number of protons and/or the number of neutrons is an even number. Second, we can look at the atomic mass on the periodic table. This mass is computed from the average of the stable nuclei. If the nuclide mass does not differ much from the average atomic mass, the chances are good (but not absolute) that that nuclide will be stable.

a. ${}^{79}_{35}\text{Br}$ has an even number of protons and is very close to the periodic table mass of 79.9. We would predict that it is stable.

b. ${}^{101}_{44}\text{Ru}$ has an even number of protons and is very close to the periodic table mass of 101.07. We would predict that it is stable.

c. ${}^{136}_{56}\text{Ba}$ has an even number of protons and neutrons and is very close to the periodic table mass of 137.33. We would predict that it is stable.

d. ${}^{180}_{73}\text{Ta}$ has an odd number of protons and neutrons but is very close to the periodic table mass of 180.9. With an odd number of protons, this element is not likely to have many stable isotopes. We could justify either argument.

Using a table of the isotopes, we find that all four of these nuclides are stable.

21.7. The key to the problem is that only a single neutron be allowed to encounter another fissionable nucleus. The excess neutrons would need to be absorbed or lost from the system.

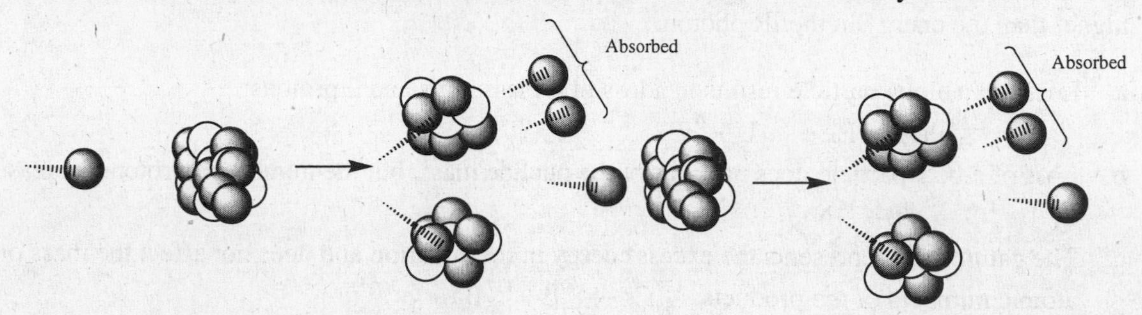

21.8. The structure on the left substitutes tritium for one hydrogen in the molecule. However, the substitution is at the acidic proton of the carboxylic acid and will probably be removed or exchanged for a nonlabeled hydrogen atom. There is no guarantee that the tritium will remain with the molecule. The structure on the right, with the iodine substitution, is not structurally the same as the salicylic acid and may behave differently. The middle structure is the same as unlabeled salicylic acid and is unlikely to lose the carbon atom. It is the best choice.

Solutions to Student Problems

1. Hydrogen, ${}^{1}_{1}\text{H}$, has the smallest atomic number (1) and the smallest mass number (1).

3. Yes, an atom can have no neutrons. Hydrogen, ${}^{1}_{1}\text{H}$, has no neutrons.

5. Yes, as long as there are two protons in the nucleus, we have an atom of helium regardless of the mass. One example would be helium-2, ${}^{2}_{2}\text{He}$, versus hydrogen-3, ${}^{3}_{1}\text{H}$. The helium-2 nuclide is not stable.

7. a. ${}^{12}_{7}\text{N}$ has 7 protons and electrons with 12 – 7 = 5 neutrons.
b. ${}^{124}_{51}\text{Sb}$ has 51 protons and electrons with 124 – 51 = 73 neutrons.
c. ${}^{152}_{63}\text{Eu}$ has 63 protons and electrons with 152 – 63 = 89 neutrons.
d. ${}^{9}_{4}\text{Be}$ has 4 protons and electrons with 9 – 4 = 5 neutrons.

9. $120\,\text{g K} \times \dfrac{0.0118\,\text{g}\,{}^{40}_{19}\text{K}}{100\,\text{g K}} = 0.014\,\text{g}\,{}^{40}_{19}\text{K}$

11. a. Strontium is an alkaline earth metal. Alkaline earth metals are not found in their elemental forms, so we will find strontium in the biosphere as Sr^{2+}, not Sr.
b. Strontium will have chemical reactivity very similar to that of calcium, which is present in large amounts in most soils and plants. The strontium will be incorporated into the soils and also into the plants, making it difficult to remove.

13. These two forms of electromagnetic radiation differ only in the amounts of energy carried in each photon (and therefore in wavelength and frequency), with the energy in the gamma ray much higher than the energy in the IR photon.

15. a. Loss of an alpha particle results in a loss of 4 mass units and 2 protons:

$$^{239}_{94}Pu \rightarrow {}^{4}_{2}He + {}^{235}_{92}U + {}^{0}_{0}\gamma$$

b. Loss of a beta particle does not change the nuclide mass, but the number of protons increases by 1: $^{14}_{6}C \rightarrow {}^{0}_{-1}\beta + {}^{14}_{7}N$.

c. The gamma ray represents the excess energy in the equation and does not affect the mass or atomic numbers of the products: $^{137}_{55}Cs \rightarrow {}^{0}_{-1}\beta + {}^{137}_{56}Ba + {}^{0}_{0}\gamma$.

17. a. The reactant will have an atomic number that is 2 higher, and a mass number that is 4 higher, than polonium-215 has: $^{219}_{86}Rn \rightarrow {}^{4}_{2}He + {}^{215}_{84}Po + {}^{0}_{0}\gamma$.

b. $^{90}_{38}Sr \rightarrow {}^{0}_{-1}\beta + {}^{90}_{39}Y$

c. $^{99}_{43}Tc \rightarrow {}^{0}_{-1}\beta + {}^{99}_{44}Ru + {}^{0}_{0}\gamma$

19. $^{146}_{62}Sm \rightarrow {}^{4}_{2}He + {}^{142}_{60}Nd$

21. Pu-244 must differ by an exact multiple of 4 mass units from one of the decay series elements, because alpha decay is the only method to lower the mass of the nuclide. (Beta decay does not affect the mass.) Pu-244 is 12 mass units (3 alpha particles) heavier than Th-232 and fits into its decay series.

23. Ionizing radiation must contain a lot of energy. Cosmic rays, gamma rays, and X rays are all ionizing radiation.

25. You could use a dose no higher than absolutely needed, wear lead-lined covers (lead blocks gamma radiation), and do the diagnosis from another room (concrete blocks gamma radiation) after the dose was administered.

27. a. Both the curie and the becquerel measure the number of disintegrations (activity) of a radioactive sample. They differ only in the amount. (1 curie = 3.7×10^{10} Bq)

b. Both the rad and the rem measure an absorbed dose of radiation. A rem is adjusted by the biological effectiveness of the radiation and is then a dose equivalent.

29. $35\times10^3\,\text{Bq}\times\dfrac{1\,\text{disintegrations/s}}{1\,\text{Bq}} \doteq 35\times10^3\,\text{disintegrations/s} = 3.5\times10^4\,\text{disintegrations/s}$

31. $4.0\times10^{-6}\,\text{Sv}\times\dfrac{100\,\text{rem}}{1\,\text{Sv}} = 4.0\times10^{-4}\,\text{rem} = 0.40\,\text{millirem}$. Consider that the EPA estimates the background radiation exposure as 300 millirem. The above dose is not likely to have serious effects.

33. a. For shielding purposes, the density of the material is most important.

b. Some additional considerations would include whether absorption of radiation weakens the material or causes the material to undergo a secondary radiation emission that would escape the shielding. The chemical reactivity of the material would also be very important.

35. a. A has the longer half-life. It takes 10 min for A to drop to 50,000 from 100,000, while B takes only 5 min.

b. The nuclide indicated by the red line drops to 50,000 counts (half the original 100,000 counts) after 10 min. The half-life is 10 min.

c. Even though B has a faster decay rate (giving it a higher activity on an equal molar basis), A has a higher activity because there must be more moles of A in the sample. More decays take place in the same period.

d. The total count for A is always higher than that for B, and it lasts for a longer time. The total radiation absorbed by A will be higher and more dangerous.

37. If 75% is gone, 25% remains after 30 days. This is two half-lives. A single half-life would be 15 days.

39. After 12.3 years (one half-life), 50% will remain. After 24.6 years (two half-lives), 50% × 50% = 25% remains.

41. We first need to solve for the rate constant using the half-life formula:

$$k = \frac{0.693}{t_{1/2}} = \frac{0.693}{28.9\text{ yr}} = 0.0240\text{ yr}^{-1}$$

The fraction of the original remaining is 0.125. We can rearrange the integrated first-order rate law to solve for the decay time:

$$\ln\frac{[A]_t}{[A]_0} = -kt$$

$$t = -\frac{\ln\frac{[A]_t}{[A]_0}}{k} = \frac{\ln\frac{0.125}{1.00}}{0.0240\text{ yr}^{-1}} = 86.7\text{ yr}$$

Alternatively, we could realize that the condition of 12.5% remaining happens after 3 half-lives: 3 × 28.9 yr = 86.7 yr.

43. We can use the integrated first-order rate law to solve for the rate constant and then convert the rate constant into a half-life to solve the problem.

$$\ln\frac{[A]_t}{[A]_0} = -kt$$

$$k = -\frac{\ln\frac{[A]_t}{[A]_0}}{t} = \frac{\ln\frac{0.01}{1.00}}{3\text{days}} = 1.5\text{day}^{-1}$$

$$t_{1/2} = \frac{0.693}{k} = \frac{0.693}{1.5\text{day}^{-1}} = 0.45\text{days}$$

45. The amount of radiation, in curies, is proportional to the amount of sample present because of the first-order kinetics. If we assume that the sample needs to retain about 5% of its original strength to be useful as a demonstration, we can determine how long it would take for the sample to be reduced to this level of radiation. We first need to solve for the first-order rate constant:

$$k = \frac{0.693}{t_{1/2}} = \frac{0.693}{138\,\text{days}} = 0.00502\,\text{day}^{-1}$$

$$\ln\frac{[\text{A}]_t}{[\text{A}]_0} = -kt$$

$$t = -\frac{\ln\frac{[\text{A}]_t}{[\text{A}]_0}}{k} = \frac{\ln\frac{0.05}{1.00}}{0.00502\,\text{day}^{-1}} = 597\,\text{days}$$

This would require a new sample about every other year.

47. We first need to solve for the rate constant using the half-life formula:

$$k = \frac{0.693}{t_{1/2}} = \frac{0.693}{24000\,\text{yr}} = 2.89\times10^{-5}\,\text{yr}^{-1}$$

The fraction of the original that remains is 0.01. We can rearrange the integrated first-order rate law to solve for the decay time:

$$\ln\frac{[\text{A}]_t}{[\text{A}]_0} = -kt$$

$$t = -\frac{\ln\frac{[\text{A}]_t}{[\text{A}]_0}}{k} = \frac{\ln\frac{0.01}{1.00}}{2.89\times10^{-5}\,\text{yr}^{-1}} = 1.6\times10^{5}\,\text{years}$$

The only way naturally occurring plutonium could exist would be through the radioactive decay of longer-lived elements, such as uranium. Much of the plutonium used today is created as a by-product of nuclear fission reactors.

49. a. When protons and neutrons come together to create an nucleus, some energy is lost. This binding energy appears as a loss of mass in the nucleus relative to the separate nucleons.
b. For the nucleus to stay together, there needs to be some binding energy. The loss of mass should be true of all nuclides.

51. The binding energy, ΔE, is related to the mass defect, Δm, via one version of Einstein's equation:
$\Delta\text{E} = |\Delta\text{m}|\,\text{c}^2$

53. Because the atomic numbers of Ir (Z = 77) and Re (Z = 75) differ by 2, the reaction must be an alpha decay: ${}^{170}_{77}\text{Ir} \rightarrow {}^{166}_{75}\text{Re} + {}^{4}_{2}\text{He} + {}^{0}_{0}\gamma$. The mass defect is the difference between the product mass and the reactant mass. Because there are the same number of protons (and therefore of electrons around the neutral atoms) on both sides of the equation, we can use the atomic masses for the neutral atoms instead of calculating the mass of each bare nucleus:

$$\Delta\text{m} = (165.965740\,\text{g/mol} + 4.002603\,\text{g/mol}) - 169.974970\,\text{g/mol} = -0.006627\,\text{g/mol}$$

The energy equivalent of this mass is

$$\Delta E = |\Delta m|\, c^2 = \left|-0.006627\,\text{g/mol} \times \frac{1\,\text{kg}}{1000\,\text{g}}\right| \left(2.9979\times10^8\,\text{m/s}\right)^2$$

$$= 5.956\times10^{11}\,\text{J/mol} = 5.956\times10^{8}\,\text{kJ/mol}$$

55. a. For a spontaneous process such as the decay of radioactive U-238, we would expect the process to release energy, resulting in a loss of mass. (See Problem 52.) U-238 has the higher mass.

b. The result of the loss of mass is the release of energy and additional binding energy in the products.

57. Both processes involve the conversion between a neutron and proton. However, the beta-minus decay converts a neutron into proton, while the beta-plus decay converts a proton into a neutron.

59. a. "Doubly magic" means that a nuclide has both a number of protons and a number of neutrons that is one of the most stable numbers (2, 8, 20, 28, 50, 82, or 114).

b. Oxygen-16 has 8 each of protons and neutrons, and 8 is one of the magic numbers. Helium-4 has 2 each of protons and neutrons, matching the magic number 2. Carbon-12 has 6 each, and nitrogen-14 has 7 each. These two nuclides do not have a magic number of either protons or neutrons. In general, those nuclei without the magic numbers of protons or neutrons are more likely to be radioactive.

61. Element 114 would have a magic number of protons and would probably be more stable and have a longer half-life before decaying. A longer half-life would make the new element easier to detect.

63. In order to reach the nearest stable nuclide, lead-206 or other stable nuclide nearby, the heavy radioactive nuclides such as uranium and plutonium need to shed at lot of mass. They can do so by emitting alpha particles. For nuclides of lighter elements, there are nearby stable nuclides that can be created by converting protons and neutrons by beta decay; these nuvlides need not lose mass to achieve stability.

65. Iron and cobalt have the most stable nuclei, with the largest binding energy per nucleon. They cannot be made more stable (and actually would create less stable products) through a fission process. Elements heavier than these will undergo fission to create more stable nuclei; elements lighter than these can achieve more stability through fusion.

67. a. $^{235}_{92}\text{U} + ^{1}_{0}\text{n} \rightarrow ^{139}_{56}\text{Ba} + ^{94}_{36}\text{Kr} + 3^{1}_{0}\text{n} + \text{energy}$. The number of neutrons that appear as products is determined by the difference in mass number between the uranium plus neutron and the given products.

b. $^{235}_{92}\text{U} + ^{1}_{0}\text{n} \rightarrow ^{80}_{38}\text{Sr} + ^{153}_{54}\text{Xe} + 3^{1}_{0}\text{n} + \text{energy}$

69. a. Because both are isotopes of uranium, they must have the same number of protons; they differ only in the number of neutrons. Uranium-238 has three more neutrons than uranium-235.

b. Uranium-235 accounts for only 0.72% of all uranium atoms; the vast majority of the rest (about 99%) are uranium-238.

c. Because these two nuclides are both uranium, they are chemically very much alike. The isotopes differ mainly in their mass. The mass differs by only about 1.3%, so separation by physical means is possible, but difficult.

71. Through a fission process, only smaller nuclides result. Americium, californium, and berkelium are all heavier than uranium.

73. Some of the most important questions would involve the total radioactive dose expected and the form of the radiation emitted. What is the half-life of the nuclide, and what is its expected persistence in the body?

75. Tc-99m has a half-life of 6 hr. Over the course of the night (12 hr), only oone-quarter of the original will remain.

77. ${}^{18}_{9}\text{F} \rightarrow {}^{0}_{+1}\beta + {}^{18}_{8}\text{O}$

79. If the half-life of the nuclide is 60 hr, then 2.5 weeks (420 hr) is equivalent to 7 half-lives. The amount of a nuclide left after 7 half-lives is only $(0.5)^7 = 0.8$ %. These two figures are consistent.

81. The medical reference to see is T. H. Winters and J. R. Franza, 1982, Radioactivity in Cigarette Smoke, *New England Journal of Medicine* 306(6): 364–365. The radioactive isotopes lead-210 and polonium-210 are found in the tobacco leaves and cigarettes. The major source of the Po-210 appears to be the fertilizer applied to grow the tobacco. The isotopes are deposited in the lungs, where the emitted radiation is thought to cause cancer.

83. The gamma ray represents the energy that is given off in the reaction. If there is energy due to a negative mass defect (reactants weigh more than products) in the reaction, there will be gamma particles emitted. There are other considerations, but a full understanding of particle physics is necessary.

85. Normal radioactive decay series include alpha decay, which decreases the mass by 4 and the atomic number by 2, and beta decay, which only increases the atomic number by 1. The decay series in the order given is alpha, beta, beta, alpha, alpha, alpha, alpha, beta, beta, alpha.

87. Rn is radon, an inert gas, and Ra is radium, a reactive metal solid.

89. The main reason why Rn-222 is the most dangerous is that it is the only species with a long half-life. The other isotopes decay much more rapidly into solid species. Rn-222 has a half-life of 3.8 days, which is enough time for the gas to seep into houses and be inhaled by people inside. Once Rn-222 does decay, it decays in a matter of minutes through several steps until it creates Pb-210, which is a long-lived ($t_{1/2} = 22.3$ yr) beta emitter. Conversion of Rn-222 into Pb-210 in the lungs creates a long-term radiation source in the body that could cause problems in large enough doses.

91. The volume of the carbon nucleus is $V = \frac{4}{3}\pi r^3 = \frac{4}{3}\pi\left(1.2\times10^{-13}\ \text{cm}\right)^3 = 7.2\times10^{-39}\ \text{cm}^3$. The density is $d = \frac{m}{V} = \frac{1.98718\times10^{-23}\ \text{g}}{7.2\times10^{-39}\ \text{cm}^3} = 2.8\times10^{15}\ \text{g/cm}^3$, which makes carbon much more dense than elemental lead.

93. Websites may vary. The answer is NO!

95. A nice site with clickable names and more information can be found at **http://www.slac.stanford.edu/library/nobel/**.

96. a. If 1 g could make 5000 detectors, the amount per detector is 1/5000 g = 0.2 mg.

b. Am-240 undergoes electron capture with a half-life of 51 hr. Not only does it not emit a particle, but nearly all would be gone very quickly, which is not useful for a smoke detector that should last several years. Am-242 does emit beta particles 83% of the time, but it has a half-life of only 16 hr. Again, there would be little left in a very short time, making it also impractical for use in a smoke detector.

c. $^{241}_{95}\text{Am} \longrightarrow {}^{4}_{2}\text{He} + {}^{237}_{93}\text{Np} + \gamma$

d. The exact time to drop to exactly 1% of the original would be

$$k = \frac{0.693}{t_{1/2}} = \frac{0.693}{432.2\,\text{yr}} = 0.001603\,\text{yr}^{-1}$$

$$\ln\frac{[\text{A}]_t}{[\text{A}]_0} = -kt$$

$$t = -\frac{\ln\frac{[\text{A}]_t}{[\text{A}]_0}}{k} = \frac{\ln\frac{0.01}{1}}{0.001603\,\text{yr}^{-1}} = 2873\,\text{yr}$$

e. $^{242}_{95}\text{Am} \longrightarrow {}^{0}_{-1}\beta + {}^{242}_{96}\text{Cm} + {}^{0}_{0}\gamma$

$^{242}_{95}\text{Am} + {}^{0}_{-1}\beta \longrightarrow + {}^{242}_{94}\text{Pu} + {}^{0}_{0}\gamma$

Chapter 22: The Chemistry of Life

The Bottom Line

- DNA is a polymer of nucleotides made from two complementary single strands wrapped around each other to form a double helix. (Section 22.1)
- Proteins are polymers of amino acids linked through an amide bond (Section 22.2). They contain primary, secondary, tertiary, and sometimes quaternary structure. (Sections 22.2, 22.3)
- Proteins are made by translating the genetic code from mRNA. tRNA supplies the specific amino acids to the growing polypeptide chain. The construction occurs at the ribosomes within the cytoplasm of a cell. (Section 22.3)
- Enzymes are proteins that contain an active site and catalyze biochemical reactions. (Section 22.4)
- Carbohydrates function as structural features such as cell walls in plants and as energy storage molecules in all living organisms. (Section 22.6)
- Lipids are a diverse class of molecules with a variety of functions. (Section 22.7)
- Biological systems exploit the use of enantiomers. Most amino acids and carbohydrates in living organisms are only one of two possible enantiomers. (Section 22.9)
- The promising future activities in the chemistry of medicines include development of new antibiotics, new drugs to treat neurological disorders and cancer, and gene therapy. (Section 22.10)

Solutions to Practice Problems

22.1. The structure of the new base, [structure], is very similar to that of thymine, [structure], and should bind well with adenine.

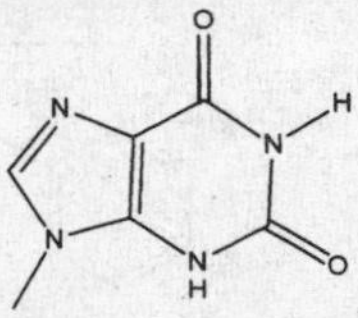

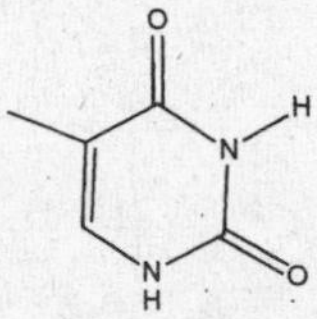

22.2. In order to get a good match, we need to find a sequence that contains the base pairs of our sequence: GATACG, *or* we need to reverse the smaller strand (CGTATC), which would then bind with a sequence of GCATAG in the longer strand:

AAAATGCTGGCATAGCGTTCCAGATACGGACTGACTGC....
CGTATC CTATGC

22.3. We break the sequence into three-letter codons: AUG|UGG|CCA|AAA|UUG|GAC|AUG|UUC|GAC|UAG, which gives a sequence of methionine-tryptophan-proline-lysine-leucine-aspartic acid-methionine-phenylalanine-aspartic acid. [The last codon is the stop codon, and no amino acid is added.]

22.4. The first reaction is the removal of hydrogen from the molecule and the formation of an alkene; this reaction is catalyzed by lyase. The second reaction involves the transfer of the acetyl group from a substrate to the molecule; this reaction is catalyzed by transferase. The third reaction is the joining of two substrates with a new bond and is catalyzed by ligase.

Solutions to Student Problems

1. DNA is a huge nucleotide polymer that has a double-helical structure. Each strand of the DNA polymer complements the other by forming base pairs. DNA provides instructions for the synthesis of proteins and enzymes that carry out the biological activities of the cell

3. We create the complementary strand by pairing A with T, T with A, C with G, and G with C: TAATTTTTCCCTGAT.

5. The cleavage will happen at each ATTA site, but also at each TAAT site because it will be opposite the ATTA on the complementary strand. There are four sites that are either ATTA or TAAT in the given strand: AT|TAAAGCCTA|AT|TACCATA|AT. (The vertical lines show the cleavage sites.)

7. Valine and leucine differ by only one carbon in the side group. Glycine and alanine also differ by only one carbon (glycine has none and alanine has one).

9. The two possible reactions follow. The amide bonds are within the dashed boxes.

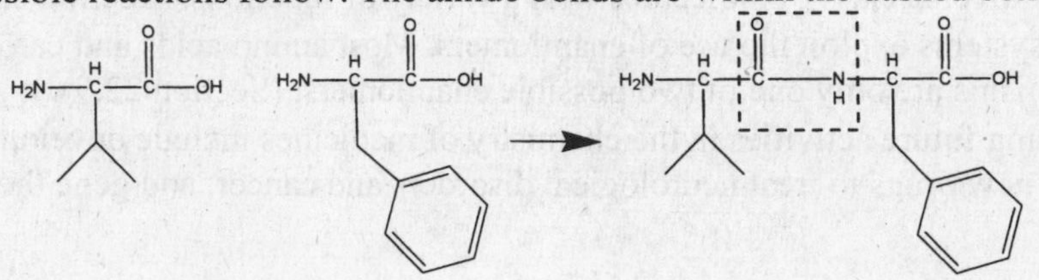

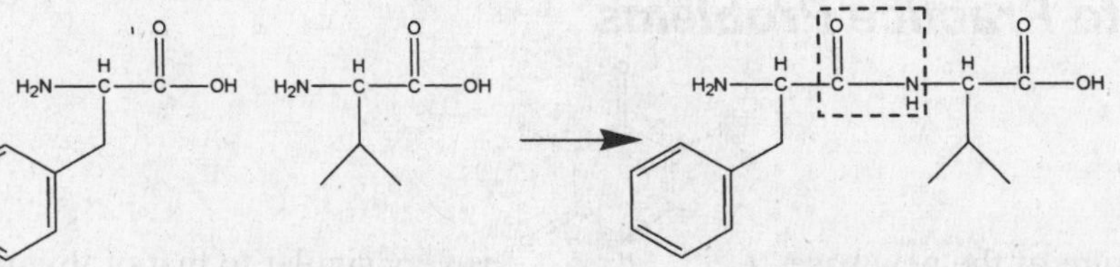

11. A codon is a sequence of three nucleic acids on an m-RNA that is translated into a particular amino acid during protein synthesis.

13. If the wrong nucleotide is used in the production of the m-RNA or an extra nucleotide is added to or removed from the m-RNA, a different three-nucleotide codon may result, which could specify a different amino acid.

15. Breaking the sequence into codons, we get
UUU|CGA|AGU|AUG|GGU|GGA|GAU|UCU|CCC|GCG
Although the AUG codon that specifies the start of the sequence is present, it also represents methionine. Furthermore, there is no UAA, UAG, or UGA that would specify the end of the sequence. We have to conclude that this fragment is in the middle of the sequence.

17. Actin's protein structure looks like a ball, so actin is a globular protein.

19. In the course of this reaction, the alcoholic carbon is oxidized; therefore, the appropriate enzyme would be in the oxidoreductase class.

21. By binding to the enzyme and inhibiting further production, the product itself can ensure that the concentration of the product does not rise too high and that the product is created only as needed (that is, when all others have been used.)

23. There are four carbons in the carbohydrate, making it a tetrose.

25. The CHO group at the right of the structure is an aldehyde functional group; the molecule is an aldose.

27. All three are made from glucose units.

29. Fructose is one of the two sugars that results from the hydrolysis of sucrose. Fructose ($C_6H_{12}O_6$) has six carbons and is a hexose sugar. The fructose structure (see Figure 22.24 of the text) has a ketone functional group, so fructose is a ketose.

31. The distinction depends merely on whether the lipid is solid at room temperature (a fat) or is liquid at room temperature (an oil).

33. The long, straight carbon tails in the molecule would be expected to pack very tightly with a large amount of interaction between molecules. This tight association leads to the formation of a solid—a fat.

35. Because the phospholipid has many polar groups and even an ionic portion (the ammonium group), it should interact well with water, while still using the nonpolar regions to form the cell wall. The ester portion of the triacylglycerol (the part with all the oxygen atoms) is polar and should interact well with water, while still using the nonpolar regions to form the cell wall.

37. The two enantiomers of serine (left) and phenylalanine (right) are shown below. Note that the enantiomers are mirror images of each other.

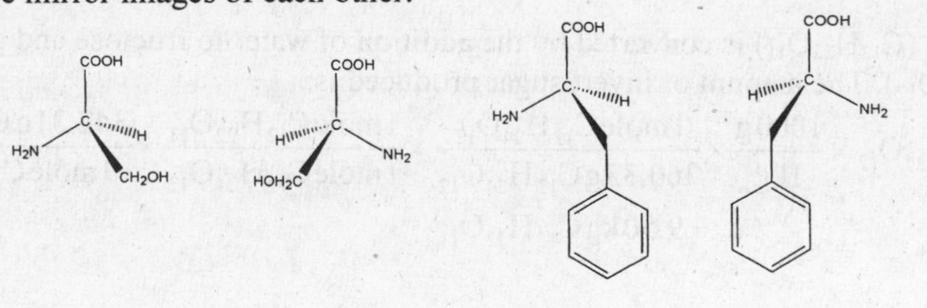

COOH COOH COOH COOH
H₂N C H H C NH₂ H₂N C H H C NH₂
CH₂OH HOH₂C

39. Because the body uses principally a single type of enantiomer, having chiral enzymes will make sure that only the proper enantiomer is used in a protein or other product. The cell will not waste resources on a molecule it cannot use.

41. The residue on tryptophan (left), along with part of the amino acid backbone, is used to create serotonin (right). There are no chiral atoms in serotonin; therefore, it has no enantiomer.

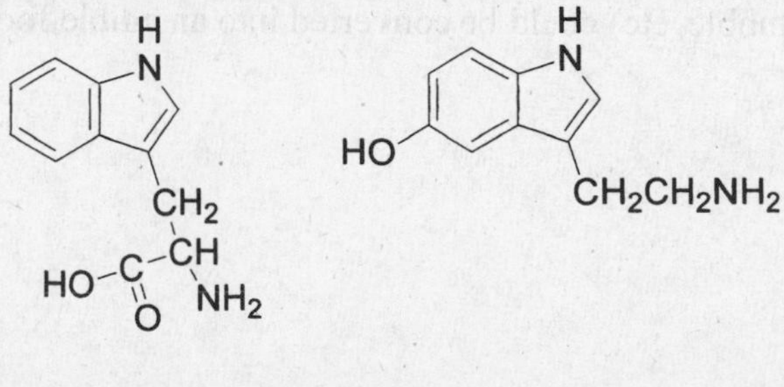

H N CH₂ CH HO C O NH₂ H N HO CH₂CH₂NH₂

43. The RNA and protein from the following DNA sequence are

DNA: TAC AGC GCT TAA ATT CCG ACG AAT AA

RNA: AUG|UCG|CGA|AUU|UAA|GGC|UGC|UUA|UU

Protein: Methionine-Serine-Arginine-Isoleucine (The fifth codon stops the build.)

45. Each of the molecules is saturated. As the chain length increases, the strength of the dispersion forces increases (recall that dispersion forces are proportional to the size or number of electrons in the molecule). The stronger the forces, the higher the temperature required to overcome those intermolecular attractions. The longer chains therefore have higher melting points.

47. If the receptors that are responsible for detecting the sweetness of a food are sensitive only to a particular enantiomeric form, the other enantiomer will not be sensed as sweet. If the receptors are *not* sensitive to just one enantiomeric form, the food could taste sweet but might not be metabolized.

49. a. The reaction is

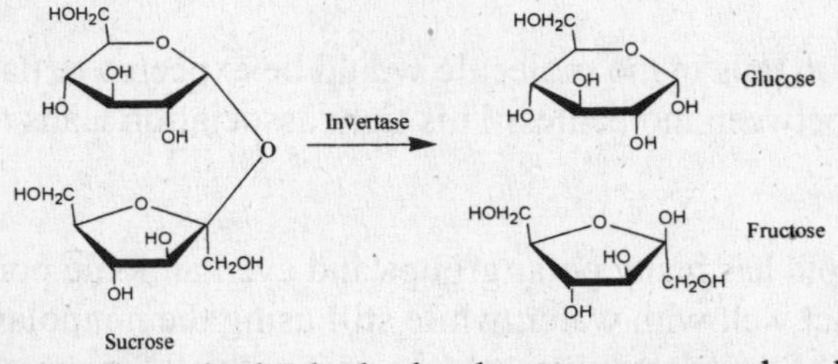

HOH₂C OH HO OH O HOH₂C HO CH₂OH OH Sucrose — Invertase → Glucose + Fructose

b. Since both reactions involves the hydrolysis, the enzyme may be able to catalyze the reaction; however, many enzymes have a very specific active site and therefore would not be able to catalyze the reaction.

c. Lysine can be coded by the AAA or AAG codon on m-RNA and UUU or UUC anticodon on t-RNA.

d. Sucrose ($C_{12}H_{22}O_{11}$) is converted by the addition of water to fructose and glucose (total = $C_{12}H_{24}O_{12}$). The amount of invert sugar produced is:

$$10\,\text{kg}\,C_{12}H_{24}O_{12} \times \frac{1000\,\text{g}}{1\,\text{kg}} \times \frac{1\,\text{mole}\,C_{12}H_{24}O_{12}}{360.33\,\text{g}\,C_{12}H_{24}O_{12}} \times \frac{1\,\text{mole}\,C_{12}H_{22}O_{11}}{1\,\text{mole}\,C_{12}H_{24}O_{12}} \times \frac{342.31\,\text{g}\,C_{12}H_{22}O_{11}}{1\,\text{mole}\,C_{12}H_{22}O_{11}} \times \frac{1\,\text{kg}}{1000\,\text{g}}$$

$$= 9.50\,\text{kg}\,C_{12}H_{22}O_{11}$$

e.

$$15.5\,\text{kg}\,C_{12}H_{22}O_{11} \times \frac{1000\,\text{g}}{1\,\text{kg}} \times \frac{1\,\text{mole}\,C_{12}H_{22}O_{11}}{342.31\,\text{g}\,C_{12}H_{22}O_{11}} \times \frac{1\,\text{mole}\,C_6H_{12}O_6}{1\,\text{mole}\,C_{12}H_{22}O_{11}} \times \frac{180.16\,\text{g}\,C_6H_{12}O_6}{1\,\text{mole}\,C_6H_{12}O_6} \times \frac{1\,\text{kg}}{1000\,\text{g}}$$

$$= 8.16\,\text{kg fructose}$$

f. If a hydrolase could be found that would convert cellulose to glucose, plant fibers (wood, wheat and corn stubble, etc) could be converted into an edible food.